ALLE ZEIT WACH 1842

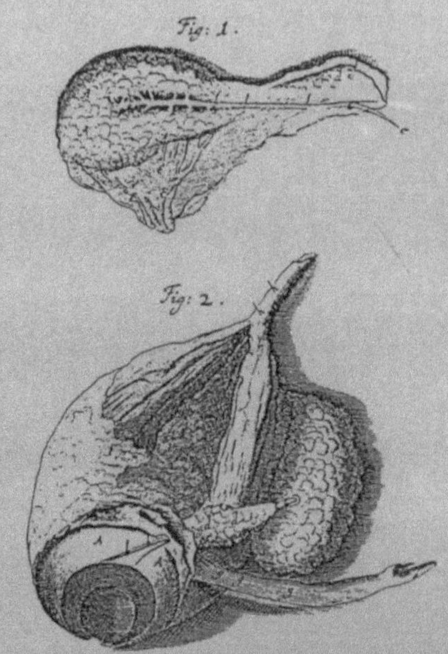

N. II. 49

ACTA ERUDITORUM

publicata Lipſiæ

Calendis Februarii, Anno M DC XCIV.

GLANDULA NOVA LACHRYMALIS una cum ductu excretorio in cervis & damis a D. Joh. Jacobo Hardero P. P. Baſileenſi anno proxime elapſo detecta & in binis literis ad --- deſcripta.

Copy of the title and figures from the first published paper on the Harderian glands by Jacob Harder, dated 1694.

S.M. Webb · R.A. Hoffman
M.L. Puig-Domingo · R.J. Reiter (Eds.)

Harderian Glands

Porphyrin Metabolism, Behavioral and Endocrine Effects

With 117 Figures

Springer-Verlag
Berlin Heidelberg NewYork
London Paris Tokyo
Hong Kong Barcelona
Budapest

Prof. Susan M. Webb

Dept. of Endocrinology
Hospital de Sant Pau. Autonomous University
Padre Claret 167
08025 Barcelona, Spain

Dr. Roger A. Hoffman

Dept. of Biology
Colgate University
Hamilton, NY 13346, USA

Dr. Manuel L. Puig-Domingo

Dept. of Endocrinology
Hospital Santa Creu de Sant Pau
University Autonoma de Barcelona
08025 Barcelona, Spain

Prof. Russel J. Reiter

Dept. of Cellular & Structural Biology
The University of Texas Health Science Center
San Antonio, TX 78284, USA

Library of Congress Cataloging-in-Publication Data
Harderian glands: porphyrin metabolism, behavioral, and endocrine effects/S.M. Webb . . . [et al.], (eds). p. cm.
Includes bibliographical references.

ISBN-13: 978-3-642-76687-9 e-ISBN-13: 978-3-642-76685-5
DOI: 10.1007/978-3-642-76685-5

1. Harderian gland. 2. Porphyrin and porphyrin compounds—Metabolism. I. Webb, S.M. (Susan M.), 1952–. [DNLM: 1. Harderian Gland—physiology. 2. Porphyrins—metabolism.

Softcover reprint of the hardcover 1st edition 1992

Typesetting: Macmillan India Ltd, Bangalore, India; Printing: Saladruck, Steinkopf & Sohn, Berlin;
Binding: Lüderitz & Bauer, Berlin
31/3020-5 4 3 2 1 0—Printed on acid-free paper

Foreword

Harderian glands, first described by Jacob Harder in two species of deer in 1694, are large, intraorbital glands which, with but few exceptions, are found in all land vertebrates. Certainly, their relatively large size, their phylogenetic age, and persistent conservation in all groups as they emerged from an aqueous to an air/land environment lend strong support to the logical deduction that they probably play an important role in the physiological adaptation to terrestrial life.

Yet, few biologists know even what or where these glands are. For a variety of reasons, the Harderian glands have not received the attention they deserve and, as a result, the published works available have been scarce and scattered throughout the world literature.

The current situation is remarkably similar to that which existed in regard to the pineal gland prior to the mid-1960s, i.e., scattered literature, unknown function, few investigators, and little interest. Yet, following a few key papers, interest in the pineal gland expanded and soon an explosive increase in the knowledge and understanding of the pineal gland took place and continues today. Will history repeat itself?

Originally, a few of us discussed the desirability of an informal Symposium on the Harderian glands. As such things are wont to do, we soon began to expand our ideas: why not solicit review chapters by known experts, and while we were at it, why not develop a complete bibliography of all published works on the glands. Why not bring together all that was known about these enigmatic glands under a single cover to serve as the "source" as of 1990, a stepping stone for new and established investigators alike. With such a volume available, biologists, biochemists, and physiologists would have the world literature at their fingertips, along with all the investigations currently under way.

Cautious inquiries to known investigators confirmed our suspicions: there was indeed an interest in such an undertaking. As a result of further inquiries, Springer-Verlag agreed to publish such a volume. For that, we are taking this opportunity to collectively thank them for their willingness to gamble and their unusual foresightedness. Such corporate behavior is rare indeed!

Originally, this book was to be made up of three parts: (1) review chapters; (2) Proceedings of the First International Symposium on the Harderian glands, held in Barcelona, Spain, on 11–13 Oct, 1990 and supported by FISS (90/8074), CIRIT (ARC550) and DGICHT (C0-86-90), (3) an accessory bibliography containing all other known papers not specifically referred to in the other two sections. Unfortunately, space limitations prevented the inclusion of Part 2. These papers, representing contemporary work on the Harderian glands, will appear in the November and December 1991 issues of Endocrinologia (Doyma, J.A., Barcelona).

It is pertinent that a complete "master" bibliography is now on floppy disc and is being continually updated by Dr. Jan Hugo, Faculty of Medicine, Department of Pathology and Medicine, Charles University, U Nemocnice, 5 CS-128 53 Prague 2, Czechoslovakia. All investigators of Harderian gland biology are urged to automatically forward a copy of their published works to him.

It is our fervent hope that this beginning is indeed just that, and that, henceforth, the mention of Harderian glands will be common knowledge and not evoke surprise or embarrassment.

February, 1992 The Editors

Table of Contents

The Past, Present, and Future of the Harderian Gland

JAMES MICHAEL OLCESE

Institute for Hormone and Fertility Research, Grandweg 64, D-2000 Hamburg 54, FRG

It is continually surprising how few people know what or where the Harderian glands are. Even among successful veterinarians, endocrinologists, and opthalmologists, there are many who have never heard of them! This lack of recognition is probably due to two major factors, namely, that it was long considered an accessory lacrimal gland of little importance generally, and secondly, it appears to be absent in the adult human. Notwithstanding this neglect over many centuries, the fact that the First International Symposium on the Harderian Gland has taken place in 1990 demonstrates the rapidly emerging interest and concomitant progress in our understanding of this peculiar tissue. One can clearly expect many new insights and important developments in the coming years.

The man whose name will be associated forever with the retro-orbital "glandula palpebrae tertiae profunda" was born in 1656 into the Age of Enlightenment. Having received his doctorate degree from the University of Basel at the age of 19, Johann Jacob Harder rose quickly through many ranks to become professor there in 1678. Throughout his career, Harder was a highly respected physician of many royal families, as well as a prominent scientist. In addition to the Harderian gland, which he first named the "glandula nova lachrymalis" in deer (Harder 1694), he was also the actual discoverer of the so-called Pacchionian glands around the sagittal sinus in the dura mater (Nouvelle Biographie Generale 1858).

In the 280 years since Harder's death in 1711, many investigators have reported the occurrence of Harderian glands not only in mammals but in amphibians, reptiles, and birds as well (cf. Miessner 1900). During most of these years, the Harderian glands were frequently presumed to be little more than special tear glands associated with the nictitating membrane. However, as has often been noted (Kennedy 1970; Sakai 1981), the glands associated with the third eyelid are not always Harderian glands, and the Harderian glands are not always associated with the third eyelid. Furthermore, the secretions from the mammalian Harderian glands are not watery like tears, but rather sebaceous. Perhaps the best criterion for characterizing the mammalian Harderian glands may be that proposed by Sakai (1981), i.e., tubuloalveolar, ocular glands which secrete lipids by a merocrine mechanism.

Webb et al. (Eds).
Harderian Glands

The Harderian glands probably occur in most mammalian orders, although absolute histochemical proof in some orders is still lacking. The *absence* of Harderian glands has been reported for the Chiroptera (bats), the terrestrial Carnivora (canines, felines), the Perissodactyla (equines) and the Primates (cf. Sakai 1981). As many of the the reported absences of Harderian glands derive from earlier publications, these reports need to be confirmed. Whether particular losses of Harderian glands across diverse mammalian orders represent negative selection or merely random genetic drift remains uncertain.

Far less is known about the Harderian glands of non-mammalian vertebrates, even in those orders where it is the sole occupant of the orbit, e.g., anuran amphibians (Minucci et al. 1990), or where it is extremely large, e.g., marine birds (Kennedy 1970). From what is known, it would appear that there are numerous similarities between the Harderian glands of all terrestrial vertebrates. For example, as in mammals, both thyroid hormones and sex steroids have important effects on the development and physiology of nonmammalian Harderian glands (Shirama et al. 1982; Gupta and Maiti 1983). Furthermore, a circadian rhythm of melatonin has been reported for pigeon Harderian glands (Vakkuri et al. 1985a, b), and is well-known for rodents. A thorough comparative analysis of the morphology, biochemistry, and physiology of the vertebrate Harderian glands would be quite valuable, but has yet to be attempted.

Only in mammals, and in particular, the rodents, have researchers begun to establish a sufficient database for the identification of significant patterns in Harderian gland biology. Among the tendencies that one finds in surveying the literature on mammalian Harderian glands are (1) high porphyrin concentrations, often of a unique nature, especially in the female rodent; (2) unusual mixtures of lipids, some of which may be components of cell membranes; (3) a morphological, and often biochemical, sexual dimorphism, at least for some rodents; and (4) synthesis of the indoleamine hormone, melatonin.

The first investigators to describe the presence of porphyrins (by virtue of the auto-fluorescence of the Harderian gland under ultraviolet light) were Derrien and Turchini in 1924. Some 18 years later, the porphyrins were chemically verified in the mouse Harderian gland (Strong 1942). A new porphyrin, Hardero-porphyrin, was isolated and characterized from the rat Harderian gland by Kennedy (1970) and subsequently shown to occur in the female hamster as well (Spike et al. 1990). Variations in Harderian gland porphyrin content with age, across the seasons, during the estrus cycle, and in response to thyroxine and steroids have been reported also (Wooley and Worley 1954; Wetterberg et al. 1970; Payne et al. 1977; Payne et al. 1979). It is noteworthy that only rodents appear to show the tendency toward a high porphyrin content in Harderian glands. Kennedy (1970) has suggested that rodent Harderian glands may lack ferrochelatase, the mitochondrial enzyme responsible for incorporating ferrous iron into protoporphyrin (to form heme), which might account for the accumulation of porphyrins. This ought to be examined in a variety of species. Perhaps such a tendency toward porphyria in diurnal species would have proven to be maladaptive. Inherited porphyria in the human is often associated with extreme sensitivity to light (i.e., photic rashes), or hepatic and neurological disorders (Martin 1985).

What is most lacking in our knowledge about Harderian gland porphyrins is that of their function. The required enzymes for porphyrin synthesis occur in the gland and are influenced by hormones, especially gonadal steroids (cf. Olcese and Wesche 1989); hence it might be expected that these porphyrins serve a physiological role. I would suggest, as a testable hypothesis, that porphyrins are secreted around the eye and hence onto the cornea. The absorbance by porphyrins of near-ultraviolet light, which the nocturnal rodent encounters normally only around the twilight hours, may generate sufficient energies of visible light to stimulate the photoreceptors. These retinal signals could be expected to influence the circadian system (Moore and Card 1985). This hypothesis might explain the reported effects of near-ultraviolet radiation on pineal melatonin production in hamsters (Podolin et al. 1987).

Another interesting proposal in this regard is that of Kennedy (1970), who, like Strong and Figge (1941) before him, noted an apparent link between Harderian gland porphyrin content and cancer susceptibility in rodents. Recently, it was shown (Sinn et al. 1987) that transgenic mice carrying the v-Ha-ras oncogene selectively develop a substantial proliferation of the Harderian gland epithelium, but not of other glandular epithelia. In view of the recent characterization of a Harderian gland-derived polypeptide growth factor in rodents (Yokoyama et al. 1989), it might prove extremely insightful to examine porphyrins and the differentiation of the Harderian gland in a cell culture system.

A unique fatty acid synthetase has been reported to occur in guinea pig Harderian glands (Seyama et al. 1981) and unusual lipids (glyceryl ether diesters, especially 1-alkyl-2,3-diacyl-glycerol) have been found in many mammalian Harderian glands. Interestingly, these lipid components also appear in both the mammalian sebum of hair follicles and in the preputial gland (Sakai 1981). The latter gland is responsible for female-attractant odors in the male urine. The Harderian glands of Syrian hamsters are reported to be sexually different in terms of the composition of these lipids (Lin and Nadakavukaren 1981), which could suggest a pheromonal function. Although the precise role for these lipids is not yet known, Thiessen (1988) has proposed that the lipid secretions of the gerbil Harderian glands are (rather like sebum) used for thermoregulation, inasmuch as these lipids (1) are released during cold-induced self-grooming episodes, (2) serve to insulate the animal against wetness, and (3) increase the absorption of radiant heat.

The sexual dimorphism of the hamster Harderian gland has been reviewed earlier (Sakai 1981; Olcese and Wesche 1989). This phenomenon is expressed not only morphologically in terms of different cell types, but also in terms of porphyrin content and composition (Spike et al. 1990), lipid composition (Lin and Nadakavukaren 1981), the concentrations of various metals (Hoffman and Jones 1981), and melatonin synthesis (Hoffman et al. 1985). The lower levels of porphyrins and melatonin synthesis in male Harderian glands can be elevated by gonadectomy, suggesting steroidal involvement. In fact, testosterone administration prevents these changes (McMasters and Hoffman 1984) and identical cytosolic androgen receptors have been identified in the Harderian glands of both male and female hamsters (Vilchis and Pérez-Palacios 1989).

Nonetheless, the mechanisms through which androgens act on Harderian gland physiology remain obscure. Likewise, a role for melatonin in the Harderian gland has yet to be demonstrated. In this regard it should not be overlooked that the melatonin-synthesizing enzyme hydroxyindole-O-methyltransferase (HIOMT), although present in the Harderian glands of rats, appears to be quite different from the pineal and retinal HIOMT enzymes, particularly with respect to substrate utilization (Cardinali and Wurtman 1972). A similarly thorough biochemical analysis of the serotonin N-aceyltransferase enzyme in the Harderian glands would be interesting, but has yet to be undertaken. Likewise, investigations into the presence or absence of specific melatonin-binding sites in this tissue could be quite useful.

In conclusion, Harderian gland research at present is replete with challenging and certainly interesting phenomena that need careful examination by morphologists, biochemists, and comparative physiologists alike. Among the many questions that demand attention in the years ahead are (1) are the Harderian glands exocrine glands only, or are certain hormonal factors produced there which might influence adjacent structures such as the eye or even the brain? (2) What is the role, if any, for the numerous "Harderian-specific" molecules, such as Harderoporphyrin, Harderian-derived growth factor, or the unique branched chain fatty acids? (3) Are thyroid hormone receptors, steroid receptors, melatonin receptors, or other receptors found in the Harderian glands of all, or only some vertebrate animals, and what is their function?

The future of the Harderian gland as an object of considerable scientific fascination seems assured. Given the ever-accelerating interest in this heretofore neglected area, research into the next century will undoubtedly force us to look on the Harderian glands in new ways. With due respect to Johann Jacob Harder, whose discovery nearly 300 years ago of the "glandula nova lachrymalis" has finally gained international attention, it seems not inappropriate to modify Harder's moniker and admit that, for many modern biologists, the Harderian gland is indeed a new gland – "glandula *nova*" – worth exploring further.

References

Cardinali DP, Wurtman RJ (1972) Hydroxyindole-O-methyltransferases in rat pineal, retina and Harderian gland. Endocrinology 91:247–252

Derrien E, Turchini J (1924) Sur l'accumulation d'une porphyrine dans la glande de Harder des rongeurs du genre *Mus* et sur sa mode d'excrétion. C R Soc Biol 91:637–639

Gupta SK, Maiti BR (1983) Influence of sex hormones on the Harderian gland activity in female domestic duckling. Mikroskopie 40:132–137

Harder JJ (1694) Glandula nova lachrymalis una cum ductu excretorio in cervis et damis. Acta Erudit Publ Lipsiae 49–52

Hoffman RA, Jones JW (1981) Concentrations of metals in the Harderian glands of male and female hamsters. Comp Biochem Physiol 69A:153–156

Hoffman RA, Johnson LB, Reiter RJ (1985) Harderian glands of golden hamsters: temporal and sexual differences in immunoreactive melatonin. J Pineal Res 2:161–168

Kennedy GY (1970) Harderoporphyrin: a new porphyrin from the Harderian glands of the rat. Comp Biochem Physiol 36:21–36

Lin WL, Nadakavukaren MJ (1981) Harderian gland lipids of male and female golden hamsters, *Mesocricetus auratus*. Comp Biochem Physiol 70B:627–630

Martin DW (1985) Porphyrins and bile pigments. In: Martin DW, Mayes PA, Rodwell VW, Granner DK (eds) Harper's review of biochemistry, 20th edn. Lange Med Publ, Los Altos, pp 331–347

McMasters K, Hoffman RA (1984) Harderian gland: regulation of sexual "type" by gonads and pineal gland. Biol Reprod 31:579–586

Miessner H (1900) Die Drüsen des dritten Augenlides einiger Säugetiere. Arch Wiss Prakt Tierkd 26:122–154

Minucci S, Chieffi Baccari G, Di Matteo L, Marmorino C, d'Istria M, Chieffi G (1990) Influence of light and temperature on the secretory activity of the Harderian gland of the green frog, *Rana esculenta*. Comp Biochem Physiol 95A:249–252

Moore RY, Card JP (1985) Visual pathways and the entrainment of circadian rhythms. Ann NY Acad Sci 453:123–133

Nouvelle Biographie Generale (1958) Didot, Paris, p 349

Olcese J, Wesche A (1989) The Harderian gland. Comp Biochem Physiol 93A:655–665

Payne AP, McGadey J, Moore MR, Thompson G (1977) Cyclic and seasonal changes in Harderian gland activity in the female golden hamster. J Endocrinol 72:41P

Payne AP, McGadey J, Moore MR, Thompson GG (1979) Changes in Harderian gland activity in the female golden hamster during the oestrous cycle, pregnancy and lactation. Biochem J 178:597–604

Podolin PL, Rollag MD, Brainard GC (1987) The suppression of nocturnal pineal melatonin in the Syrian hamster: dose-response curves at 500 and 360 nm. Endocrinology 121:266–270

Sakai T (1981) The mammalian Harderian gland: morphology, biochemistry, function and phylogeny. Arch Histol Jpn 44:299–333

Seyama Y, Otsuka H, Kawaguchi A, Yamakawa T (1981) Fatty acid synthetase from the Harderian gland of guinea pig: biosynthesis of methyl-branched fatty acids. J Biochem (Tokyo) 90:789–798

Shirama K, Kikuyama S, Takeo Y, Shimizu K, Maekawa K (1982) Development of Harderian gland during metamorphosis in anurans. Anat Rec 202:371–378

Sinn E, Muller W, Pattengale P, Tepler I, Wallace R, Leder P (1987) Coexpression of MMTV/v-Ha-ras and MMTV/c-myc genes in transgenic mice: synergistic action of oncogenes in vivo. Cell 49:465–475

Spike RC, Payne AP, Thompson GG, Moore MR (1990) High-performance liquid chromatographic analysis of porphyrins in hamster Harderian glands. Biochim Biophys Acta 1034:1–3

Strong LC (1942) Sex differences in pigment content of Harderian glands of mice. Proc Soc Exp Biol Med 50:123–125

Strong LC, Figge FHJ (1941) Fluorescence of Harderian glands in mice of cancer-susceptible and cancer-resistant strains. Science 94:331

Thiessen DD (1988) Body temperature and grooming in the Mongolian gerbil. Ann NY Acad Sci 525:27–39

Vakkuri O, Rintamäki H, Leppäluoto J (1985a) Plasma and tissue concentrations of melatonin after midnight light exposure and pinealectomy in the pigeon, *Columbia livia*. J Endocrinol 105:263–268

Vakkuri O, Rintamäki H, Leppäluoto J (1985b) Presence of immunoreactive melatonin in different tissues of the pigeon (*Columba livia*). Gen Comp Endocrinol 58:69–75

Vilchis F, Pérez-Palacios G (1989) Steroid hormone receptors and the sexual phenotype of the Harderian gland in hamsters. J Endocrinol 121:149–156

Wetterberg L, Yuwiler A, Geller E, Schapiro S (1970) Harderian gland: development and influence of early hormonal treatment on porphyrin content. Science 168:996–998
Wooley G, Worley J (1954) Sexual dimorphism in the Harderian gland of the hamster. Anat Rec 118:416–417
Yokoyama Y, Kano K, Kaji K, Seyama K (1989) Purification and characterization of a growth factor from guinea pig Harderian gland. J. Biol Chem 264:17058–17063

Comparative Anatomy of Mammalian Harderian Glands

TATSUO SAKAI

Department of Anatomy, School of Medicine, Juntendo University, 2-1-1 Hongo, Bunkyo-ku, Tokyo 113, Japan

1 Introduction

The mammalian eye possesses two sets of major exocrine glands: one in the inner canthus and the other in the outer canthus. The former, which may be named the medial ocular glands, includes the Harderian gland and the nictitans gland. These two glands are distinguished on the basis of the nature of their secretions. The Harderian gland secretes lipids by a merocrine mechanism whereas the nictitans gland secretes glycoproteins (Sakai 1981). The latter set, which may be called the lateral ocular glands, is regarded generally as the lacrimal gland, and is divided into two groups on the basis of their gross anatomy (Sakai 1989).

In recent years, the Harderian glands have attracted more and more attention, since they are quite unique among structures consituting the mammalian body. Morphologically, the mammalian Harderian glands are the sole exocrine glands which secrete lipids by exocytosis (Wooding 1980; Sakai and Yohro 1981; Winterhager and Kühnel 1983). Biochemically, the secretion of the glands contains unusual substances such as 1-alkyl-2,3-diacylglycerols (Otsuru et al. 1983) and porphyrins (Jackson et al. 1976; Akagi et al. 1987). Physiologically, several functions of the gland have been proposed including pheromone secretion (Payne 1977, 1979; Harriman and Thiessen 1985; Johnston 1986), thermoregulation (Thiessen et al. 1977; Thiessen and Kittrell 1980; Thiessen 1988), and photoreception (Wetterberg et al. 1970; Pavet et al. 1984), which have almost nothing to do with the eye. Thus, the mammalian Harderian gland is one of the most fascinating objects to those who want to uncover the secret of the nature behind its diverse and complex appearance.

The distinction between the Harderian gland and the nictitans gland had not been clearly understood until Sakai (1981) introduced the above-mentioned definition of the mammalian Harderian gland based on secretion. The distinction between the Harderian and nictitans glands is valid only in the mammals. The medial ocular glands in lower vertebrates so far investigated secrete glycoproteins and no lipids, and are called the Harderian gland regardless of their secretion (Sakai 1981).

Webb et al. (Eds).
Harderian Glands

Confusion between the Harderian and nictitans gland in mammals renders anatomical descriptions of the Harderian gland in the literature difficult to interpret. One cannot judge whether given anatomical descriptions concern the Harderian gland or the nictitans gland until appropriate histochemical or ultrastructural descriptions of the gland are available. Frequent lack of macroscopic descriptions, including opening location and number of ducts, size, and shape of the gland and topographical relations to the surrounding structures, sometimes makes interpretation of description in the literature difficult.

The present chapter on the anatomy of the Harderian gland, will critically interpret the literature information that is arranged in taxonomical order in Table 1. On the basis of these data as well as the author's own observations, some aspects of the anatomy of the gland will be reviewed, as in a previous review (Sakai 1981); (1) the Harderian gland in individual mammals, (2) anatomy of the mammalian Harderian gland including its innervation, and functional significance of the orbital venous sinus, and finally, (3) significance of the mammlian Harderian gland from the viewpoint of comparative anatomy.

2 Harderian Gland in Individual Mammalian Species

The concept of the Harderian gland had not been clear until the first review on the mammalian Harderian gland (Sakai 1981) was published. In this article, Sakai (1981) reviewed the historical change of the concept and introduced a new definition of the mammalian Harderian gland. "The mammalian Harderian gland is those ocular glands that have tubuloalveolar endpieces (tubular alveoli) and secrete lipid by a merocrine mechanism." Based on this criterion, Sakai (1981) surveyed literature on the Harderian gland and summarized its distribution among mammalian species. In the 10-year interim, numerous reports have been published on the gland and the summary of the distribution of the mammalian Harderian gland must now be somewhat revised (Tables 1, 2).

In monotremes, Griffiths (1978) mentioned "Harderian glands" in platypus and echidna, but did not give any descriptions of their anatomy or histology. Until descriptions of the structure are available, one cannot decide whether it is a Harderian or a nictitans gland.

The medial ocular glands in three South American marsupials examined by Paule (1957) are histochemically all positive for lipids and negative for sugars, indicating that these glands are Harderian glands. Sakai and van Lennep (1984) studied the medial ocular glands of the Australian marsupials and identified Harderian glands in some species and nictitans glands in others based on histology and histochemistry. Distribution of the Harderian and nictitans glands among Australian marsupials is interesting; the Harderian and nictitans glands may occur in two different species in the same genus (*Antechinus*), and

Table 1. Literature of anatomical descriptions on the mammalian Harderian and nictitans gland

Order *Species* (common name) Type of gland Description[a]	Author (year)
Monotremata	
Ornithorhynchus sp. (platypus)	
Tachyglossus sp. (echidna)	
Harderian gland?	
Present	Griffiths (1978)
Marsupialia	
Didelphis marsupialis (Virginia opossum)	
Caluromys philander (woolly opossum)	
Marmosa mexicana (mouse opossum)	
Harderian gland	
Histochem (sugar, lipid)	Paule (1957)
Perameles nasuta (long-nosed bandicoot)	
Antechinus stuartii (brown antechinus)	
Antechinus swainsonii (dusky antechinus)	
Harderian gland	
Histol, histochem (sugar)	Sakai and van Lennep (1984)
Isoodon macrourus (short-nosed bandicoot)	
Antechinus flavipes (yellow-footed antechinus)	
Dasyurus viverrinus (eastern native cat)	
Dasyurus maculatus (tiger cat)	
Nictitans gland	
Histol, histochem (sugar)	Sakai and van Lennep (1984)
Petauroides volans (greater glider)	
Gland with Harderian and nictitans tissues	
Histol, histochem (sugar)	Sakai and van Lennep (1984)
Sminthopsis murina (common dunnart)	
Sminthopsis crassicaudata (fat-tailed dunnart)	
Trichosurus vulpecula (brush possum)	
Macropus eugenii (tammar wallaby)	
No gland	
(Absent)	Sakai and van Lennep (1984)
Insectivora	
Erinaceus europaeus (hedgehog)	
Gland with Harderian and nictitans tissues	
Histol	Loewenthal (1892a)
Sorex unguiculatus (big-clawed shrew)	
Harderian gland	
Anat, histol, cytol	Sakai (1981, unpublished)
Blarina brevicauda (short-tailed shrew)	
Harderian gland	
Histochem (sugar, lipid)	Paule (1957)
Condylura cristata (star-nosed mole)	
No gland	
(Absent)	Paule (1957)

Table 1 (*continued*)

Insectivora (*continued*)	
Suncus murinus (musk shrew)	
Harderian gland	
Anat, histol, cytol	Sakai (1989)
Chiroptera	
Myotis leucifugus (little brown bat)	
Dasmondus rufus (vampire rat)	
No gland	
(Absent)	Paule (1957)
Primates	
Generally absent	Wiedersheim (1909) Prince (1956)
Edentata	
Dasypus novemcinctus (nine-banded armadillo)	
Gland with two portions (nictitans gland, Harderian gland)	
Histol, cytol	Weaker (1981)
Lagomorpha	
Oryctolagus cuniculus (rabbit)	
Harderian gland with two portions (white and red parts)	
Histol	Loewenthal (1892b, 1896)
Histochem	Paule (1957)
Histol, cytol	Björkman et al. (1960)
Histochem (lipid)	Kühnel and Wrobel (1968)
Histol, histochem (lipid), cytol	Kühnel (1971)
Cytol	Jost et al. (1974)
Cytol	Wooding (1980)
Cytol	Winterhager and Kühnel (1983)
Nictitans gland	
Histol	Loewenthal (1892b, 1896)
Histol	Shinoda (1958)
Sylvaligus floridanus (cottontail rabbit)	
Harderian gland	
Histochem (sugar, lipid)	Paule (1957)
Ochotona rufescens (Afghan pika)	
Harderian gland	
Histol, cytol	Sakai (1981, unpublished)
Rodentia	
Mus musculus (house mouse)	
Harderian gland	
Histol	Buschke (1933)
Histochem (sugar, lipid)	Paule (1957)
Histol, cytol	Woodhouse and Rhodin (1963)
Development	Michael et al. (1988)
Histol, cytol	Shirama et al. (1988)

Table 1 (*continued*)

Rodentia (*continued*)	
Rattus rattus (rat)	
Harderian gland	
Histol	Loewenthal (1892b, 1896)
Histol	Walter (1925)
Histol	Buschke (1933)
Histol, histochem (sugar, lipid)	Kanwar (1960)
Anat	Akiyama (1961)
Cytol	Kelényi and Orbán (1964)
Histochem, cytol	Tsutsumi et al. (1966)
Development	Müller (1969a,b 1970)
Development	Vianna et al. (1975a,b)
Innervation	Huhtala et al. (1977)
Cytol	Brownscheidle and Niewenhuis (1978)
Innervation	Shirama et al. (1979)
Innervation	Tsukahara and Jacobowitz (1987)
Cytol	Abe et al. (1980, 1981)
Mesocricetus auratus (golden hamster)	
Harderian gland	
Cytol	Bucana and Nadakavukaren (1972a)
Innervation	Bucana and Nadakavukaren (1972b)
Innervation	Norvell and Clabough (1972)
Development	Bucana and Nadakavukaren (1972c, 1973)
Cytol	Sun and Nadakavukaren (1980)
Cytol	Nadakavukaren and Lin (1983)
Histol, cytol	Payne et al. (1982, 1985)
Innervation	Vega-Alvarez et al. (1989)
Meriones unguiculatus (Mongolian gerbil)	
Harderian gland	
Anat, histol, histochem (lipid), cytol	Sakai and Yohro (1981)
Cytol	Kaiho and Ichikawa (1982)
Histol, cytol	Johnston et al. (1983)
Nictitans gland	
Histol, cytol	Sakai and Yohro (1981)
Apodemus sylvaticus (woodmouse)	
Harderian gland	
Histol, cytol	Johnston et al. (1987)
Pseudomys australis (Plains mouse)	
Harderian gland	
Histol, cytol	Johnston et al. (1985)
Cavia aperea (guinea-pig)	
Harderian gland	
Histochem (sugar, lipid)	Paule (1957)
Histol, cytol	Sakai (unpublished)

Table 1 (*continued*)

Rodentia (*continued*)	
Tamiasciurus hudsonicus (red squirrel)	
Sciurus niger (fox squirrel)	
S. carolinensis (gray squirrel)	
Citellus tridecemlineatus (13-lined squirrel)	
Tamias striatus (chipmunk)	
Peromyscus leucopus (deer mouse)	
Harderian gland	
Histochem (sugar, lipid)	Paule (1957)
Cetacea	
Balaenoptera sp., (mysticete)	
Megaptera sp., (mysticete)	
Balaena sp., (mysticete)	
Delphinus sp., (odontocete)	
Phocaena sp., (odontocete)	
Delphinapterus sp., (odontocete)	
Hyperodon sp., (odontocete)	
Harderian gland? (possibly nictitans gland)	
Histol	Pütter (1903)
Platanista gangetica (Ganges dolphin)	
Pontoporia sp. (La Plata dolphin)	
Harderian gland? (possibly nictitans gland)	
Histol	Sakai (1981)
Carnivora	
(Fissipedia)	
Canis lupus familiaris (house dog)	
Felis catus (house cat)	
No Harderian gland	
(Absent)	Loewenthal (1892b, 1896)
(Absent)	Paule (1957)
(Absent)	Shinoda (1958)
Nictitans gland	
Histol	Loewenthal (1892b, 1896)
Anat	Shinoda (1958)
(Pinnipedia)	
Odobenus sp., (walrus)	
Phoca sp., (seal)	
Harderian gland? (possibly nictitans gland)	
Present	Pütter (1903)
Sirenia	
Dugong dugong (dugong)	
Harderian gland? (possibly nictitans gland)	
Histol	Sakai (1981)
Perisodactyla	
Equus cavarus (horse)	
No Harderian gland	
(Absent)	Loewenthal (1896)
Nictitans gland	
Histol	Loewenthal (1896)

Table 1 (*continued*)

Artiodactyla	
Sus scropha (pig)	
Harderian gland	
Histol	Loewenthal (1896)
Histol	Miessner (1900)
Anat, histol, histochem (sugar, lipid), cytol	Kühnel (1974)
Nictitans gland	
Histol	Loewenthal (1892b, 1896)
Bos primigenius (house cow)	
Ovis ammon (house sheep)	
No Harderian gland	
(Absent)	Loewenthal (1892b, 1896)
Nictitans gland	
Histol	Loewenthal (1892b, 1896)
Cervus elaphus (red deer)	
Dama dama (fallow deer)	
Harderian gland	
Anat, histol	Miessner (1900)

[a] The field "description" describes type of description in literature: anat, macroscopic anatomy; histol, light microscopic anatomy; histochem, histochemistry (sugar, reaction for sugar residues such as PAS; lipid, reaction for lipids such as Sudan black); cytol, electron microscopy; present, (absent) mention of presence or absence.

furthermore, Harderian and nictitans tissues may be found mixed in one gland (*Petauroides volans*), as in the Harderian gland of the hedgehog.

In insectivores, the medial ocular glands are found in hedgehogs and shrews, and not in moles. Based on the available descriptions, the glands are found to be the Harderian gland. Interestingly, the Harderian gland in hedgehogs contains two tissues: Harderian tissue with wide lumina and nictitans tissue with narrow lumina (Loewenthal 1892a). The Harderian gland in the big-clawed shrew and the musk shrew (Sakai 1981, 1989) is elongated caudally even beyond the ear (Fig. 1).

In chiropterans and primates, no medial ocular glands have been reported.

In edentates, Weaker (1981) reported that the glands in the nine-banded armadillo contain both Harderian and nictitans tissues and thus resemble the Harderian glands of the hedgehog in this respect.

The Harderian gland is best developed both in lagomorphs and rodents. All the species so far investigated in these two orders possess large Harderian glands located behind the eyeball. Nictitans glands have been reported in the rabbit and the gerbil, but not in the other species. Macroscopic anatomy of the Harderian gland is discussed in detail later in this chapter, microscopic anatomy in other chapters of this book.

Sea mammals including cetaceans, pinnipeds, and dugongs possess well-developed ocular glands with very wide lumina in both canthi of the eye. The histological appearance of these glands in conventional paraffin sections is very

Table 2. The occurrence of the medial ocular glands and their tissue type among mammals

	Gland[a]	Type of tissue[b]
Monotremata	+	?
Marsupialia	+ / −	Harderian or nictitans
Insectivora	+ / −	Harderian
Dermoptera	?	
Chiroptera	−	
Primates	−	
Edentata	+	Harderian with nictitans
Pholidota	?	
Lagomorpha	+	Harderian (and nictitans)
Rodentia	+	Harderian (and nictitans)
Cetacea	+	Harderian? (possibly nictitans)
Carnivora		
[Fissipedia]	+	Nictitans
[Pinnipedia]	+	Harderian? (possibly nictitans)
Tubulidentata	?	
Proboscidea	?	
Hyracoidea	?	
Sirenia	+	Harderian? (possibly nictitans)
Perissodactyla	+	Nictitans
Artiodactyla	+	Harderian and/or nictitans

[a] +, and − indicate presence and absence of the medial ocular glands, respectively. + / − means that the gland is present in some species and absent in others; ?, unknown.
[b] "Harderian" and "nictitans" indicate lipid-secreting and glycoprotein-secreting tissue in the medial ocular gland, respectively. Descriptions in parentheses concern properties in some species.

similar to that of typical Harderian glands in rodents. However, no histochemical studies have been done on these glands. Observations of Waller and Harrison (1978), who reported positive reactions for sugar residues of eyelid glands of whales may indicate that the medial ocular glands in sea mammals are the nictitans gland rather than the Harderian gland.

The medial ocular glands in fissiped carnivores such as dogs and cats are the nictitans gland because of their histochemical reactions as reported by Loewenthal (1892b, 1896) and Paule (1957). Those of the perissodactyls such as horses are also the nictitans gland, judged from their histological structure (Loewenthal 1892b, 1896).

The original description of the Harderian gland by Harder (1694) was made in artiodactyls. However, anatomical studies on their Harderian glands are rare except for those of Loewenthal (1892b, 1896), Miessner (1900), and Kühnel (1974). According to these studies, the glands in deer appear to be the Harderian gland and those in the calf and sheep, the nictitans gland. The pig possesses apparently both Harderian and nictitans glands.

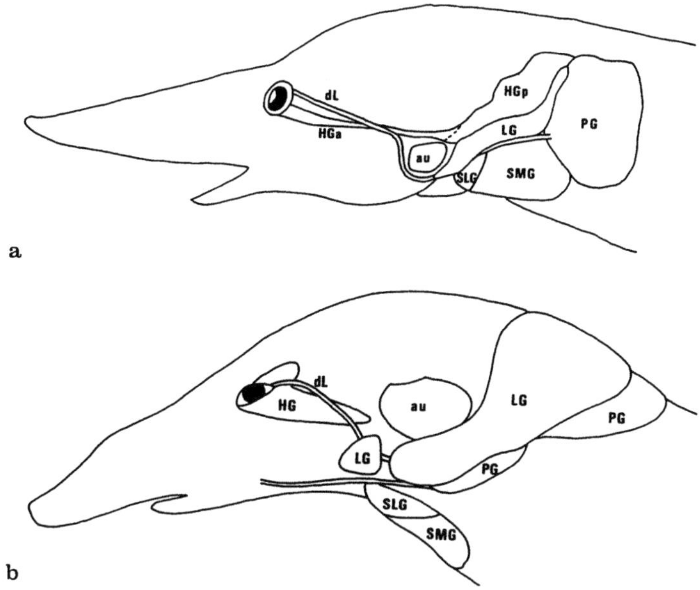

Fig. 1a,b. Schematic drawings of the head of two insectivores showing the major ocular and salivary glands. **a** Musk shrew, *Suncus murinus* and **b** big-clawed shrew, *Sorex unguiculatus*. Abbreviations: *au* auricle; *dL* lacrimal duct; *HG*, *HGa*, *HGp* Harderian gland and its anterior and posterior portions; *LG* lacrimal gland; *PG* parotid gland; *SLG* sublingual gland; *SMG* submandibular gland

3 Macroscopic Anatomy of the Mammalian Harderian Gland

3.1 Location and Shape of the Harderian Gland

Macroscopic anatomy of the Harderian gland is very variable among mammals so far investigated. The Harderian gland is best developed in rodents and lagomorphs. In these animals, the gland may be larger than the eyeball and located behind it. These animals also possess orbital venous sinuses which cover all the orbital contents including the Harderian gland (Figs. 2, 3). The significance of the orbital sinus is discussed later.

In insectivores such as the big-clawed shrew and the musk shrew, the Harderian gland is very elongated posteriorly. In these animals the orbital venous sinus is absent (Figs. 1–3).

3.2 Duct of the Harderian Gland

Loewenthal (1896), who studied the Harderian gland in several species, argued that each Harderian gland has a single duct while the nictitans gland has several. Miessner (1900) stated that the duct of the Harderian gland in deer bifurcates to make two openings. Shinoda (1958) examined the macroscopic structure of the

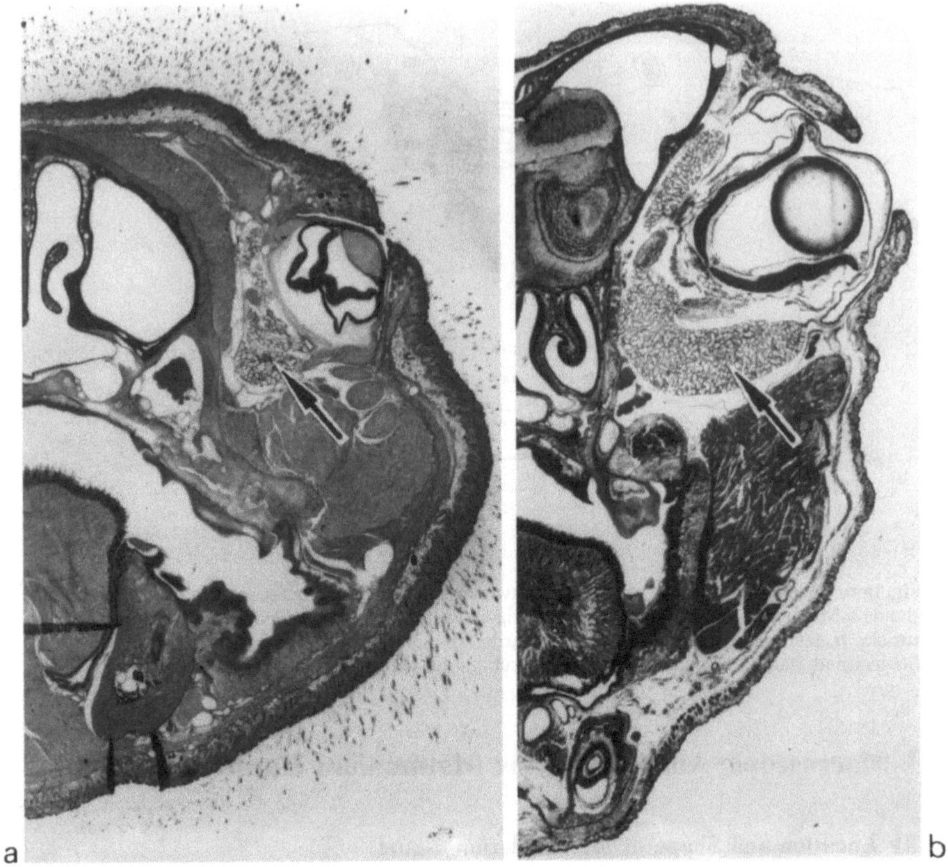

Fig. 2a,b. Frontal sections of the head of the musk shrew (**a**), *Suncus murinus* and the mouse (**b**), *Mus musculus*. The eye of the musk shrew is poorly developed compared with that of the mouse. The Harderian glands (*arrows*) are much smaller in the musk shrew than in the mouse since a major part of the musk shrew Harderian gland is located posteriorly to the eye

Harderian and nictitans glands in cats, dogs, rabbits, and mice, and confirmed fundamentally the observations of Loewenthal (1896). According to the author's own observations, the Harderian gland in insectivores, lagomorphs, and rodents has a single duct. Ductal anatomy of the Harderian and nictitans gland in other species is unknown.

The eye of sea mammals such as whales, pinnipeds, and sirenians is reported to possess Harderian glands with multiple excretory ducts (Pütter 1903). Histologically, these glands appear to be Harderian glands because of their wide glandular lumina. However, whether these glands are actually the Harderian or the nictitans glands remains to be determined by means of histochemical analysis of these glands, as mentioned above.

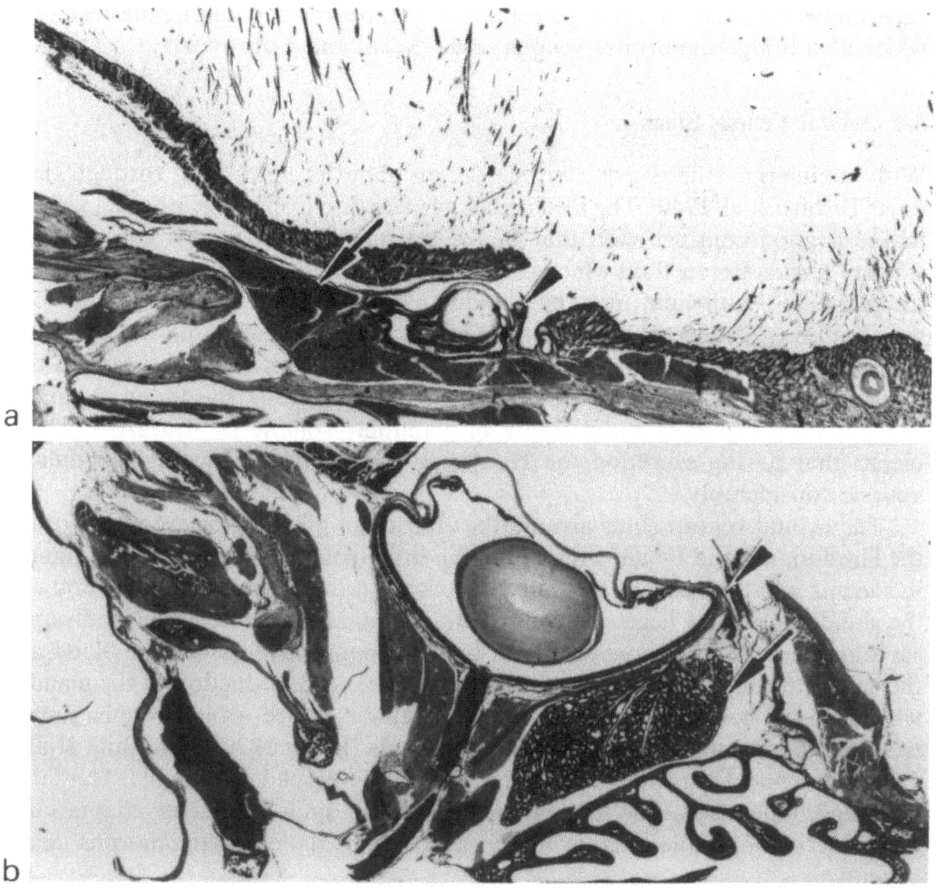

Fig. 3a,b. Horizontal sections of the head of the big-clawed shrew (**a**), *Sorex unguiculatus* and the Mongolian gerbil (**b**), *Meriones unguiculatus*. The eye of the big-clawed shrew is very poorly developed compared with that of the Mongolian gerbil. The Harderian glands (*arrows*) are elongated posteriorly in the big-clawed shrew and located behind the eye in the Mongolian gerbil. The third eyelid is indicated by *arrowheads*

3.3 Innervation

Innervation of the Harderian gland has been studied mainly in rats, hamsters, and rabbits (Bucana and Nadakavukaren 1972a; Norvell and Clabough 1972; Huhtala et al. 1977; Butler et al. 1984; Tsukahara and Jacobowitz 1987; Vega-Alvarez et al. 1989). Autonomic nerve varicosities have been observed in the interstitium of these glands. The parasympathetic nerve fibers come from the facial nerve, change their neuron in the pterygopalatine ganglion (Butler et al. 1984), and enter the gland from behind through fine strands bridging the orbital venous sinus (Sakai and Yohro 1981). The sympathetic fibers coming from the superior cervical ganglion (Pangerl et al. 1989) appear to enter along the arteries, which also bridge the orbital venous sinus (Sakai and Yohro 1981).

3.4 Orbital Venous Sinus

With cholinergic stimulation, the Harderian gland excretes lipid through the duct (Tashiro et al. 1940). The mechanism of excretion from the Harderian gland makes a good contrast with that of the salivary glands. Serous cells of the salivary glands secrete fluid which is transported from the capillary blood across the glandular epithelium into the glandular lumina. Therefore, the volume of secretion may exceed the mass of the salivary gland in a short period. On the other hand, the Harderian gland squeezes its luminal content by contraction of myoepithelial cells. Thus, the volume of excretion from the Harderian gland is thought to be equal to the decrease of glandular volume by squeezing. This means that during excretion the Harderian gland must reduce its glandular volume considerably.

The orbital venous sinus around the Harderian gland helps excretion from the Harderian gland by allowing reduction in its volume. The Harderian gland in rodents and lagomorphs is located deep behind the eyeball. The reduction of the glandular volume necessary for excretion would be difficult if the gland were surrounded not by a venous sinus but by usual connective tissue. The blood in the sinus can easily change its volume to allow volume reduction of the gland, whereas usual connective tissue cannot. A well-developed orbital venous sinus has been found only in lagomorphs and rodents (Davis 1929; Yamashita et al. 1980; Sakai and Yohro 1981), which have a well-developed retro-ocular Harderian gland. One may argue that the orbital venous sinus represents adaptation for excretion by squeezing from the Harderian gland in lagomorphs and rodents.

4 Comparative Anatomy of the Harderian Gland

4.1 Mammalian Harderian Gland

In comparative anatomy of the exocrine glands, the location of the ductal opening is far more important than that of the gland itself. While in most exocrine glands including the Harderian, lacrimal, and salivary glands, the location of the ductal opening is almost constant among species for each gland, its glandular location is quite variable, as is seen in the Harderian gland of insectivores and rodents (Fig. 1). Fahrenholz (1937) argued that the constancy and variation of ductal opening result from developmental processes. The ductal opening is thought to represent the developmental origin of its primordium, whereas the glandular location may change dramatically during the course of development.

Regarding the importance of the ductal opening, division of the medial and lateral ocular glands appears reasonable. The lateral ocular glands are further divided anatomically into superior and inferior groups (glandula lacrimalis superior and inferior) on the basis of their ductal anatomy (Sakai 1989). Glands in the superior group have multiple ducts opening into the superior conjuctival sac, and are found in larger mammals, whereas those in the inferior group have single ducts opening at the outer edge of the conjunctival sac, and are found in smaller mammals. Thus, distinction based on ductal anatomy is valid also in the lacrimal glands.

On the other hand, division of the mammalian medial ocular glands into the Harderian and nictitans glands is mainly based on the histological and histochemical properties and not on the ductal anatomy. Loewenthal (1896) and Shinoda (1958) studied the medial ocular glands of several species and argued that the Harderian gland has a single duct and the nictitans gland multiple ones. However, observation of the marsupial Harderian glands contradicts this generalization. The medial ocular glands in the genus *Antechinus* are quite similar in macroscopic appearance, but are either the Harderian or the nictitans gland depending upon species. Furthermore, the gland in the marsupial greater glider, as well as those in the hedgehog, nine-banded armadillo, and Mongolian gerbil, may contain both Harderian and nictitans tissues in the same gland. These observations indicate that the lipid-secreting tissue which characterizes the Harderian gland has no strict correlation with the ductal anatomy.

4.2 Harderian Gland in Mammals and Lower Vertebrates

The eye of land vertebrates is furnished with exocrine glands to lubricate the eyelids. The glandular tissue is generally well developed at the inner and the outer canthus, forming a medial and a lateral ocular gland. As mentioned above, the medial ocular glands are called the Harderian glands in lower vertebrates and either the Harderian or the nictitans glands in mammals, whereas the lateral ocular glands are called the lacrimal glands.

The medial and the lateral ocular glands exhibit unique histological specializations in individual classes. In marine turtles, the Harderian gland is mucous-secreting, and the lacrimal gland serves as a salt gland to excrete excess salt. The Harderian glands of lizards contain two portions with different types of secretory cells, whereas the lacrimal gland has only one cell type (Schwarz-Karsten 1937; Baccari et al. 1990). The Harderian gland in birds contains numerous plasma cells and is regarded as an accessory immune organ.

The Harderian gland in mammals is characterized by its histological structure with lipid-secreting epithelial cells. The histological differentiation of the medial ocular glands into the Harderian gland appears to be a secondary feature, considering the distribution of Harderian gland tissue among mammals. Harderian glands of the greater glider (marsupial), of the hedgehog (insectivore), the nine-banded armadillo (edentate), of the Mongolian gerbil (rodent) contain both Harderian and nictitans tissues (Loewenthal 1892b, 1896; Weaker 1981; Sakai and van Lennep 1984). Furthermore, the medial ocular glands of the marsupial genus *Antechinus* vary in their histological structure depending upon species, being Harderian in *A. stuartii* and *swainsonii*, and nictitans in *A. flavipes* (Sakai and van Lennep 1984).

As regards the Harderian gland tissue in the outer ocular glands, Sashima and co-workers (1989) reported that the rat lacrimal gland may contain lipid-secreting tissue which appears similar to that of the Harderian gland. They called this phenomena "Harderianization" after Coujard and Coujard (1974). Furthermore, demonstration of porphyrin in the lacrimal gland (Hugo et al. 1989) may indicate additional similarities between the lacrimal and the Harderian gland tissues. These reports indicate that the glandular tissues of both the medial and lateral ocular glands have a potential to develop tissue of Harderian gland type with lipid and porphyrin secretion.

References

Abe J, Sugita A, Katsume Y, Yoshizuka M, Tamura N, Iwanaga S, Nishida T (1980) Scanning electron microscopic observations of the Harderian gland in rat. Kurume Med J 27:239–246

Abe J, Sugita A, Hamasaki M, Nakamura K, Iwanaga S, Nagae K, Atsuji K, Tsunawaki A, Abe T, Matsumoto T, Yo S, Murakami M (1981) Scanning electron microscopic observations of the myoepithelial cells of normal and contracting status in the rat Harderian gland. Kurume Med J 28:103–112

Akagi N, Morimoto T, Watanabe M (1987) The Harderian gland and the porphyrins. Bull Osaka Med Sch 33:97–115

Akiyama K (1961) Anatomical study of the orbit and its contents in albino rat (in Japanese). Acta Soc Ophthalmol Jpn 65:1718–1735

Baccari G, Minucci S, Di Matteo L, Chieffi G (1990) Harderian gland and the lacrimal gland of the lizard *Podarcis s. sicula*: histology, histochemistry, and ultrastructure. Anat Rec 226:269–278

Björkman B, Nicander L, Schantz B (1960) On the histology and ultrastructure of the Harderian gland in rabbits. Z Zellforsch Mikrosk Anat 52:93–104

Brownscheidle CM, Niewenhuis RJ (1978) Ultrastructure of the Harderian gland in male albino rats. Anat Rec 190:735–753

Bucana CD, Nadakavukaren MJ (1972a) Innervation of the hamster Harderian gland. Science 175:205–206

Bucana CD, Nadakavukaren MJ (1972b) Fine structure of the hamster Harderian gland. Z Zellforsch Mikrosk Anat 129:178–187

Bucana CD, Nadakavukaren MJ (1972c) Ultrastructural investigation of the postnatal development of the hamster Harderian gland. I. Cell division. Z Zellforsch Mikrosk Anat 135:149–153

Bucana CD, Nadakavukaren MJ (1973) Ultrastructural investigation of the postnatal development of the hamster Harderian gland. II. Male and female. Z Zellforsch Mikrosk Anat 142:1–12

Buschke W (1933) Die Hautdrüsenorgane (Hardersche Drüsen, Inguinaldrüsen, Präputialdrüsen, Analdrüsen, Kaudaldrüsen, Kieferdrüsen) der Laboratoriumsnagetiere und die Frage ihrer Abhängigkeit von den Geschlechtsdrüsen. Z Zellforsch Mikrosk Anat 18:217–243

Butler JM, Ruskell GL, Cole DF, Unger WG, Zhang SQ, Blank MA, McGregor GP, Bloom SR (1984) Effects of VIIth (facial) nerve degeneration on vasoactive intestinal polypeptide and substance P levels in ocular and orbital tissues of the rabbit. Exp Eye Res 39:523–532

Coujard M (1974) La ⟨Hardérisation⟩ de la glande de Loewenthal chez le Rat blanc. C R Seances Soc Biol 168:231–233

Davis FA (1929) The anatomy and histology of the eye and orbit of the rabbit. Trans Am Ophthalmol Soc 27:401–441

Fahrenholz C (1937) Drüsen der Mundhöhle. In: Bolk L, Göppert E, Kallius E, Lubosch W (eds) Handbuch der vergleichenden Anatomie der Wirbeltiere vol 3. Urban & Schwarzenberg, Berlin, pp 115–210

Griffiths M (1978) The biology of the monotremes. Academic Press, New York

Harder JJ (1694) Glandula nova lachrymalis una cum ductu excretorio in cervis et damis. Acta-Eruditorium-Lipsiae 1694:49–52

Harriman AE, Thiessen DD (1985) Harderian letdown in male Mongolian gerbils (*Meriones unguiculatus*) contributes to proceptive behavior. Horm Behav 19:213–219

Hugo J, Vokuraka M, Sanitrák J, Melková Z, Janousek V (1989) First demonstration of porphyrins in the lacrimal glands: similarity to the Harderian gland. Biomed Biochim Acta 48:K1–K4

Huhtala A, Huikuri KT, Palkama A, Tervo T (1977) Innervation of the rat Harderian gland by adrenergic and cholinergic nerve fibres. Anat Rec 188:263–271

Jackson AHD, Kenner GW, Smith KM, Suckling CJ (1976) Pyrroles and related compounds – XXXVII Harderoporphyrin. Tetrahedron 32:2757–2762

Johnston HS, McGadey J, Thompson GG, Moore MR, Payne AP (1983) The Harderian gland, its secretory duct and porphyrin content in the Mongolian gerbil (*Meriones unguiculatus*). J Anat 137:615–630

Johnston HS, McGadey J, Thompson GG, Moore MR, Breed WG, Payne AP (1985) The Harderian gland, its secretory duct and porphyrin content in the Plains mouse (*Pseudomys australis*). J Anat 140:337–350

Johnston HS, McGadey J, Payne AP, Thompson GG, Moore MR (1987) The Harderian gland, its secretory duct and porphyrin content in the woodmouse (*Apodemus sylvaticus*). J. Anat 153:17–30

Johnston RE (1986) Effects of female odors on the sexual behavior of male hamsters. Behav Neural Biol 46:168–188

Jost UL, Kühnel W, Schimassek H (1974) A morphological and histochemical analysis of the Harderian gland in the rabbit. Cytobiologie 8:440–456

Kaiho M, Ichikawa A (1982) An application of the ferrocyanide-reduced osmium tetroxide for the ultrastructural analysis of gerbil, *Meriones unguiculatus* Harderian gland. J Electron Microsc 31:410–414

Kanwar KC (1960) Morphological and cytochemical studies on the Harderian glands of rats. Cellule 61:129–143

Kelényi G, Orbán S (1964) Electron microscopy of the Harderian gland of the rat: maturation of the acinar cells and genesis of the secretory droplets. Acta Morphol Acad Sci Hung 13:155–166

Kühnel W (1971) Struktur und Cytochemie der Harderschen Drüse von Kaninchen. Z Zellforsch Mikrosk Anat 119:384–404

Kühnel W (1974) Zur funktionellen Morphologie der Harderschen Drüse des Schweins (*Sus scropha* L.). Mit Bemerkungen zur Frage ihrer Innervation. Berl Münch Tieraurztl Wochenschr 87:9–15, 21–25

Kühnel W, Wrobel KH (1968) Histochemische Studien an der Harderschen Drüse des Kaninchens. Morphol Jahrb 111:493–500

Loewenthal N (1892a) Notiz über die Harder'schen Drüse des Igels. Anat Anz 7:48–54

Loewenthal N (1892b) Beitrag zur Kenntnis der Harder'schen Drüse bei den Säugetieren. Anat Anz 7:546–556

Loewenthal N (1896) Drüsenstudien. I. Die Harder'schen Drüse. Int Mschr Anat Physiol 13:27–65

Michael MI, Khalil SH, Matta CA, Rizk TA (1988) Normal development of the ocular glands of the mouse. Folia Morphol (Prague) 36:47–52

Miessner H (1900) Die Drüsen des dritten Augenlides einiger Säugetiere. Arch Wiss Prakt Tierheilkd 26:122–154

Müller HB (1969a) Die postnatale Entwicklung der Harderschen Drüse der weissen Ratte. I. Lichtmikroskopische Befunde. Z Zellforsch Mikrosk Anat 100:421–438

Müller HB (1969b) Die postnatale Entwicklung der Harderschen Drüse der weissen Ratte. II. Fermenthistochemische Befunde. Histochemie 20:181–196

Müller HB (1970) Untersuchungen zur postnatalen Entwicklung der Harderschen Drüse der Ratte. Verh Anat Ges 64:503–509

Nadakavukaren MJ, Lin WL (1983) Influence of short photoperiods on the ultrastructure of hamster Harderian glands. Cell Tissue Res 233:209–213

Norvell JE, Clabough JW (1972) Adrenergic and cholinergic innervation of the hamster Harderian gland. Science 178:1102–1103

Otsuru O, Otsuka H, Kasama T, Seyama Y, Sakai T, Yohro T (1983) The characterization of 2,3-alkanediol diacyl esters obtained from the Harderian glands of Mongolian gerbil (*Meriones unguiculatus*). J Biochem 94:2049–2054

Pangerl A, Pangerl B, Buzzell GR, Jones DJ, Reiter RJ (1989) Characterization of beta-adrenoceptors in the Syrian hamster Harderian gland: sexual differences and effects of either castration or superior cervical ganglionectomy. J Neurosci Res 22:456–460

Paule WJ (1957) The comparative histochemistry of the Harderian gland. Dissertation, Ohio State University, Columbus, OH

Payne AP (1977) Pheromonal effects of Harderian gland homogenates on aggressive behaviour in the hamster. J Endocrinol 73:191–192

Payne AP (1979) The attractiveness of Harderian gland smears to sexually naive and experienced male golden hamsters. Anim Behav 27:897–904

Payne AP, McGadey J, Johnston HS, Moore MR, Thompson GG (1982) Mast cells in the hamster Harderian gland: sex differences, hormonal control and relationship to porphyrin. J Anat 135:451–461

Payne AP, McGadey J, Johnston HS (1985) Interstitial porphyrins and tubule degeneration in the hamster Harderian gland. J Anat 140:25–36

Pévet P, Heth G, Hiam A, Nevo E (1984) Photoperiod perception in the blind mole rat (*Spalax ehrenbergi*, Nehring): involvement of the Harderian gland, atrophied eyes, and melatonin. J Exp Zool 232:41–50

Prince JH (1956) Comparative anatomy of the eye. Charles C Thomas, Springfield, pp 298–305

Pütter A (1903) Die Augen der Wassersäugethiere. Zool Jahrb Abt Anat Ontogen 17:99–402

Sakai T (1981) The mammalian Harderian gland: morphology, biochemistry, function and phylogeny. Arch Histol Jpn 44:299–333

Sakai T (1989) Major ocular glands (Harderian gland and lacrimal gland) of the musk shrew (*Suncus murinus*) with a review on the comparative anatomy and histology of the mammalian lacrimal glands. J Morphol 201:39–57

Sakai T, van Lennep EW (1984) The Harderian gland in Australian marsupials. J Mammal 65:159–162

Sakai T, Yohro T (1981) A histological study of the Harderian gland of Mongolian gerbils, *Meriones meridianus*. Anat Rec 200:259–270

Sashima M, Hatakeyama S, Satoh M, Suzuki A (1989) Harderianization is another sexual dimorphism of rat exorbital lacrimal gland. Acta Anat 135:303–306

Schwarz-Karsten H (1937) Über die Orbitaldrüsen von *Lacerta agilis, Lacerta muralis, Ophisops elegans, Tarentola mauretanica* und *Tropidonotus natrix*. Morphol Jahrb 80:248–279

Shinoda S (1958) Harder's gland in some mammals. Igaku Kenkyu (Acta Medica) 28:4623–4635 (in Japanese)

Shirama K, Furuya T, Takeo Y, Shimizu K, Maekawa K (1979) Neural control of the Harderian gland in the rat. Zool Mag 88:149–154 (in Japanese)

Shirama K, Harada T, Kohda M, Hokano M (1988) Fine structure of melanocytes and macrophages in the Harderian gland of the mouse. Acta Anat 131:192–199

Sun CY, Nadakavukaren MJ (1980) Effect of testosterone on the female hamster Harderian gland pigmentation and ultrastructure. Cell Tissue Res 207:511–517

Tashiro S, Smith CC, Badgen E, Kazur E (1940) Chromodacryorrhea, a new criterion for biological assay of acetylcholine. Proc Sci Exp Biol Med 44:658–661

Thiessen DD (1988) Body temperature and grooming in the Mongolian gerbil. Ann NY Acad Sci 525:27–39

Thiessen DD, Kitterell EMW (1980) The Harderian gland and thermoregulation in the gerbil (*Meriones unguiculatus*). Physiol Behav 24:417–424

Thiessen DD, Graham M, Perkins J, Marcks S (1977) Temperature regulation and social grooming in the Mongolian gerbil (*Meriones unguiculatus*). Behav Biol 19:279–288

Tsukahara S, Jacobowitz DM (1987) Peptidergic innervation of the rat Harderian gland. Histochemistry 87:233–236

Tsutsumi A, Iwata K, Ogawa K, Matsuura K (1966) Histochemical and electron microscopic observations on the Harderian gland of the albino rat. Arch Histol Jpn 27:553–567 (in Japanese)
Vega-Alvarez JA, Menendez-Pelaez A, Coalla-Perez C, Martinez-Telleria A (1989) Sexual dimorphism in the cholinergic innervation of the hamster (*Mesocricetus auratus*) Harderian glands. Z Mikrosk Anat Forsch 103:166–172
Vianna GF, Cruz AR, Azoubel R (1975a) Mitotic activity and nuclear/cytoplasmatic ratio of the lachrymal and Harderian glands of the rat during postnatal life. Acta Anat 92:1–7
Vianna GF, Cruz AR, Azoubel R (1975b) Allometric study of the lachrymal and Harderian glands of the rat during postnatal life. Acta Anat 92:161–170
Waller GH, Harrison RJ (1978) The significance of eyelid glands in delphinids. Aquat Mam 6:1–9
Walter A (1925) Über die Hautdrüsen mit Lipoidsekretion bei Nagern. Beitr Pathol Anat 73:142–167
Weaker FJ (1981) Light microscopic and ultrastructural features of the Harderian gland of the nine-banded armadillo. J Anat 133:49–65
Wetterberg L, Geller E, Yuwiler A (1970) Harderian gland: an extraretinal photoreceptor influencing the pineal gland in neonatal rats? Science 167:884–885
Wiedersheim R (1909) Vergleichende Anatomie der Wirbeltiere. Gustav Fischer, Jena
Winterhager E, Kühnel W (1983) Membrane specializations of the cells of the Harderian gland of the rabbit with particular reference to the mechanism of exocytosis. Cell Tissue Res 231:623–636
Woodhouse MA, Rhodin JAG (1963) The ultrastructure of the Harderian gland of the mouse with particular reference to the formation of its secretory product. J Ultrastruct Res 9:76–98
Wooding FBP (1980) Lipid droplet secretion by the rabbit Harderian gland. J Ultrastruct Res 71:68–78
Yamashita T, Takahashi A, Honjin R (1980) The spatial aspect of the mouse orbital venous sinus. Okajimas Folia Anat Jpn 56:329–336

Harderian Glands and Their Development in Laboratory Rats and Mice

KAZUHIKO SHIRAMA and MASAMI HOKANO

Department of Anatomy, Tokyo Medical College, Shinjuku-ku, Shinjuku, Tokyo 160, Japan

1 Introduction

The Harderian glands of the rat and mouse lie, for the most part, deep within the orbit. However, the external rim of the gland extends forward into the space between the bulbus oculi and the orbital rim, and can be seen as soon as the palpebral fascia has been removed from its orbital attachment. The gland consists of two portions, a smaller lobule occupying the upper portion of the eye socket, and a lower, larger and more diffuse lobule occupying the floor of the optic cavity. As there is no marked histological difference between the lobules, they can be considered together as a single entity. The glands of rats and mice are characterized by the production of porphyrins and its accumulation within the glandular lumina. The function of the gland remains unknown, although it is usually regarded as a source of lubrication and protection for the cornea. The purpose of this section is to summarize what has been reported to date on the Harderian glands of rats and mice.

2 Histological Organization

The Harderian gland contains several lobules which vary from one species to the next. In rats and mice, the lobules form a single excretory duct which originates at the hilus through the loose pedicular connective tissue. The duct, which generally opens on the outer side of the third eyelid, is seldom seen in sections of the gland because the gland contains no branched duct system (Sakai and Yohro 1981). The duct is separated by interlobular connective tissue. Previously, some authors considered the Harderian gland to be a compound tubuloalveolar gland. However, it is now considered to be a simple branched tubuloalveolar gland in rats and mice.

The basic morphology of the rat and mouse Harderian gland is that of a typical exocrine gland with alveoli which lead to tubules. The terminal portions of the gland are composed of alveoli separated by small amounts of

Webb et al. (Eds).
Harderian Glands

loose connective tissue located in the interstices of the gland. These alveoli are lined with acinar cells. Each alveolous contains one or two types of secretory and myoepithelial cells (Fig. 1).

2.1 Secretory Epithelium

Light microscopic observations on the Harderian gland have been reported in both rats (Duthie 1934; Grafflin 1942; Arvy 1959; Kanwar 1960; Rapin 1962; Tilgner 1968; Müller 1969a b, 1970) and mice (Derrien and Turchini 1924; Cohn 1955; Kittel 1972). In rats, two types of secretory cells (Types A and B) comprise the glandular epithelium (Grafflin 1942; Kanwar 1960; Rapin 1962; Müller 1969a, b, 1970). The type A cell contains large cytoplasmic vacuoles and has very little intervacuolar cytoplasm whereas the type B cell contains small unstained vacuoles and larger amounts of basophilic intervacuolar cytoplasm. The number and size of vacuoles are greater in type A cells and therefore these cells are more numerous than the type B in most terminal portions of the gland. In mice,

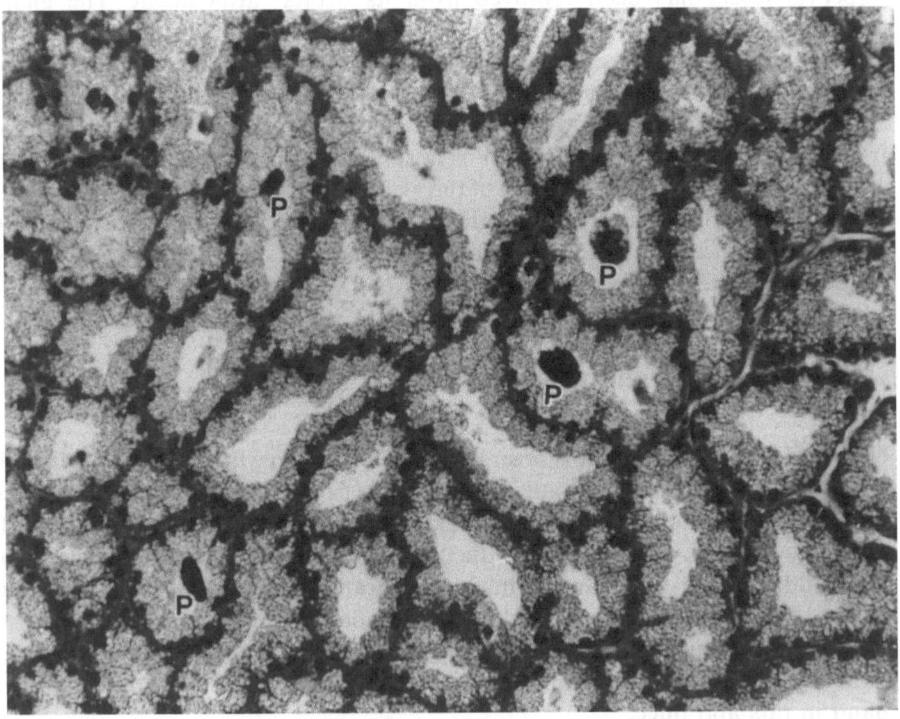

Fig. 1. Light micrograph of the Harderian gland from adult female mouse, showing the alveoli. Some of the alveoli contain intraluminal accretions of porphyrin (porphyrin-lipid complex: *P*). H & E, ×320

it has been difficult to distinguish between the two types of secretory cells in the terminal portions under light microscopy, but the two cell types can easily be recognized using the electron microscope (Woodhouse and Rhodin 1963; Watanabe 1980; Strum and Shear 1982a, b; Fig. 2).

In both rat and mouse, pigment is found in some of the glandular lumina but is not observed within the grandular epithelium. The quantity of pigments within the lumina of the terminal portions varies considerably. In some terminal portions, the lumina contain a few pale gold-yellow strands of pigment; other lumina contain large masses of reddish brown pigment, whereas in some terminal portions, pigment is absent. The pigment is thought to consist mainly of porphyrins because of their red fluorescence (Derrien and Turchini 1924) and their chemical nature (Kennedy 1970).

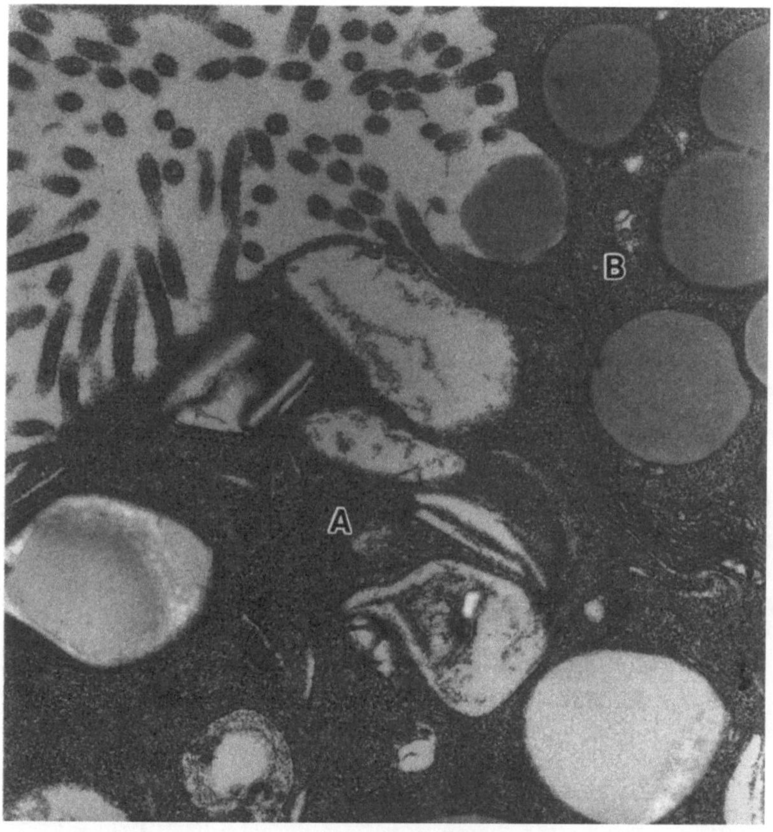

Fig. 2. Electron micrograph of the Harderian gland of mouse, showing the apical portion of two cell types, type *A* and type *B*. Note the differences in the lipid droplets. Lipid droplets from type A cells have irregular shapes and contain lipid-like materials. Most of the lipid droplets from type B cells contain densely stained materials. ×29000

Some conventional ultrastructural studies on the Harderian gland have been reported in both rats (Orbán and Kelényi 1962; Kelényi and Orbán 1984; Tsutsumi et al. 1966; Brownscheidle and Niewenhuis 1978; Abe et al. 1980; Carriere 1985) and mice (Woodhouse and Rhodin 1963; Watanabe 1980; Strum and Shear 1982b). The glandular epithelial cells in both species are pyramidal in shape and extend from the basement membrane to the lumen of an individual secretory unit. Their nuclei are located basally and nucleoli are frequently present. Binucleate cells and mitotic figures are not found in the glandular epithelium. Junctional complexes and microvilli of variable length and number are present at the apices of the cells.

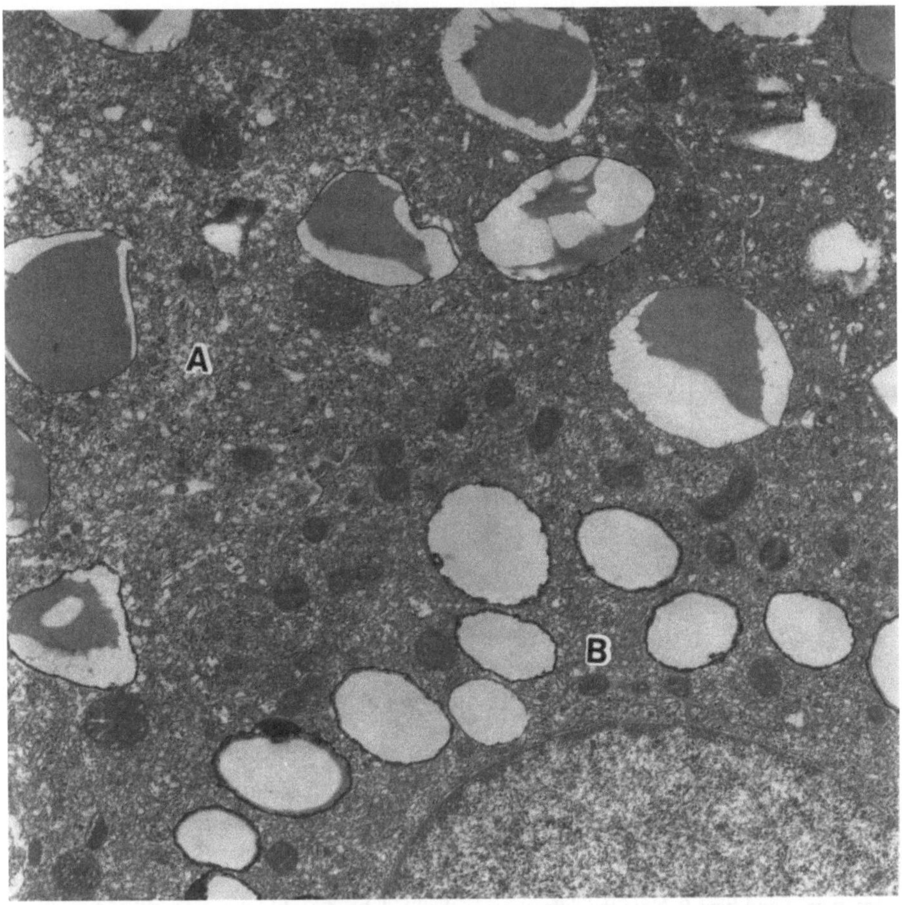

Fig. 3. Electron micrograph of the Harderian gland from adult male rat showing two cell types, type A (*A*) and type B (*B*), in the glandular epithelium. ×12000

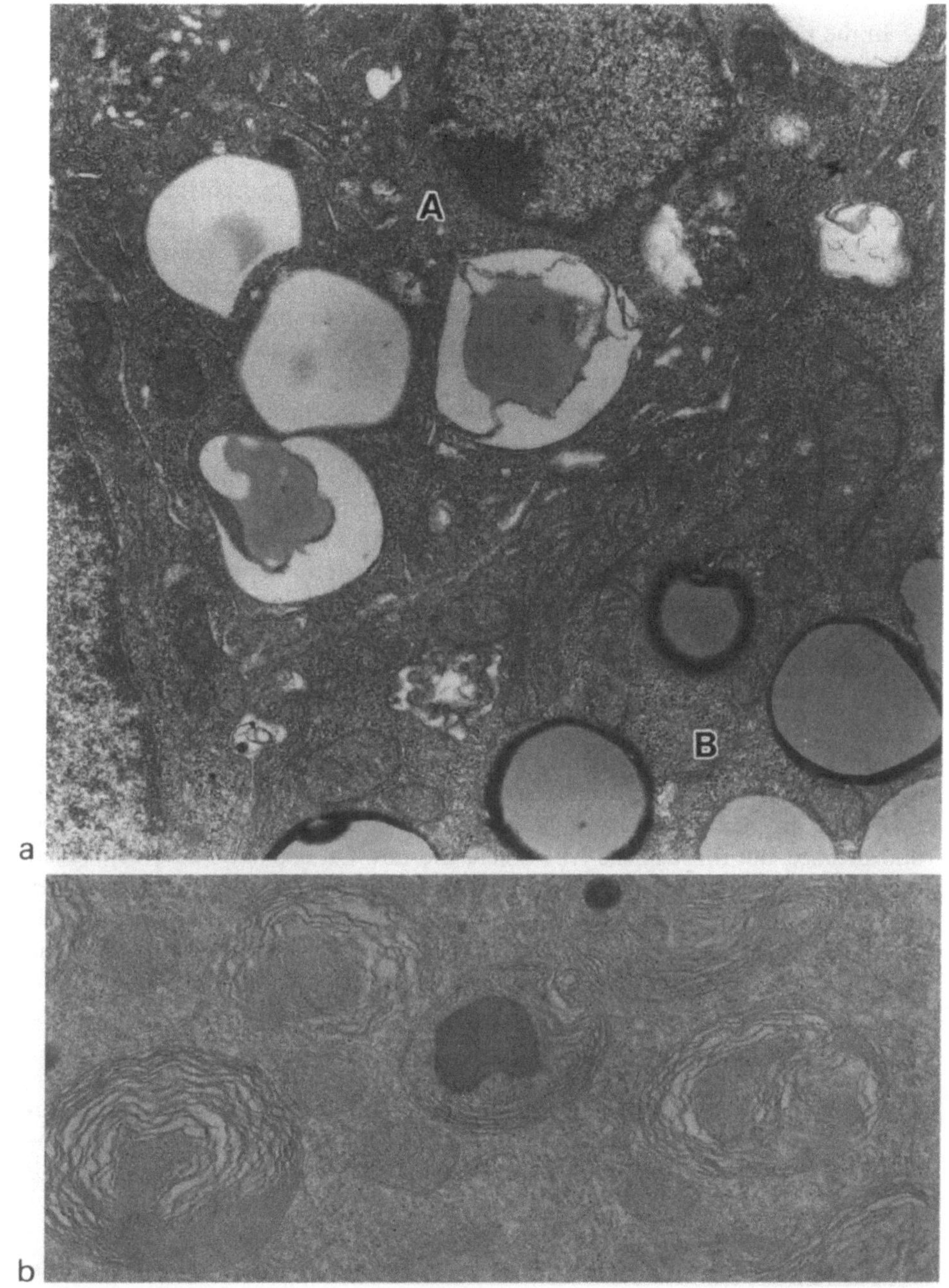

Fig. 4a. Electron micrograph of the Harderian gland of mouse, showing the basal portion of the two cell types, type *A* and type *B*. Note the differences in the lipid droplets. Lipid droplets from type A cells have irregular shapes and contain lipid-like materials. Most of the lipid droplets from type B cells contain a peripheral ring of densely stained material. **b** With unstained section, the lipid droplets appear to be lamellated membranes. ×29000

In the rat, the type A cell is characterized by the large number of secretory vacuoles present throughout most of the cytoplasm. The vacuoles have a maximum diameter of about 2.5 μm and their shape is variable. Most of the secretory vacuoles contain a lipid-like substance and granular osmophilic material (Brownscheidle and Niewenhuis 1978; Carriere 1985). Type B cells occur less frequently than type A cells (approximately 1:4). The cytoplasmic secretory vacuoles are less numerous than in those of the type A cell and smaller, with a maximum diameter of about 1.7 μm. The vacuoles of this type of cell often possess border laminations (Fig. 3). Large numbers of mitochondria, extensive endoplasmic reticulum, and a prominent Golgi apparatus are seen within the cytoplasm of this type of cell (Brownscheidle and Niewenhuis 1978).

In the mouse, the differences between A and B secretory cell types are best seen by the electron microscope (Woodhouse and Rhodin 1963; Watanabe 1980; Strum and Shear 1982b). The type A cells are more numerous than type B cells in the secretory epithelium (approximately 1:10). The type A cells contain large numbers of vacuoles which tend to aggregate at the apex of the cell. The less numerous type B cells are surrounded by densely stained materials or densely stained rings (Fig. 4a). In unstained sections, the materials appear to be lamellae composed of fibro-membranous layers (Fig. 4b). There are profiles suggesting coalescence of secretory vacuoles in both types of cells. The gland (particularly in female mice) contains intraluminar fibro-membranaous masses corresponding to the site of brown pigment accumulation seen by light microscopy (Woodhouse and Rhodin 1963; Fig. 5). However, it is still unknown whether an intraluminal component which may consist of porphyrin compounds is produced by the type A or type B cells. Strum and Shear (1982b) have demonstrated that lipid droplets of the type A cells in the gland of mice displayed a red fluorescence, but those in the type B cells did not. On this basis, they concluded that porphyrins are present in the lipid droplets of the type A cells. On the other hand, this component seems to be identical with lamellated fibrous membranes observed in the secretory vacuoles of the type B cells, suggesting that the porphyrins may also be secreted by the type B cells (Watanabe 1980).

2.2 Myoepithelial Cells

The presence of myoepithelial cells has been demonstrated in the Harderian gland of the mouse (Chiquoine 1958; Woodhouse and Rhodin 1963; Watanabe et al. 1981; Abe et al. 1981; Strum and Shear 1982b) and rat (Brownscheidle and

Fig. 5a. Light micrograph of the Harderian gland from female mouse (C3H/He), showing the pigment accumulations (*P*) in the alveolar lumina. A mast cell is seen in the interstitial tissue (*arrow*). A 1-μm-thick section stained with toluidine blue. × 320. **b** Electron micrograph of the same pigment shown in **a**. The substance is composed primarily of fibro-membranous material. Apical portion of secretory cell with microvilli (*MV*) is seen in a lower part of the picture. × 12000

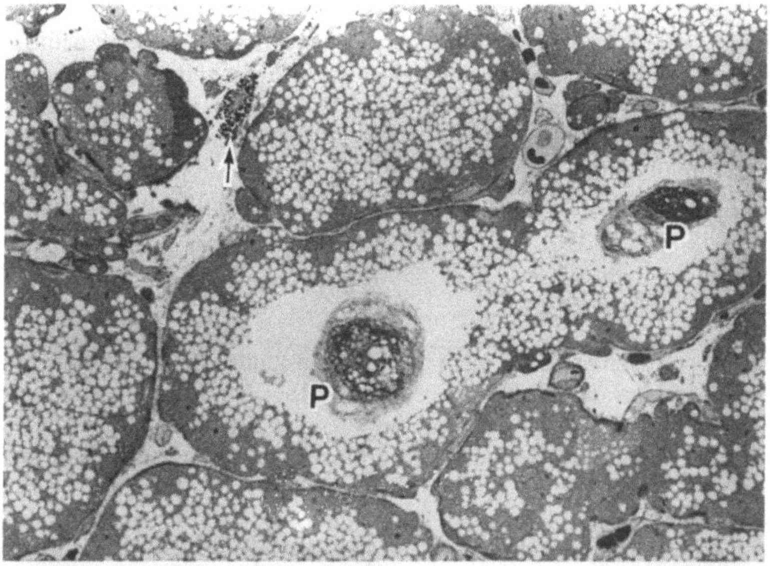
a
P
P

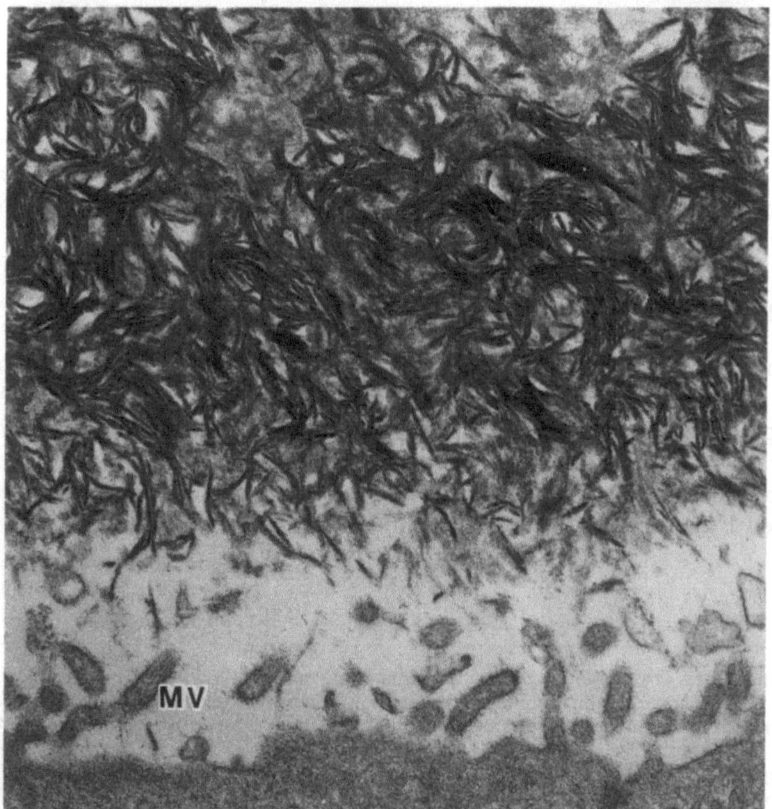
b
MV

Niewenhuis 1978). Myoepithelial cells, located between the basement membrane and secretory epithelium, are characterized by their shape and by the concentration of organelles at the poles of the nuclei (Chiquoine 1958). These cells also contain two populations of filaments. The thick filaments measured 12–17 nm in diameter and thin filaments 6–7 nm. Based on the size and appearance of similar filaments in smooth muscle cells, it is suggested that the thick filaments correspond to myosin and the thin to actin (Strum and Shear 1982b).

2.3 Interstitial Connective Tissue

The interstitial connective tissue of the Harderian gland contains small blood vessels, fibroblasts, melanocytes, mast cells, and other connective tissue cellular elements (Fig. 6a). In mice, there is marked sexual difference in the Harderian gland, with the females having a larger number of mast cells than males (Shirama et al. 1988b). Studies on the male showed that the mast cell count greatly increases after castration plus adrenalectomy, reaching levels comparable to those found in the female. This postoperative rise in mast cell numbers is completely prevented by administration of androgen (Shirama et al. 1988b). The presence of dentritic cells containing melanin granules has been demonstrated in the interstitial tissue of the mouse Harderian gland, employing a silver impregnation technique and electron microscopy (Fig. 6b). Furthermore, two types of melanocytes, either with or without the various developmental stages of melanin granules, are found in the gland (Shirama et al. 1988a). In the rat, plasma cells and eosinophils are also found in the connective tissue septae of the gland (Brownscheidle and Niewenhuis 1978).

2.4 Cytology

The grandular epithelium of the Harderian gland contains clear secretory vacuoles. It has been demonstrated histochemically that their contents are mainly lipids. The mode of secretion had been thought to be of the apocrine type (Cohn 1955; Tsutsumi et al. 1966) based on the light microscopic observations. In electron microscopic observations, the secretory vacuoles were observed to be released by exocytosis (Woodhouse and Rhodin 1963; Brownscheidle and

Fig. 6a. Melanocytes (*arrows*) in the interstitial tissue of a female mouse Harderian gland. Several alveoli can be seen, two of which contain intraluminal accretions of porphyrin (lipid-porphyrin complex: *P*). Masson-Fontana silver impregnation, lightly counterstained with Kernechtrot. ×400. **b** Electron micrograph of a macrophage in the interstitial tissue of a mouse Harderian gland, showing aggregated melanin granules that were taken up as multiples from the melanocytes. The surface is highly irregular with many projections and lamellipodia, and the cytoplasm contains numerous lysosomes (*L*), a prominent Golgi apparatus (*G*) and a moderate number of mitochondria (*M*). ×12000

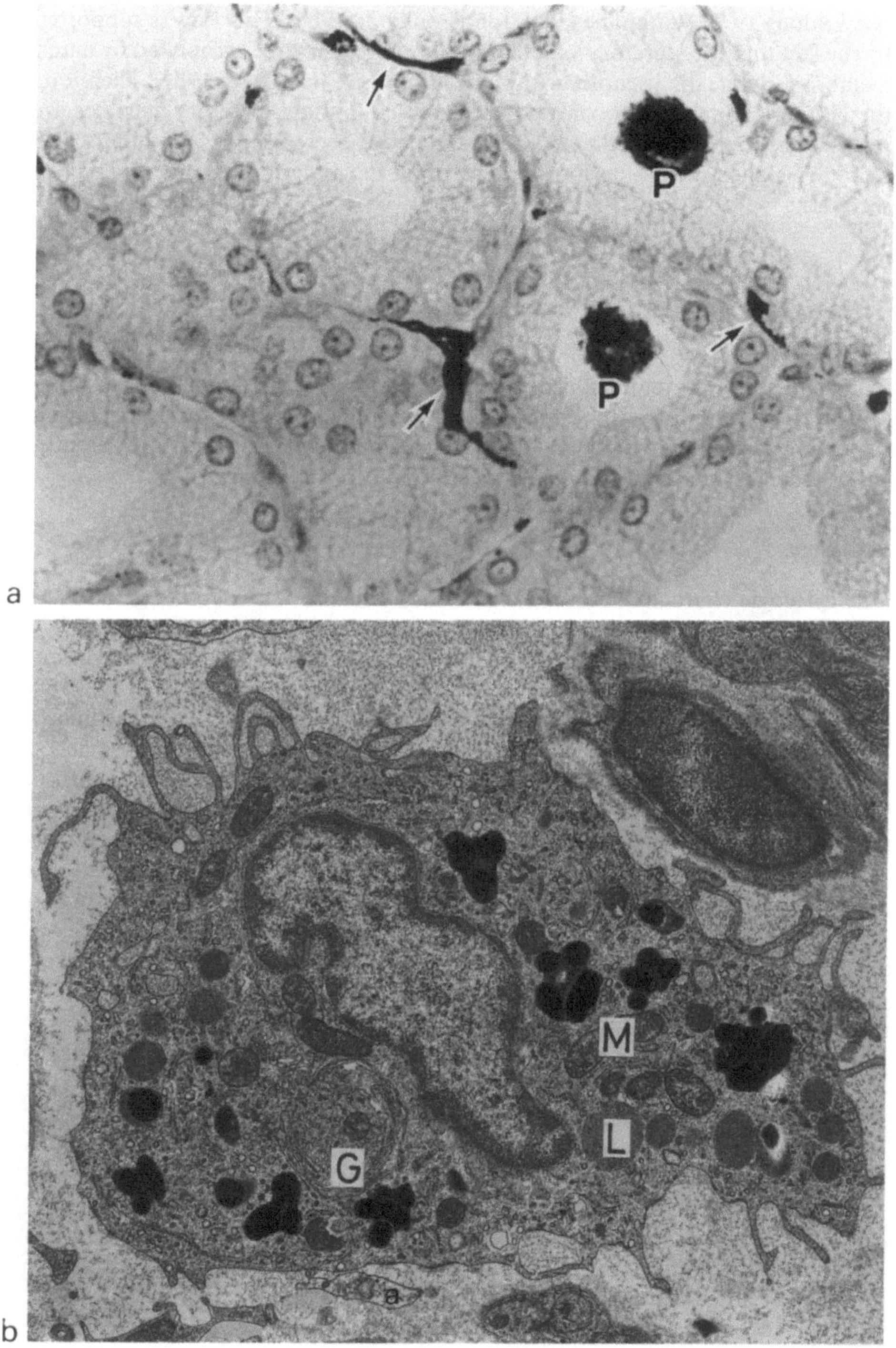
P
P
a
M
L
G
a
b

Niewenhuis 1978; Watanabe 1980; Johnston et al. 1985). This view is supported by the fact that the secretory vacuoles in the cytoplasm are surrounded by a unit membrane, while the vacuoles within the lumina are not (Sakai 1981). Therefore, the Harderian gland is the only organ in which lipoidal secretory vacuoles are released by exocytosis (the merocrine type of secretion), while other lipid secretory glands such as dermal glands exhibit the apocrine or holocrine type of secretion.

3 Secretory Products

The main secretory materials of the rat and mouse Harderian glands are lipids which are visible in frozen sections as sudanophilic materials and porphyrins which are seen in the glandular lumina as pigment concretions. More recently, other secretory materials have also been found in the gland.

3.1 Lipids

The cytologic characteristics of the secretory cells in the Harderian gland have been studied chemically and histochemically. These studies have established that the vacuoles within the secretory cell cytoplasm contain various types of lipids (Grafflin 1942; Paule 1957; Kanwar 1960; Rapin 1962; Hais et al. 1968; Kasama et al. 1970, 1989; Murawski and Jost 1974; Watanabe 1980). The lipid is composed of unsaturated fats, phospholipids, cholesterol esters, and neutral lipids or triglycerides. The major lipid class of the rat Harderian gland consists of unsaturated alchols and monoenoic fatty acids in the wax ester (Murawski and Jost 1974). In infantile rats, the glandular lipid is 1-alkyl 2,3-diacylglycerols (glycerol ether diester = GEDE) which can change into wax esters in adult rats (Zacker et al. 1976). GEDE also forms the major lipid fraction in the Harderian gland of the mouse (Kasama et al. 1970, 1989; Watanabe 1980).

3.2 Porphyrins

The Harderian glands of rodents are characterized by the presence of pigment concretions in the alveolar lumina (Grafflin 1941). The presence of porphyrin in the Harderian gland was first reported by Derrien and Turchini (1924) and later by Towbin et al. (1945) and Rohoyni and Kelényi (1962). Pigments are visible as solid intraluminal accretions in histological preparations only when the amount of porphyrin becomes excessive. In some strains of mice, sparse pigment concretions in alveolar lumina are associated with low concentrations of porphyrin (Shirama et al. 1980, 1981b; Fig. 7). Porphyrin pigments typically fluoresce bright red under ultraviolet light (Grafflin 1942). The porphyrin in mice was first determined to be protoporphyrin by Bittner and Watson (1946). This was

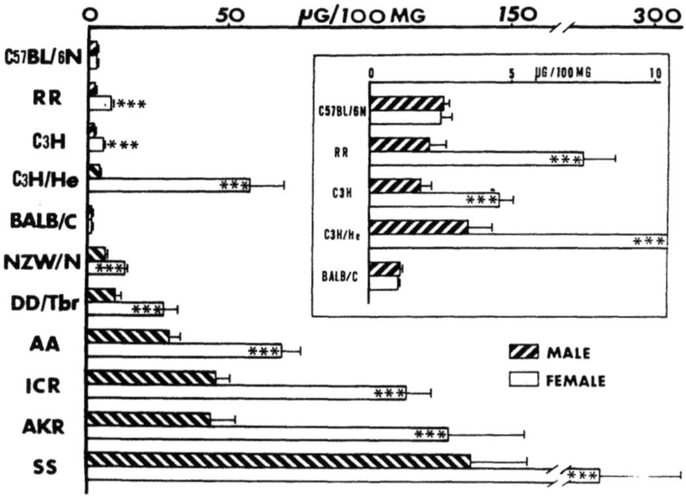

Fig. 7. Strain and sexual differences in porphyrin concentration of the Harderian gland of mice. Porphyrins are expressed as µg/100 mg wet wt. tissue. Values are mean ± SEM of five to eight animals. ***$p < 0.001$ compared with male (Student's *t* test)

confirmed by Johnston et al. (1985), who found protoporphyrin to be the predominent porphyrin form (approximately 95%) in the Australian plains mouse, although copro- and hexa-porphyrins were also detected. Tomio and Grinstein (1968) found that the enzymatic system which catalyzes the total synthesis of protoporphyrin IX is present in the Harderian glands of the rat. They measured glandular porphyrins and found protoporphyrin IX: 87%, coproporphyrin III: 11%, tricarboxylic porphyrin: 3%.

Most of the porphyrins soon incorporate metallic ions which usually lead to hemoglobin and myoglobin synthesis. The porphyrins from the Harderian gland, however, are largely metal-free, and, therefore, characteristically fluoresce bright (Jackson et al. 1976). The accumulation of metal-free porphyrin in the gland is attributed to the lack of a metal-chelating enzyme. The porphyrin in the gland may enter the circulation via interstitial blood vessels or lymphatics and in this way contribute to metal-complex formation, e.g., hemoglobin.

It is widely accepted that three enzymes, δ-aminolevulinate synthetase (δ-ALA synthetase), δ-aminolevulinate dehydratase (δ-ALA dehydratase), and uroporphyrinogen synthetase occur first in the porphyrin biosnthetic pathway (Fig. 8). The rate-limiting enzyme in the porphyrin biosynthetic pathway is δ-ALA synthetase whose synthetic activity is inhibited by the presence of protohemin IX. The amount of porphyrin is usually regulated by a feedback mechanism. Although it was assumed that the Harderian gland served as a passive storage organ for this pigment prior to the studies of Cohn (1955) and

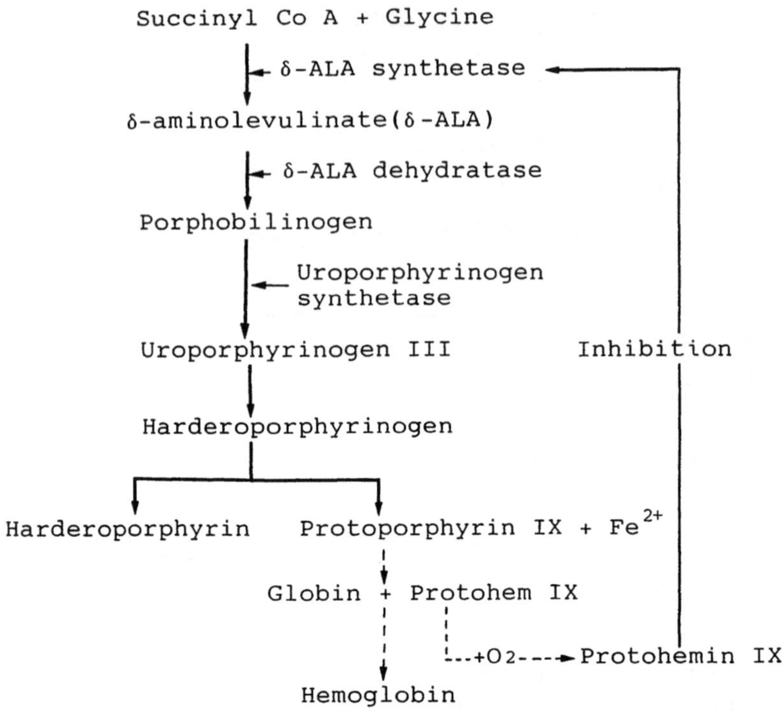

Fig. 8. The biosynthetic pathway of porphyrin. The reactions indicated by *short continuous arrows* are present in the Harderian gland, those by *broken arrows* are not. The other *long continuous arrow* (from protohemin IX to δ-ALA synthetase) indicates the end-product inhibition

Davidheiser and Figge (1958), the presence of these enzymes suggests that the Harderian gland may be responsible for the biosynthesis of porphyrin. Actually, upon incubation of Harderian gland homogenates in the presence of the porphyrin precursor, δ-ALA, porphyrin is formed (Davidheiser and Figge 1955; Figge and Davidheiser 1957; Tomio and Grinstein 1968). The δ-ALA dehydratase which catalyzes the second step has been purified from Harderian gland homogenates of rats and shown to be a sulfydryl cytoplasmic enzyme (Tomio et al. 1968).

There are remarkable differences between the Harderian gland and other porphyrin-producing organs such as the liver in porphyrin metabolism. Porphyrin compounds of the Harderian gland actually accumulate within alveolar lumina as pigment concretions. This accumulation may be explained by overproduction due to the absence of protohemin IX which inhibits the activity of δ-ALA synthetase. On the other hand, the Harderian gland is resistant to the induction of δ-ALA synthetase by hepatic porphyria-inducing chemicals, such as chlorinated benzenes, suggesting that there is a difference between the gland and the liver in the responsiveness of this enzyme to certain chemicals (Eida et al. 1977).

3.3 Melatonin

It has been thought that among mammalian tissues, only the pineal gland contained the enzyme hydroxyindole-O-methyltransferase (HIOMT), which catalyzes the terminal step in melatonin synthesis. The first finding indicating the possible presence of melatonin in the Harderian gland was a discovery of HIOMT in this tissue (Cardinali and Wurtman 1972). According to the work of Vlahakas and Wurtman (1972), Harderian gland HIOMT activity differs from the enzyme in the pineal gland and retina in being Mg^{2+}-dependent and in having a lesser affinity for N-acetylserotonin. In spite of this difference, they all produce methylated N-acetylindoles that react in an identical fashion with specific antimelatonin antibodies (Bubenik et al. 1974). In fact, melatonin has been assayed in the Harderian gland as well as in the pineal gland and retina using radioimmunoassay (Pang et al. 1976) and immunohistochemical studies (Bubenik et al. 1976a, b), suggesting that the gland contains melatonin in its secretory cells. Moreover, rat Harderian gland tissue in culture can acetylate tryptamine and serotonin and O-methylate the N-acetylserotonin to form melatonin (Brammer et al. 1978; Balemans et al. 1980). Melatonin levels in the Harderian gland show diurnal variations (Bubenik et al. 1976b; Reiter et al. 1983), while enzymatic activity related to melatonin biosynthesis does not (Brammer et al. 1978).

The appearance of a correlation between the localization of melatonin and of porphyrins in the Harderian gland is of particular interest. In the rat, the amount of porphyrin in the Harderian gland was doubled 20 min after intraperitoneal administration of melatonin (Joó and Kahán 1975; Shirama et al. 1979), while the removal of the pineal gland, which contains melatonin, induces high amounts of porphyrins (Reiter et al. 1983; Niwano et al. 1984). On the other hand, melanocyte-stimulating hormone (MSH) substantially decreased the porphyrin content of the gland (Joó and Kahán 1975).

The function of indoleamines within the Harderian gland remains unknown. Wetterberg et al. (1970a, b) expressed the view that porphyrins in the Harderian gland may serve as the convertor of ultraviolet frequencies that are below the perceptual threshold to frequencies within visual range. In such a case, melatonin may act as a substance inhibiting or stimulating the visualization process by a modification of porphyrin content. Exposure of rats to continuous light or darkness for 35 days significantly decreased Harderian gland porphyrin levels (Ulrich et al. 1974). These findings were verified by Shirama et al. (1978a, b) and Strum and Shear (1982b). The relation of melatonin to light perception seems to be more general in character. Despite numerous reports on this subject, the actual mechanism of melatonin effect on the visual system and the other organs is not yet known.

3.4 Other Products

More recently, the presence of arginine vasopressin, oxytocin, and neurophysins have been found in the rat Harderian gland as well as in the pineal gland and retina using radioimmunoassay (Gauquelin et al. 1988). Immunoreactive somatostatin has also been identified in the male rat Harderian gland by radioimmunoassay (Puig-Domingo et al. 1988). They demonstrated that growth hormone treatment leads to a dramatic increase in the levels of somatostatin in the Harderian gland, similar to its effect on somatostatin in other organs. Growth hormone administration also has a slight restorative effect on the Harderian gland in hypophysectomized rats whose glands has considerably diminished (Ebling et al. 1975a, b).

4 Innervations

Autonomic nerve fibers have been found in the Harderian gland of the rat (Arvy 1959; Tervo et al. 1968; Huhtala et al. 1977; Brownscheidle and Niewenhuis 1978) and mouse (Watanabe 1980; Strum and Shear 1982b). The Harderian gland receives cholinergic innervation with clear vesicle containing nerve terminals near the myoepithelial cells and the basal lamina of the secretory cells. Tashiro et al. (1940) has reported that injection of rats with acetylcholine resulted in secretion of "bloody tears" (porphyrin pigment) from the Harderian gland. The morphological basis of this phenomenon seems to be the cholinergic innervation of the myoepithelial cells whose contraction causes expulsion of the secretory product. Huhtala et al. (1977) reported that the cholinergic fibers to the rat Harderian gland derive from the zygomatic nerve. Furthermore, zygomatectomy significantly reduced porphyrin contents in the rat Harderian gland (Shirama et al. 1979). Adrenergic terminals which contain dense-cored vesicles have been found in association with blood vessels in the Harderian gland of rats (Huhtala et al. 1977) and mice (Watanabe 1980; Strum and Shear 1982b). Based on the site of terminals, adrenergic innervation is probably associated with vascular function. Huhtala et al. (1977) showed that adrenergic nerve fibers originate from the superior cervical ganglion via the trigeminal branches using the technique of selective denervations.

In addition to the cholinergic and adrenergic innervations, peptidergic innervations of the rat Harderian gland have been demonstrated. Using immunohistochemical techniques, Tsukahara and Jacobowitz (1987) showed positive reactions to neuropeptide-Y, neurotensin, calcitonin-gene-related peptide and especially to vasoactive intestinal polypeptide (VIP) and cholesystokinin. Some VIP-like immunoreactive nerves seem to be co-localized with acetylcholinesterase-positive nerves. Ruehle (1981) showed that radioactive oxytocin and lysine vasopressin are taken up in the rat Harderian gland. Subsequently, the presence of oxytocin, arginine vasopressin, and neurophysins in the Harderian gland was demonstrated by Gauquelin et al. (1988).

5 Development

Studies on the development of the Harderian gland have been undertaken by many authors (Kelényi and Orbán 1964; Müller 1969a, b; Shirama et al. 1977a; Gray et al. 1982; Michael et al. 1988). According to Michael et al. (1988), the Harderian glands of the mouse first appear in the posterior part of the region of the eye in the form of solid tubules on the 16th and 18th day of gestation. A few days after birth, the gland is attached to the orbital sinus, at the root of the nictitating membrane by a pedical of loose connective tissue. The anteroposterior extent of the gland increases and presents tubules as groups. Each group consists of many tubules, which vary in shape and are often still without lumina. Further development of the gland occurs around the time at which the eyelids open (Fig. 9). At this stage, the tubules have a lumen. The wall of the tubules consist of columnar cells containing large numbers of secretory granules. In rats, the gland develops in the same manner.

Vianna et al. (1975a, b) have studied the mitotic activity in the Harderian glands of the rat during postnatal life in an attempt to evaluate the structural alterations which might justify the modification in observed growth rhythm. After birth, the glands exhibit a high mitotic activity which decreases gradually until the 20th day. They concluded that the differential growth of the gland occurs in two stages; from birth until the 15th day, and from the 15th day until adult life, and that the development is characterized by a high rate of growth immediately before and after eyelid disjunction.

The enzymes related to the synthesis of glandular products are first noted histochemically at the 6th day of age (Müller 1969a). The porphyrins are first detectable at around the 7th day of age and thereafter, the porphyrin level shows a marked rise during the closed-eye period, reaching a peak around the time of eyelid disjunction, and then decreases gradually until the 25th day (Shirama et al. 1981a; Fig. 10).

The development of the Harderian gland has also been examined by electron microscopy in the rat (Kelényi and Orbán 1965) and mouse (Shirama and Hokano unpublished). Kelényi and Orbán (1964) showed, in infantile rats, a maturation process within the glandular epithelium which involved progressive differentiation of the secretory cells. Four cell types were demonstrated in the process. Cell types 1 and 2 disappear during the developmental process, so that finally the gland consists of only two cell types. In the mouse, although two types of secretory cells, type A and type B, also comprise the glandular epithelium in fully developed glands, the time of neonatal appearance is different between the two. Type A cells first appear at the 5th day, while type B at around

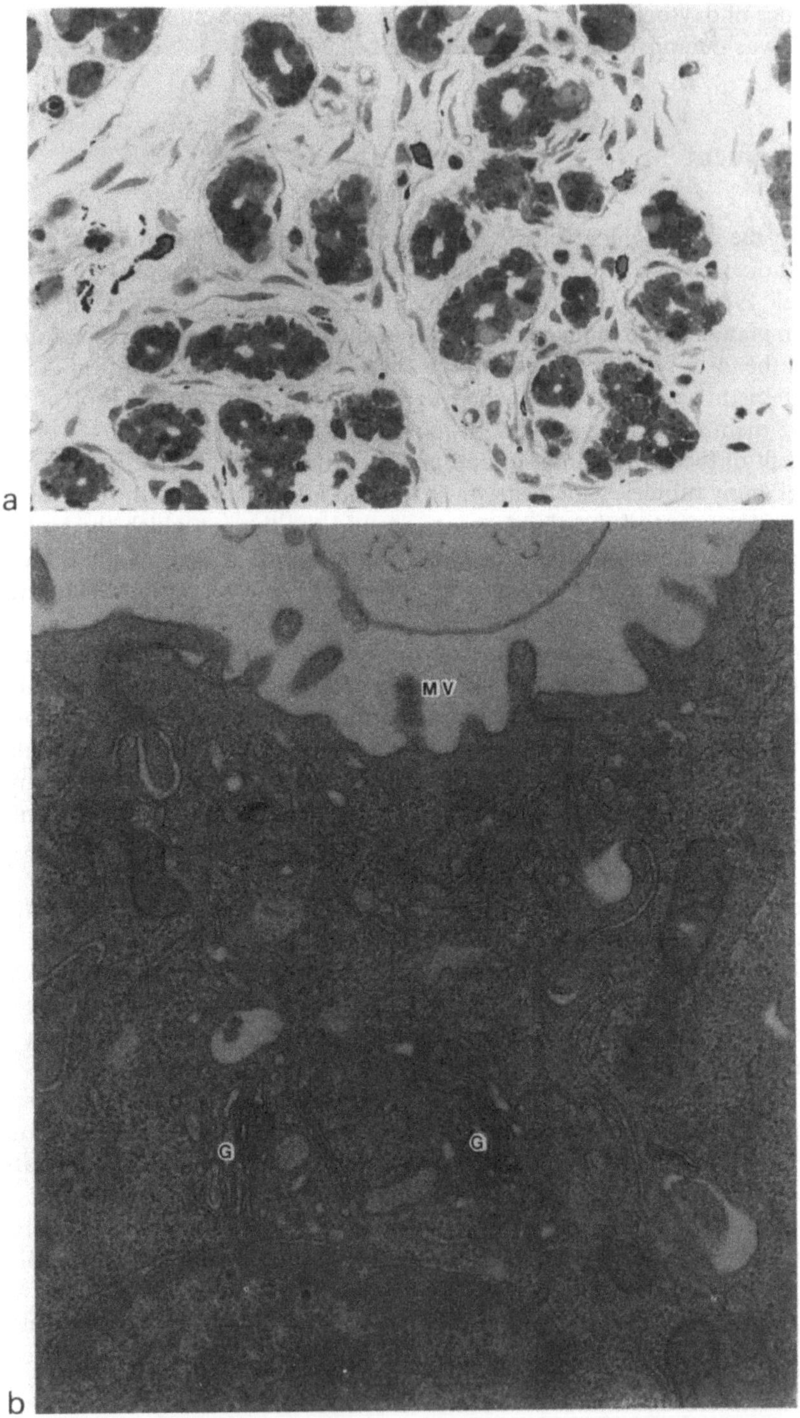
a
MV
G
G
b

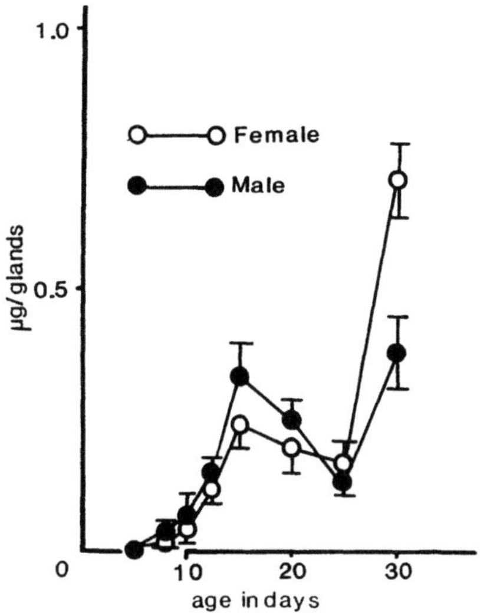

Fig. 10. Content of porphyrins in the Harderian gland of mice during early postnatal life is expressed as μg/pair glands. Each *point* (*open circles* female; *closed circles* male) represents the mean value, and the *bars* show the standard error of the mean. Five animals were used for each point

the 7th day, corresponding to the time at which porphyrins are first detected (Fig. 11).

6 Hormones

6.1 Gonadal Steroids

The Harderian gland may be a target organ for gonadal steroids (Brooksbank et al. 1973). In rats, ovariectomy significantly reduces porphyrin content of the Harderian gland, and administrations of estradiol plus progesterone exert a marked stimulatory effect on glandular porphyrins in ovariectomized females (Ulrich et al. 1974). Shirama et al. (1987) have observed a low level of porphyrin in the Harderian gland of both male and female rats after administration of

Fig. 9a. Light micrograph of the Harderian gland from 4-day-old mouse, showing Harderian gland alveoli. The lumina are considerably narrowed. H & E. ×200. **b** Electron micrograph of the Harderian gland from 4-day-old mouse, showing nonvacuolated (undifferentianted) epithelium. Note a cytoplasm of high density. The Golgi apparatus (*G*) is easily recognized, being well developed, and is located near the nucleus. The cell has short microvilli (*MV*) which extend into the llumen. ×47000

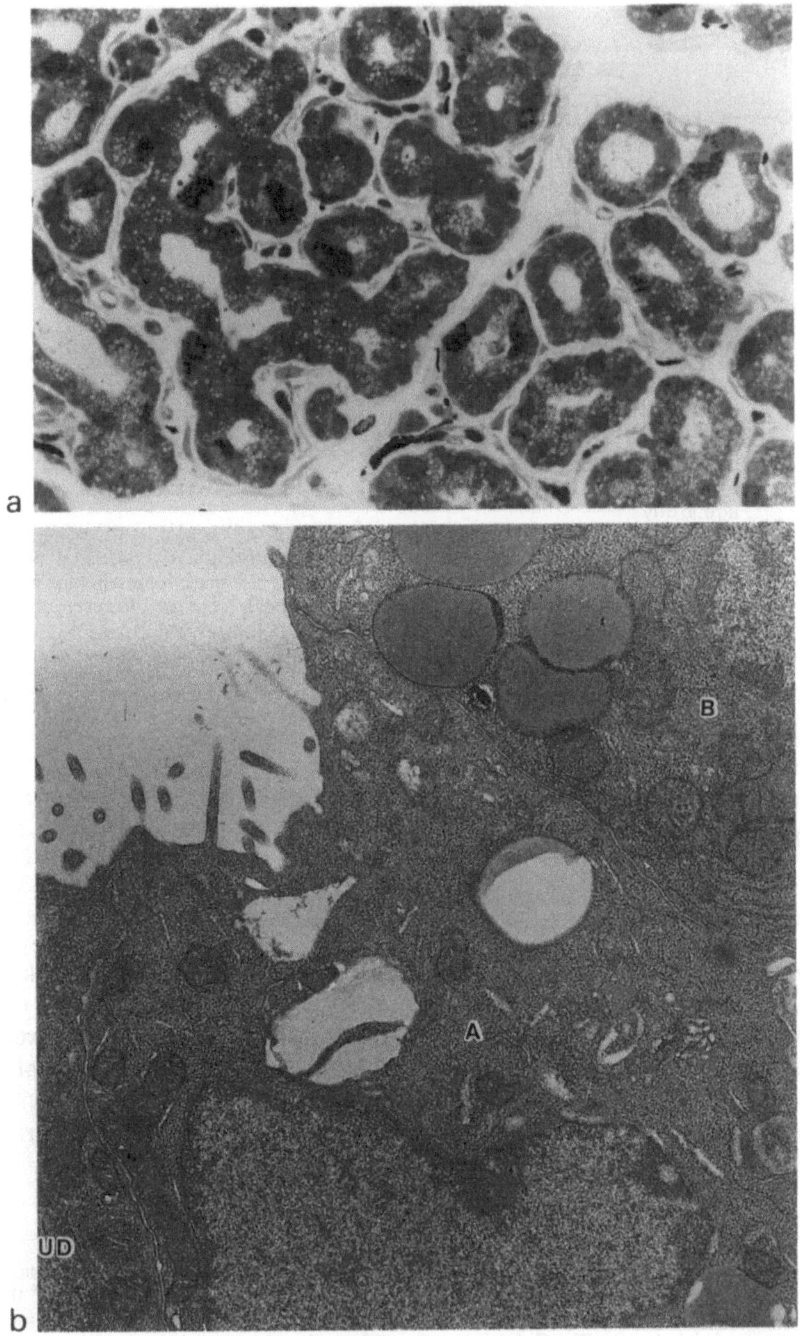
a
b
B
A
UD

testosterone, indicating that ovarian hormones, like androgens in the male, are necessary for regulation of the Harderian gland porphyrin synthesis.

In mice, the Harderian gland exhibits a marked sexual dimorphism insofar as there are higher amounts of porphyrins in the glands of females (Strong 1942; Figge and Davidheiser 1957; Margolis 1971; Shirama et al. 1981a). Margolis further demonstrated that the quantity of δ-ALA synthetase differed between males and females of the same strain. This has never been seen in the glands of rats. Furthermore, work on male mice has shown that experimental procedures such as castration plus adrenalectomy, which result in an increase of glandular porphyrin, are also associated with administered androgen doses (Shirama et al. 1981a; Fig. 12). Thus, the androgenic hormones exhibit an inhibitory effect on porphyrin synthesis in the Harderian glands of rats and mice, although they may differ in several ways in their response to other hormones such as estrogen and progesterone. It is possible that both testosterone and some adrenal substances, possibly an androgen, may inhibit the δ-ALA synthetase activity in the Harderian gland.

Harderian gland porphyrin levels have been shown also to vary with the estrous cycle. In the mouse, but not in the rat, porphyrin levels are lowest during metestrus and highest at diestrus (Shirama et al. 1981a).

6.2 Thyroid Hormones

In the rat, growth of the Harderian glands after eyelid disjunction is correlated ontogenetically with an increase in thyroxine secretion (Dohler et al. 1979). Thyroidectomy causes Harderian gland regression (Boas and Scow 1954), and treatment of neonatal animals with thyroxine accelerates development of the gland (Wetterberg et al. 1970b). The herbicide 2,4-dichlorophenyl-p-nitrophenyl ether (TOK) lowers serum thyroxine levels. Prenatal exposure of rats and mice to TOK produces severe intraorbital defects resulting in the destruction of the Harderian glands. This effect is noticeable only postnatally because the Harderian gland is not fully grown or functioning until after birth (Gray et al. 1982). The mouse (Singh and McKenzie 1969, 1970) and rat (Kok-van Alphen and Noach 1978) Harderian glands are also responsive to injection of thyrotropin. Furthermore, thyroxine 5′-deiodinase (Type 11), the enzyme which produces 3,5,3′-triiodothyronine (T3) via peripheral 5′-deiodination of thyroxine (T4), is found in the rat Harderian gland as well as in the pineal gland (Guerrero et al.

Fig. 11a. Light micrograph of the Harderian gland from 7-day-old mouse, showing the appearance of two cell types in the glandular epithelium. The cells with clear vacuoles are more numerous than cells with dark vacuoles. H & E. ×200. **b** Electron micrograph of the Harderian gland from 7-day-old mouse, showing three cell types, type A (*A*), type B (*B*) and undifferentiated cell (*UD*), in the glandular epithelium. Type B cells are first detectable at 7th day of age, while type A cells first appear at the 5th day of age. ×18000

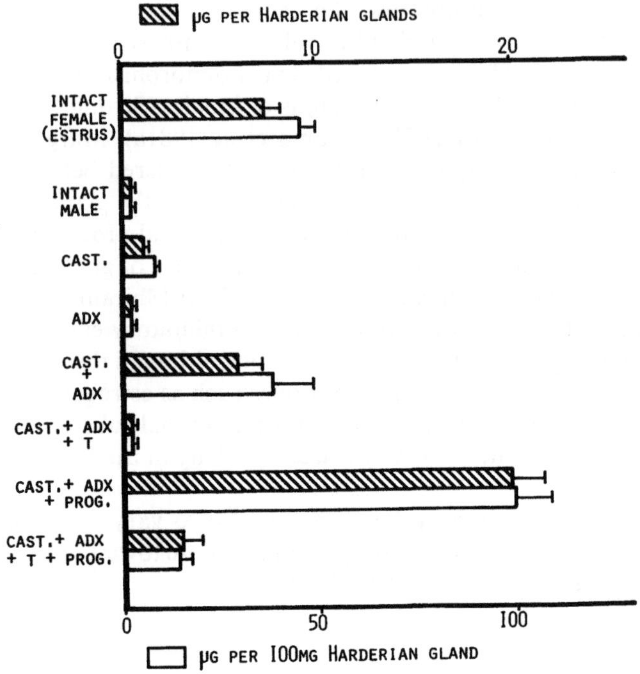

Fig. 12. Influence of the extirpation of testes (*CAST*) and/or adrenals (*ADX*) and of sex hormone replacement (*T* testosterone; *PROG* progesterone) on the content of porphyrins in male mice. Porphyrins are expressed as µg/pair gland (*hatched bar*) and µg/100 wet wt. tissue (*open bar*). Values are mean ± SEM of nine animals

1987; Tanaka et al. 1986). The activity of this enzyme exhibits a nycterochemeral rhythmicity coincident with melatonin synthetic activity both in the pineal and Harderian glands (Guerrero et al. 1987).

7 Light

The Harderian gland has been implicated in photoreception and in the control of the serotonin and HIOMT rhythms in the pineal glands of neonatal rats (Wetterberg et al. 1970a, b). In adult animals, however, it does not seem to have photoreceptor capability (Reiter and Klein 1971; Reiter 1973). The experiments of Kittel (1972) and Reiter (1973) suggested that the Harderian gland clearly reacts to continuous light and continuous darkness. Exposure of rats to continuous light for relatively long periods (65 to 100 days) results in significant regression of the Harderian glands (Reiter 1973; O'Steen et al. 1978) as well as the extraocular muscle (O'Steen et al. 1975, 1978), although there is a report suggesting that exposure to high intensity light for a short period results in a weight increase of the glands due to an increase in fluid content (Johnson et al. 1979). Moreover, the secretory activity of the Harderian gland is modified by environmental lighting conditions (Feria-Velasco et al. 1978). In rats exposed to continuous light or continuous darkness, the porphyrin contents of the glands are significantly lowered (Shirama et al. 1977a, 1987). The porphyrin levels correlate with light intensity (Shirama et al. 1977b), with the wave length of the light (Kittel 1973; Hoffman et al. 1985), and with light-dark rhythm (Hugo et al. 1987). Exposure to short photoperiod (8:16 LD) for 50 days results in significantly elevated lysosomal enzyme (acid phosphatase, hexosaminase, and β-glucuronidase) activities in the rat Harderian gland, suggesting that these enzymes are dependent upon the light-dark cycle to which the animals are exposed (Vaughan et al. 1988b). On the other hand, continuous light exposure leads to increased δ-ALA synthesase activity in the rat liver (Magnus et al. 1974), but the presence of this phenomenon in the Harderian gland remains unclear.

In the mouse, continuous light exposure damages the Harderian gland histologically. Glandular epithelial cells undergo squamous metaplasia, and the porphyrin contents drop markedly (Strum and Shear 1982a). None of these responses are affected by pineal removal (Reiter 1973), indicating that these pronounced cellular changes appear to result from the direct effect of light on the gland.

8 Tumors and Diseases

Harderian gland tumors occur spontaneously with a high incidence in animals over 24 months of age, but they are rare prior to 6 months (Rothwell and Everitt 1986; Sheldon et al. 1983). According to the observations of Sheldon et al. Harderian gland tumors could be classified into four histologic types, viz., papillary, acinar, cystic, and cystic papillary. The papillary type is the most frequently observed. These tumors can transform from one type to another. Moreover, there are marked strain variations in the incidence of tumors in the mouse.

Harderian gland tumors can be induced by carcinogenic agents in both rats (Zawirska 1965a, b, 1966) and mice (Figge 1944; Hueper and Figge 1945). In mouse Harderian glands, the susceptibility to carcinogenic inducement is closely related to Harderian gland porphyrin content, which probably reflects its metabolism and excretion (Strong and Figge 1941; Figge 1944).

The correlation between chromodacryorrhea and porphyrin formation in the Harderian gland has been observed by Eida et al. (1975). Since oversecretion of red tears (chromodacryorrhea) had occurred in pantothenic acid-deficient rats (Figge and Salomon 1942), they concluded that chromodacryorrhea is caused by secretion of excess protoporphyrin formed due to the increase in δ-aminolevalinate synthetase activity in the Harderian gland.

The presence of autoimmune mechanisms in the Harderian gland appeared from the work of Pisarev et al. (1968). They produced severe Harderian gland dacryoadenitis, characterized by an increase in the intercorneal distance, cellular damage, and lymphocytic infiltration in guinea pigs injected with crude homogenates of the Harderian gland, whereas Singh and McKenzie (1970), employing a similar approach, were unable to produce this effect in mice. Recently, Liu et al. (1986) demonstrated, however, that further purification of those crude homogenates enables the recovery of a specific antigen capable of inducing severe Harderian gland disease in the rat. This autoimmune dacryoadenitis is produced specifically in the Harderian gland.

9 Ocular Immune System

The ocular secretory immune system is important in the defense of mucosal surfaces against microbial challenge. This protective function is mediated primarily through secretory immunoglobulin A (lgA), which is the prominent antibody in most external secretions. In rats, the exorbital lacrimal gland contains larger numbers of lgA-producing cells in the interstitium than does the Harderian gland (Gudmundsson et al. 1985), though these two glands have a common embryological origin and similar blood vascular system. This suggests that the Harderian gland is not the primary site of the ocular secretory immune system in the rat. However, further studies will be needed to reach a definite conclusion.

10 Conclusion

While this chapter summarizes all information on the Harderian glands of rats and mice, we still cannot answer the question of what is their function. It would appear that the presence of characteristic porphyrins in the gland is an important clue to the eventual understanding of Harderian gland function. That this substance can be altered by hormonal manipulations and by environmental lighting would seem to provide us with a useful model for further study on the physiological role. For example, are there, in fact, gonadal steroid receptors in the gland? What is the function of glandular melatonin? Is the porphyrin-synthetizing enzyme, δ-ALA synthetase, really influenced by light exposure?

There is some evidence that the Harderian gland plays an immunological role in birds. Attempts to demonstrate a similar immunological role for the Harderian gland in rodents have been unsuccessful.

References

Abe J, Sugita A, Katsume Y, Yoshizuka M, Tamura N, Iwanaga S, Nishida T (1980) Scanning electron microscopic observations of the Harderian gland in rat. Kurume Med J 27:239–246

Abe J, Sugita A, Hamazaki M, Nakamura K, Iwanaga S, Nagae K, Atsuji K, Tsunawaki A, Abe T, Matsumoto T, Yo S, Murakami M (1981) Scanning electron microscopic observations of the myoepithelial cells of normal and contracting status in the rat Harderian gland. Krume Med J 28:103–112

Arvy L (1959) Contribution a l'histoenzymologie de la glande de Harder du rat albino CR Soc Biol 153:915–917

Balemans MGM, Pévet P, Legerstee WC, Nevo E (1980) Preliminary investigations on melatonin and 5-methoxytryptophol synthesis in the pineal, retina, and Harderian gland of the mole rat and in the pineal of the mouse "Eyeless". J Neural Transm 49:247–255

Bittner JJ, Watson CJ (1946) The possible association between porphyrins and cancer in mice. Cancer Res 6:337–343

Boas NF, Bates RW (1954) Role of the thyroid and anterior pituitary glands in the maintenance of the Harderian glands of the rat. Endocrinolgy 55:601–612

Boas NF, Scow RO (1954) Apparent exophthalmos in the rat following cortisone treatment or thyroidectomy. Endocrinology 55:148–155

Brammer GL, Yuwiler A, Wetterberg L (1978) N-acethyltransferase activity of the rat Harderian gland. Biochem Biophys Acta 526:93–99

Brooksbank BWL, Wilson DAA, Clough G (1973) The in vivo uptake of ^{3}H-androsta-4, 16-dien-3-one in tissues of the adult male rat. J Endocrinol 57:i–ii

Brownscheidle CM, Niewenhuis RJ (1978) Ultrastructure of the Harderian gland in male albino rats. Anat Rec 190:735–754

Bubenik GA, Brown GM, Uhlir I, Grota LJ (1974) Immunohistological localization of N-acetylindolealkylamine in pineal, retina and cerebellum. Brain Res 81:233–242

Bubenik GA, Brown GM, Grota LJ (1976a) Immunohistochemical localization of melatonin in the rat Harderian gland. J Histochem Cytochem 24:1173–1177

Bubenik GA, Brown GM, Grota LJ (1976b) Differential localization of N-acethylated indolealkylamines in CNS and the Harderian gland using immunohistology. Brain Res 118:417–427

Cardinali DP, Wurtman RJ (1972) Hydroxyindole-O-methyl transferases in rat pineal, retina and Harderian gland. Endocrinology 91:247–252

Carriere R (1985) Ultrastructural visualization of intracellular porphyrin in the rat Harderian gland. Anat Rec 213:496–504

Chiquoine AD (1958) The identification and electron microscopy of myoepithelial cells in the Harderian gland. Anat Rec 132:569–577

Cohn SA (1955) Histochemical observations on the Harderian gland of the albino mouse. J Histochem Cytochem 3:342–353

Davidheiser RH, Figge FHJ (1955) Enzymatic porphyrin synthesis in Harderian glands and other organs and tissues of mice. Proc Soc Exp Biol Med 90:461–462

Davidheiser RH, Figge FHJ (1958) Comparison of porphyrin producing enzyme activities in Harderian glands of mice and other rodents. Proc Soc Exp Biol Med 97:775–777

Derrien E, Turchini J (1924) Sur l'accumulation d'une porphyrine dans la glande de Harder des rongeurs du genre *Mus* et sur sa mode l'excretion. C R Soc Biol 91:637–639

Dohler KD, Wong CC, von zur Muhlen A (1979) The rat as model for the study of drug effects on thyroid function: consideration of methodological problems. Pharmacol Ther 5:305–318

Duthie ES (1934) Studies on the cytology of the Harderian gland of the rat. Q J Microsc Sci 76:549–557

Ebling FJ, Ebling E, Randall V, Skinner J (1975a) The effects of hypophysectomy and of bovine growth hormone on the responses to testosterone of prostate, preputial, Harderian and lacrymal glands, and of brown adipose tissue in the rat. J Endocrinol 66:401–406

Ebling FJ, Ebling E, Randall V, Skinner J (1975b) The synergistic action of α-melanocyte-stimulating hormone and testosterone on the sebaceous, prostate, preputial, Harderian and lacrymal glands, seminal vesicles and brown adipose tissue in the hypophysectomized-castrated rat. J Endocrinol 66:407–412

Eida K, Kubota N, Nishigaki T, Kikutani M (1975) Harderian gland. V. Effect of dietary pantothenic acid deficiency on porphyrin biosynthesis in Harderian gland of rats. Chem Pharm Bull 23:1–4

Eida K, Hasumi F, Nishimura N, Kikutani M (1977) Harderian gland. VI. Effect of chlorinated benzenes on porphyrin biosynthesis in the Harderian gland of rat. Chem Pharm Bull 25:1209–1214

Feria-Velasco A, Tapia-Arizmendi G, Malpica I (1978) Efectos de las condiciones de luz ambiental sobre la ultraestructura de la glándula de Harder de la rata recién nacida. Arch Invest Méd (Méx) 9:67–84

Figge FHJ (1944) Fluorescence studies on cancer. I. Porphyrin metabolism, Harderian gland fluorescence, and susceptibility to carcinogenic agents. Cancer Res 4:464–471

Figge FHJ, Davidheiser RH (1957) Porphyrin synthesis by mouse Harderian gland extracts: sex, age, and strain variation. Proc Soc Exp Biol Med 96:437–439

Figge FHJ, Salomon K (1942) Prevention of porphyrin incrustations on pantothenic acid-deficient rats by Harderian gland ablation. J Lab Clin Med 27:1496–1501

Gauquelin G, Gharib C, Ghaemmaghami F, Allevard A, Cherbal F, Geelen G, Bouzeghrane F, Legros J-J (1988) A day-night rhythm of vasopressin and oxytocin in rat retina, pineal and Harderian gland. Peptides 9:289–294

Grafflin AL (1941) The porphyrin-excreting Harderian gland of the albino rat. Anat Rec 79:25–26 (Abstr)

Grafflin AL (1942) Histological observations upon the porphyrin-excreting Harderian gland of the albino rat. Am J Anat 71:43–64

Gray Jr LE, Kavlock RJ, Chernoff N, Ferrell J, McLamb J, Ostby J (1982) Prenatal exposure to the herbicide 2,4-dichlorophenyl-p-nitrophenyl ether destroys the rodent Harderian gland. Science 215:293–294

Gudmundsson OG, Sullivan DA, Bloch KJ, Allansmith MR (1985) The ocular secretory immune system of the rat. Exp Eye Res 40:231–238

Guerrero JM, Puig-Domingo M, Vaughan GM, Reiter RJ (1987) Characterization of type-11 thyroxine 5'-deiodinase activity in the rat Harderian gland. Life Sci 41(9):1179–1185

Hoffman RA, Johnson LB, Corth R (1985) The effects of spectral power distribution and illuminance levels on key parameters in the male golden hamster and rat with preliminary observations on the effects of pinealectomy. J Pineal Res 2(3):217–233

Hueper WC, Figge FHJ (1945) Porphyrin excretion of Harderian glands in its relation to actinic carcinogenesis in hairless rats. Cancer Res 5:328–330

Hugo J, Krijt J, Vokurka M, Janousek V (1987) Secretory response to light in rat Harderian gland: possible photoprotective role of Harderian porphyrin. Gen Physiol Biophys 6:401–404

Huhtala A, Huikuri KT, Palkama A, Tervo T (1977) Innervation of the rat Harderian gland by adrenergic and cholinergic nerve fibers. Anat Rec 188:263–272

Jackson AH, Kenner GW, Smith KM, Suckling CJ (1976) Pyrroles and related compounds-XXXVII. Harderoporphyrin. Tetrahedron 32:2757–2762

Johnson DD, Rudeen PK, O'Steen WK (1979) Photically induced experimental exophthalmos: role of Harderian and pituitary glands. Invest Ophthalmol Visual Sci 18:1280–1284

Johnston HS, McGadey J, Thompson GG, Moore MR, Breed WG, Payne AP (1985) The Harderian gland, its secretory duct and porphyrin content in the plains mouse (*Pseudomys australis*). J Anat 140:337–350

Joó I, Kahán A (1975) The porphyrin content of Harderian glands in rats and the melatonin-melanocyte stimulating hormone system. Endokrinologie 65:308–312

Kanwar KC (1960) Morphological and cytochemical studies on the Harderian glands of rats. Cellule 61:129–143

Kasama K, Uezumi N, Itoh K (1970) Characterization and identification of glycerol ether diesters in Harderian gland tumors of mice. Biochim Biophys Acta 202:56–66

Kasama K, Blank ML, Snyder F (1989) Identification of 1-alkyl-2-acyl-3-(2',3'-diacylglycerol) glycerols, a new type of lipid class, in Harderian gland tumors of mice. J Biol Chem 264(16):9453–9461

Kelényi G, Orbán S (1964) Electron microscopy of the Harderian gland of the rat: maturation of the acinar cells and genesis of the secretory droplets. Acta Morphol Acad Sci Hung 13:155–165

Kennedy GY (1970) Harderoporphyrin: a new porphyrin from the Harderian gland of the rat. Comp Biochem Physiol 36:21–36

Kittel R (1972) Histologische Untersuchungen über den Einflu von Licht und Dunkelheit auf die Hardersche Drüse der wei en Maus. Gegenbaurs Morphol Jahrb 117:461–471
Kittel R (1973) Influence of intensity of light and wavelength on the secretory activity of Harder's gland in white mouse. Folia Morphol (Prague) 21:398–400
Kok-van Alphen CC, Noach EL (1978) Hormonal influences on ocular refraction in the rat. Nature 271:62–66
Liu SH, Sakai F, Prendergast RA, Silverstein AM (1986) Experimental autoimmune dacryoadenitis II. Harderian gland disease in the rat. Invest Ophthalmol Visual Sci 28:276–280
Magnus IA, Janousek V, Jones K (1974) The effect of environmental lighting on porphyrin metabolism in the rat. Nature 250:504
Margolis FL (1971) Regulation of porphyrin biosynthesis in the Harderian gland of inbred mouse strains. Acta Biochem Biophys 145:373–381
Michael MI, Khalil SH, Matta CA, Rizk TA (1988) Normal development of the ocular glands of the mouse. Folia Morphol (Prague) 36:47–52
Müller HB (1969a) Die postnatale Entwicklung der Harderschen Drüse der weien Ratte. I. Lichtmikroskopische Befunde. Z Zellforsch Mikrosk Anat 100:421–438
Müller HB (1969b) Die postnatale Entwicklung der Harderschen Drüse der weien Ratte. II. Fermenthistochemische Befunde. Histochemie 20:181–196
Müller HB (1970) Untersuchungen zur postnatale Entwicklung der Harderschen Drüse der Ratte. Anat Ges 64:503–509
Murawski U, Jost U (1974) Unsaturated wax esters in the Harderian gland of the rat. Chem Phys Lipids 13:155–158
Niwano Y, Ishii T, Ishibashi T (1984) Effects of pinealectomy on the porphyrin content of the Harderian gland in rats. Jpn J Zootech Sci 55:894–896
Orbán S, Kelényi G (1962) Submicroscopic structure and histophysiology of the Harderian gland of the rat. Acta Biol (Budapest) 4:433 (Astr)
O'Steen WK, Shear CR, Anderson KV (1975) Extraocular muscle degeneration after exposure of rats to incandescent radiant energy. J Cell Sci 18:157–177
O'Steen WK, Kraeer SL, Shear CR (1978) Extraocular muscle and Harderian gland degeneration and regeneration after exposure of rats to continuous fluorescent illumination. Invest Ophthalmol Visual Sci 17:847–856
Páng SF, Brown GM, Grota LJ, Rodman RL (1976) Radioimmunoassay of melatonin in pineal glands, Harderian glands, retina and sera of rats or chickens. Fed Proc 35:691 (Abstr)
Paule WJ (1957) The comparative histochemistry of the Harderian gland. Dissertation, Ohio State University, Columbus, Ohio
Paule WJ, Hayes ER (1958) Comparative histochemical studies on the Harderian gland. Anat Rec 130:436 (Abstr)
Pisarev MA, Altschuler N, Davison TA (1968) Immune exophthalmos due to Harderian gland antigen. Endocrinology 83:903–906
Puig-Domingo M, Guerrero JM, Reiter RJ, Peinado MA, Sabry I, Viader M, Webb SM (1988) Identification of immunoreactive somatostatin in the rat Harderian gland: regulation of its content by growth hormone, beta-adrenergic agonists and calcium channel blockers. Peptides 9:571–574
Rapin M (1962) Apercus sur la glande de Harder du rat. Son development sa secretion et son histologie. Soc Zool Fr 77:239 (Abstr)
Reiter RJ (1973) Comparative effects of continual lighting and pinealectomy on the eyes, the Harderian glands and reproduction in pigmented and albino rats. Comp Biochem Physiol 44:503–509
Reiter RJ, Klein DC (1971) Observations on the pineal gland, the Harderian glands, the retina and the reproductive organs of adult female rats exposed to continuous light. J Endocrinol 51:117–125
Reiter RJ, Richardson BA, Matthews SA, Lane SJ, Ferguson BN (1983) Rhythms in immunoreactive melatonin in the retina and Harderian gland of rats: persistence after pinealectomy. Life Sci 32:1229–1236
Rohonyi B, Kelényi G (1962) Porphyrin content of the Harderian gland of the rat. Acta Biol Hung 13:241–245
Rothwell TLW, Everitt AV (1986) Exophthalmos in ageing rats with Harderian gland: disease. Lab Anim 20:97–100

Ruehle HJ (1981) The Harderian gland: target for cyclic nonapeptide neurohypophyseal hormones: autoradiographic studies in the rat. Z Mikrosk Anat Forsch (Leipz) 95:459–465

Sakai T (1981) The mammalian Harderian gland: morphology, biochemistry, function and phylogeny. Arch Histol Jpn 44:299–333

Sakai T, Yohro T (1981) A histological study of the Harderian gland of Mongolian gerbils, *Meriones meridianus*. Anat Rec 200:259–270

Sheldon WG, Curtis M, Kodell RL, Weed L (1983) Primary Harderian gland neoplasms in mice. J Natl Cancer Inst 71:61–68

Shirama K (1978) Induction of persistent estrus by constant light: Effects of neonatal constant light and Harderian gland function. Neuroendocrinology 27:193–203

Shirama K, Takeo Y, Shimizu K, Maekawa K (1977a) Influence of environmental lighting and sex hormones on the development of Harderian gland in the rat. Annot Zool Jpn 50:70–80

Shirama K, Takeo Y, Shimizu K, Maekawa K (1977b) Influence of continuous lighting intensities on porphyrin content in rat Harderian gland. Zool Mag (Tokyo) 86:259–260

Shirama K, Furuya T, Takeo Y, Shimizu K, Maekawa K (1979) Neural control of the Harderian gland in the rat. Zool Mag (Tokyo) 88:149–154

Shirama K, Furuya T, Takeo Y, Shimizu K, Maekawa K (1980) Sex differences in Harderian gland porphyrins of inbred mouse strains. Acta Anat Nippon 55:623–631 (in Japanese)

Shirama K, Furuya T, Takeo Y, Shimizu K, Maekawa K (1981a) Influence of some endocrine glands and of hormone replacement on porphyrins of the Harderian glands of mice. J Endocrinol 91:305–312

Shirama K, Furuya T, Takeo Y, Shimizu K, Maekawa K (1981b) Sex and strain differences in porphyrin content of the Harderian gland of mice. Zool Mag (Tokyo) 90:247–250

Shirama K, Kohda M, Kohda M, Hokano M (1987) Effects of lighting conditions and of hormone replacement on the level of porphyrins in the rat Harderian gland. J Endocrinol Invest 10:79–82

Shirama K, Harada T, Kohda M, Hokano M (1988a) Fine structure of melanocytes and macrophages in the Harderian gland of the mouse. Acta Anat 131:192–199

Shirama K, Kohda M, Hokano M (1988b) Effects of endocrine glands and hormone replacement on the mast cell count of the Harderian gland of mice. Acta Anat 131:327–331

Singh SP, McKenzie JM (1969) Effect of thyrotropin on sulfur-35 sulfate uptake by mouse Harderian gland. Endocrinology 85:952–955

Singh SP, McKenzie JM (1970) The influence of thyrotropin on Harderian gland glycosaminoglycans of the mouse. Proc Soc Exp Biol Med 135:478–482

Strong LC (1942) Sex differences in pigment content of Harderian gland mice. Proc Soc Exp Biol Med 50:123–125

Strong LC, Figge FHJ (1941) Fluorescence of Harderian glands in mice of cancer-susceptible and cancer-resistant strain. Science 94:331 (Abstr)

Strum JM, Shear CR (1982a) Constant light exposure induces damage and squamous metaplasia in Harderian glands of albino mice. Tissue Cell 14:149–161

Strum JM, Shear CR (1982b) Harderian glands in mice: fluorescence, peroxidase activity and fine structure. Tissue Cell 14:135–148

Tanaka K, Murakami M, Greer MA (1986) Type-11 thyroxine 55-deiodinase in the rat pineal gland. Biochem Biophys Res Commun 137:863–868

Tashiro S, Smith CC, Badger E, Kezur E (1940) Chromodacryorrhea, a new criterion for biological assay of acetlycholine. Proc Soc Exp Biol Med 44:658–661

Tervo TA, Huhtala A, Palkama A (1968) Innervation of the rat Harderian gland. Exp Eye Res 22:299–300 (Abstr)

Tilgner S (1968) Zur Morphologie der Harderschen Drüse der Ratte (*Rattus norvegicus*, Berkenhout 1769). Z versuchstierk Bd 10:189–195

Tomio JM, Grinstein M (1968) Porphyrin biosynthesis 5. Biosynthesis of protoporphyrin IX in the Harderian glands. Euro J Biochem 6:80–83

Tomio JM, Tuzman V, Grinstein M (1968) σ-Aminolevulinate dehydratase from rat Harderian gland. Purification and properties. Euro J Biochem 6:84–87

Towbin EJ, Fanta PE, Hodge HC (1945) The porphyrin of Harder's gland. Proc Soc Exp Biol Med 60:228–231

Tsukahara S, Jacobowitz DM (1987) Peptidergic innervation of the rat Harderian gland. Histochemistry 87:233–236

Tsutsumi A, Iwata K, Ogawa K, Matsuura K (1966) Histochemical and electron microscopic observation on the Harderian gland of the albino rat (in Japanese). Arch Histol Jpn 27:553–567

Tvrzická E, Rezanka T, Krijt J, Janousek V (1988) Identification of very-long-chain fatty acids in rat and mouse Harderian gland lipids by capillary gas chromatography-mass spectrometry. J Chromatogr 431:231–238

Ulrich R, Yuwiler A, Geller E, Wetterberg L (1974) Effects of sex hormones and environmental lighting on rat Harderian gland porphyrin. J Endocrinol 63:99–102

Vaughan MK, Chenoweth PC, Sabry I, Chambers JP, Reiter RJ (1988a) Lysosomal enzymes in the rat Harderian gland are altered by either bromocriptine treatment or hypophysectomy and hormone replacement therapy. Proc Soc Exp Biol Med 187:469–473

Vaughan MK, Nordio M, Chenoweth PJ, Chambers JP, Reiter RJ (1988b) Underfeeding and exposure to short photoperiod alters rat pineal and Harderian gland lysosomal enzyme activities. Proc Soc Exp Biol Med 189:211–216

Vianna GF, Cruz AR, Azoubel R (1975a) Mitotic activity and nuclear/cytoplasmic ratio of the lachrymal and Harderian glands of the rat during postnatal life. Acta Anat 92:1–7

Vianna GF, Cruz AR, Azoubel R (1975b) Allometric study of the lachrymal and Harderian glands of the rat during postnatal life. Acta Anat 92:161–170

Vlahakes GJ, Wurtman RJ (1972) A Mg^{2+}-dependent hydroxyindole-O-methyl-transferase in rat Harderian gland. Biochem Biophys Acta 261:194–197

Watanabe M (1980) An autoradiographic, biochemical, and morphological study of the Harderian gland of the mouse. J Morphol 163:349–365

Watanabe M, Kihara T, Shimono R (1981) Electron microscopy of myoepithelial cells of the mouse Harderian gland. Bull Osaka Med Sch 27:1–7

Wetterberg L, Geller E, Yuwiler A (1970a) Harderian gland: An extraretinal photoreceptor influencing the pineal gland in neonatal rats? Science 167:884–885

Wetterberg L, Yuwiler A, Geller E, Schapiro S (1970b) Harderian gland: development and influence of early hormonal treatment on porphyrin content. Science 168:996–998

Woodhouse MA, Rhodin JAG (1963) The ultrastructure of the Harderian gland of the mouse with tparticular reference to the formation of its secretory product. J Ultrastr Res 9:76–98

Zacker J, Egge H, Murawski U (1976) 1-Alkyl-2,3-diacylglycerine in der Harderschen Drüse der Ratte. Hoppe-Seyler's Z Physiol Chem 357:288–289

Zawirska B (1965a) Microfluoroscopic studies on porphyrins in Harder's gland and in tumours induced in it. II. Microfluoroscopic studies on porphyrins in precancerous states and tumors induced with methylcholanthrene in Harder's gland in white rats. Arch Immunol Ther Exp 13:238–275

Zawirska B (1965b) Microfluoroscopic studies on porphyrins in Harder's gland and in tumors induced in it. III. Histotopography of porphyrins and growth of transplanted sarcoma in Harder's gland in white rats. Arch Immunol Ther Exp 13:377–390

Zawirska B (1966) Microfluoroscopic studies on porphyrins in Harder's gland and in tumors induced in it. IV. Comparison of carcinogenesis induced with methylcholanthrene in Harder's gland, the submandibular salivary gland and subcutaneous tissue in white rats. Arch Immunol Ther Exp 14:622–629

The Structure of the Harderian Gland of the Golden Hamster

A.P. PAYNE, J. MCGADEY and H. S. JOHNSTON

Department of Anatomy, Glasgow University, Glasgow G12 8QQ, UK

1 Introduction

Hamsters are a group of cricetid rodents numbering some 23 species and with a widespread distribution in eastern Europe and Asia. The golden hamster (*Mesocricetus auratus* Waterhouse) is a common laboratory species, although its distribution in the wild is difficult to assess. First described in 1833, it is not recorded again until 1930, when a nest of pups found in the desert near Aleppo, Syria was acquired by Professor Aharoni of the Zoology Department of the Hebrew University at Jerusalem; two females and a male from this litter are said to have given rise to all the present captive hamsters (Fulton 1968). Expeditions to the Middle East have not been successful in establishing the present status of the golden hamster (Murphy 1971), although the closely related *Mesocricetus brandtii* (considered by some to be the same species, e.g., Corbet and Hill 1986) has been trapped in Turkey.

The Harderian gland of the golden hamster has received more attention than that of any other species. Apart from morphological studies, it has been used in a wide variety of investigations e.g., of hormonal control, of porphyrin biosynthesis, of lipid biochemistry (Lin and Nadakavukaren 1981), of proteins (Hoh et al. 1984), of metallic ions (Hoffman and Jones 1981), of photoperiod, indoles, and monoamine oxidase (Hoffman et al. 1985; Menendez-Pelaez et al. 1988; 1989), of somatostatin (Puig-Domingo et al. 1988) and of pheromone production (Payne 1979). Many of these studies have involved combined investigations of gland structure and activity. In view of this widespread interest, it is perhaps fitting that the hamster Harderian gland should receive special attention within this volume.

2 General Features

The gross anatomy of the hamster Harderian gland has received only brief attention (Kittel 1962) but would appear to resemble the situation described for gerbils (Brownscheidle 1974) or rats (Grafflin 1942). The gland is horseshoe-

Webb et al. (Eds).
Harderian Glands

shaped and the two main lobes are firmly adherent to the nictitating membrane (third eyelid) through which the duct of the gland travels to open on its posteroinferior surface. As with other rodents, the gland is additional to two true lacrimal glands, the exorbital and infraorbital. The gland is the largest structure in the bony orbit, bigger than the eye itself, and it surrounds the posterior part of the eyeball on its inferior, medial (nasal) and superior surfaces. In this position, the gland is in contact with the optic nerve and several of the extraocular muscles; in the rat, the Harderian gland surface is irregularly shaped to interdigitate with other orbital structures but this is not conspicuous in the hamster.

Histologically, although the gland is divided into lobules, (1) there is no morphologically distinct duct system within the substance of the gland and (2) there is no evidence that particular parts of the tubule system are secretory while others are not (i.e., there is no evidence for secretory endpieces). This suggests that the gland should be thought of as compound tubular rather than tubulo-acinar.

3 Particular Features

Under this heading we wish to consider in turn the epithelial cells of the gland, the luminal contents, the myoepithelial cells, interstitial elements, and duct. This consideration will use material (both published and unpublished) obtained by the authors, together with information provided in a number of key references by others (including Hoffman 1971; Bucana and Nadakavukaren 1972a, 1973; Lin and Nadakavukaren 1979; Sun and Nadakavukaren 1980)

3.1 The Epithelial Cells

From the beginning, it was apparent to investigators that the hamster Harderian gland showed sexual dimorphisms which encompassed both its histological appearance and its secretory products (Christensen and Dam 1953; Woolley and Worley 1954). It is, therefore, necessary to consider the epithelial cells of the two sexes separately. Furthermore, it must be borne in mind that gland structure is controlled by (among other factors) gonadal steroids. This is the subject of chapters in this Volume by Payne, Spike and Moore, and by Nadakavukaren.

3.1.1 The Epithelial Cells of the Male Gland

These are shown diagrammatically in Fig. 1 and as a low power SEM in Fig. 2a. Epithelial cells are columnar, although they often appear pyramidal because of the convoluted nature of the tubules. The nucleus is basally placed, large, round, and with one or more nucleoli. Mitotic figures are rarely seen in adult glands.

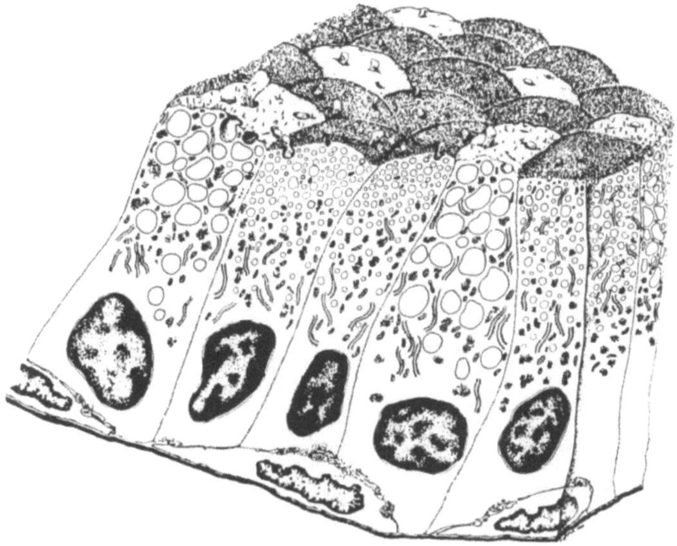

Fig 1. A diagrammatic representation of tubule cells from the male hamster Harderian gland. Epithelial cells with small lipid vacuoles (Type I cells) and large vacuoles (Type II cells) are shown. The cytoplasm of both types contains numerous ribosomes and polytubular complexes. The surface of both cell types bears microvilli but these are dispersed on the bulging surfaces of Type II cells; pitting, caused by the release of lipid vacuoles which can be seen as blebs, is evident. Contractile myoepithelial cells intervene between the basal surface of the epithelial cells and the basal lamina

The cytoplasm of the epithelial cells contains many free ribosomes together with clumps of rough endoplasmic reticulum often situated near the nucleus. There is also much smooth endoplasmic reticulum in the cytoplasm in the form of minute vesicles resulting in a very dense, homogeneous appearance to the interior of the cell. The Golgi apparatus (though mentioned by some authors, e.g., Bucana and Nadakavukaren 1972a) is not readily found. Mitochondria are numerous and not particularly large; nevertheless, many porphyrinogenic enzymes are mitochondrial and morphometric analysis of mitochondrial volumes or the surface area of cristae might in the future give useful insight into the relationships between structure and biosynthetic capacity.

All male cells contain large numbers of a unique ultrastructural feature, the polytubular complex (see Fig. 2b). This presents as a series of tubular profiles, each some 30 nm in diameter. Often, transverse and longitudinal profiles are seen within the same complex. The ultrastructural affinities of these complexes are difficult to assess, but from work on the woodmouse (*Apodemus sylvaticus*), it has been suggested that they arise from vacuolated smooth ER (Johnston et al. 1987). A variety of functions have been ascribed to them; they are packages of enzymes (Bucana and Nadakavukaren 1972a); they prevent the formation of porphyrins (Jones and Hoffman 1976), or their storage (Bucana and Nadakavukaren 1973). They occur normally only in adult males or females given androgens

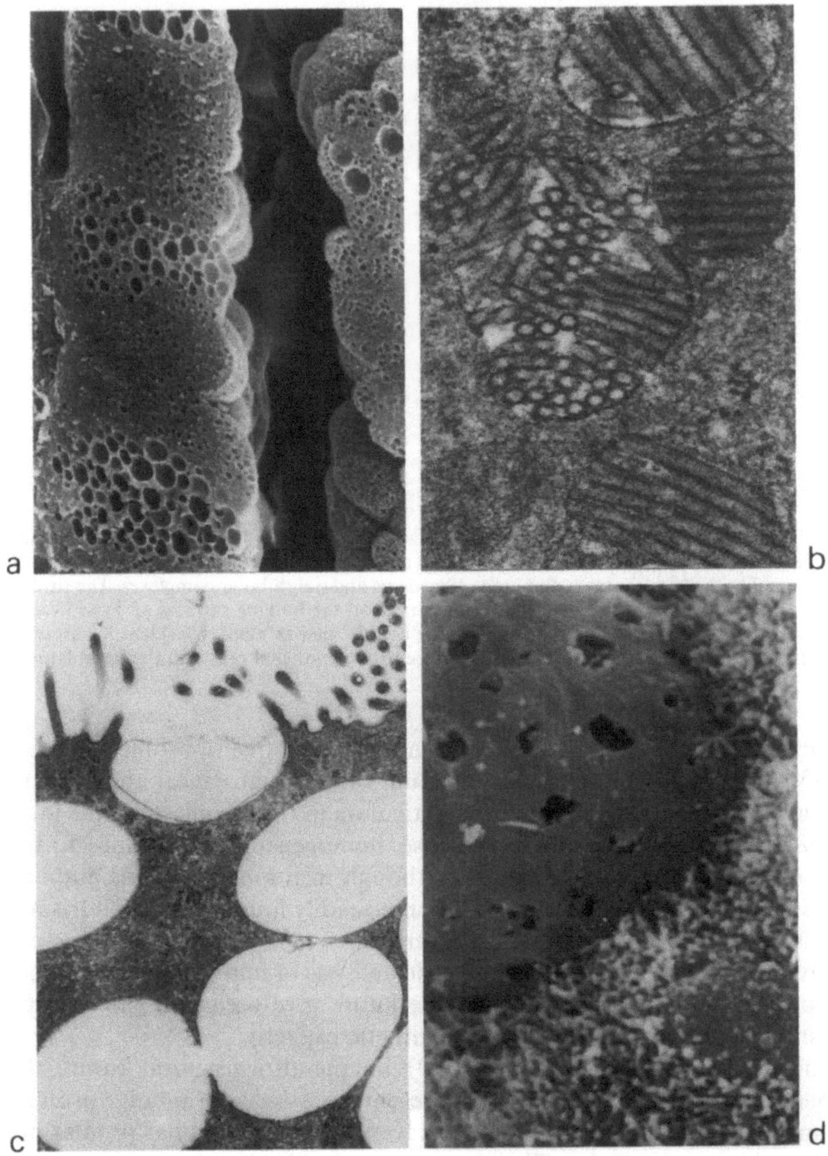

Fig 2a. SEM of the cut surface of a tubule of the male hamster Harderian gland. Type I cells (small vacuoles) and Type II cells (large vacuoles) are clearly distinguishable. (×600). **b** Polytubular complexes occur in great numbers within the cytoplasm of all male hamster Harderian gland epithelial cells.(× 64000). **c** A portion of the apical region of a hamster Harderian gland epithelial cell. A lipid vacuole (with an intact membrane surrounding it) is being released by exocytosis. (×18500). **d** SEM of male epithelial cells showing several Type I cells with numerous microvilli surrounding the bulging surface of a Type II cell. The latter is relatively devoid of microvilli and shows clear pitting where vacuoles have been released. (×6000)

(i.e., in individuals where porphyrin synthesis is suppressed). They do not occur in normal females, and they disappear in males after castration (Payne et al. 1977; Lin and Nadakavukaren 1979).

The central and apical parts of the cell are dominated by lipid vacuoles limited by a single membrane (Fig. 2c). It is the size of these vacuoles which leads to the classification of cell types in the male (see below) into Type I with small vacuoles and Type II with very large ones (Figs. 1 and 2a). The cell surfaces of Type I cells bear numerous stubby microvilli while, in general, Type II cells have a more bulging apex and microvilli are sparse: scanning views of both cell types reveal areas of pitting where apical lipid vacuoles have been released (see Fig. 2d). The method of vacuole release has long been a subject of debate. Initially thought to be an apocrine mechanism (Cohn 1955), it is now generally considered to be merocrine i.e. vacuoles are released by exocytosis (Woodhouse and Rhodin 1963); (see Fig. 2c). Nevertheless,

1. there are ultrastructural studies which suggest that apocrine secretion may occur in some species (Johnston et al. 1985);
2. almost all studies on the hamster and other rodent species have postulated an element of holocrine secretion, since whole cells can be seen in the process of extrusion into the lumen and intraluminal collections of epithelial cell nuclei often occur (Hoffman 1971; Bucana and Nadakavukaren 1973; Johnston et al. 1983; Carriere 1985). However, if the loss of the whole cells is widespread, it is interesting that mitotic figures are rarely seen in adults (Bucana and Nadakavukaren 1973).

Male epithelial cells also contain numerous membranous bodies, often seen as circular profiles, which have been described in other species (e.g., *Apodemus*, Johnston et al. 1987). There is some evidence, particularly the presence of apparently intermediate stages, to suggest a relationship between the membranous bodies and the mitochondria of the epithelial cells. The membranous bodies closely resemble forms of peroxisome reported by Gorgas (1984) and Gorgas and Zaar (1984).

Junctional complexes such as desmosomes between adjacent lateral cell walls are not frequent, nor is there pronounced interdigitation between adjacent cells. At the apices of cells, there are relatively inconspicuous tight junctions.

3.1.2 The Epithelial Cells of the Female Gland

These share the same general features of male epithelial cells but differ in the following respects:

1. They normally never contain polytubular complexes [but do so if the female is given androgens (Sun and Nadakavukaren 1980)];
2. They are homogeneous i.e. female glands possess only one cell type. This type contains numerous small lipid vacuoles and, on this basis, has been likened to the male Type I cell. Nevertheless, it must be borne in mind that

there has been no rigorous morphometric analysis of lipid vacuole size or vacuole volume/unit volume to confirm the similarity;
3. They contain concentric lamellar formations with smooth membranes (Bucana and Nadakavukaren 1972a, 1973), which are not characteristic of the male, and which may derive from juxtanuclear structures.

In addition, there is usually less rough endoplasmic reticulum evident in the epithelial cells of females than males. The smooth endoplasmic reticulum appears as a closely woven meshwork of smooth-walled tubules and vesicles.

The surface features revealed by SEM and referred to above for male cells all occur in the female gland. Thus, some cells have dense microvilli while in others the microvilli are sparse and the surface is pitted as a result of lipid vacuole release.

3.1.3 Classification of Epithelial Cell Types

The number of glandular epithelial cell types varies between species (at present, from one to three) and, in the golden hamster, between the sexes. In the hamster, classification may be based on:

1. Lipid vacuole size. Cells with numerous small vacuoles are usually referred to as Type I cells, those with large vacuoles as Type II cells (Hoffman 1971). These are shown in Figs. 1 and 2.
2. Electron density. Type II cells have been reported as being more electron dense (darker). This is said to be due to the possession of more ribosomes (Bucana and Nadakavukaren 1972a), but it is possible that, if the lipid vacuoles are larger, other cellular organelles become more densely packed.
3. The cell surface. As noted above, Type II cells have a surface relatively devoid of microvilli compared with Type I cells.

Does this mean that Type I and II are fundamentally different cell types? Clearly this cannot be the case. Castration leads to a loss of Type II cells in the male hamster Harderian gland, while androgen administration restores them; similarly, if androgens are administered to female hamsters, Type II cells appear de novo in the Harderian gland (Sun and Nadakavukaren 1980). They must therefore represent the same cell type in different stages of a secretory cycle or in different stages of activity or development. In addition, there is no evidence of separate precursor cells during early development of the gland (Bucana and Nadakavukaren 1973).

3.2 The Luminal Contents

The lumina of many tubule profiles contains large quantities of secretory material such as intact lipid vacuoles together with other cytoplasmic debris and nuclei; these often remain remarkably well preserved.

In females, the lumen may also contain solid accretions which are visible to the naked eye and which render the female gland a brown/black color compared to the pale yellow/brown of the male gland. These accretions autofluoresce red under ultraviolet light and are therefore assumed to possess a high porphyrin content; the denser ones do not fluoresce, probably because their porphyrin content is so high as to produce a quenching effect. They show a layered appearance, evidence of cumulative deposition of material, and TEM examination reveals a crystalline appearance typical of porphyrin. Since porphyrin synthesis is suppressed by androgens, it is not surprising that these accretions are also found (1) in castrated males and (2) briefly, in early postnatal males.

While lipid vacuoles are shed by exocytosis and, possibly, by other mechanisms (see above), it is still not clear where porphyrin is stored within the epithelial cell nor how it is secreted. Crystalline porphyrin can sometimes be visualised within degenerating hamster epithelial cells, while profiles which resemble porphyrin crystals occur in rat Harderian gland cells which are believed to be dying (Carriere 1985). In such cases, the porphyrin presumably leaves the cell by a holocrine mechanism. In more viable cells, it may be that porphyrin leaves by a merocrine mechanism, solubilized within the lipid vacuoles (Brownscheidle and Niewenhuis 1978) which form the bulk of the luminal contents and which may sometimes be seen "trapped" within luminal porphyrin accretions.

In some circumstances, the tubule lumen is invaded by neutrophils. This usually occurs in ovariectomised or aged post-reproductive females where tubule degeneration is also present (Payne et al. 1985; Spike et al. 1985, 1986, 1988). These neutrophils differ from those within the interstitium in having numerous vacuoles and filopodial extensions. Their appearance suggests active phagocytosis, probably of lipid or cellular debris.

3.3 Myoepithelial Cells

Myoepithelial cells occur in many glands such as breast and lacrimal, salivary, and sweat glands, as well as in the Harderian gland. They form a complex network around the tubules (Leeson 1960), in keeping with their presumed role in assisting the transport of luminal contents. In addition, scanning studies by the authors have revealed cells forming rings around tubules; if these are myoepithelial cells, an additional function may be to restrict luminal flow. The myoepithelial cells lie between the basal surface of the epithelial cells and the basal lamina (Fig. 1).

The fine structure of myoepithelial cells has been described in the Harderian gland of the rat (Chiquoine 1958; Leeson 1960) and is similar to that found in the golden hamster (Bucana and Nadakavukaren 1972a). The nucleus is centrally placed and elongated. The cytoplasm contains fine contractile filaments and pinocytotic vesicles.

Junctions between myoepithelial cells have not been described. Hemi-desmosomes are often present at the basal surface of the myoepithelial cell where it apposes the basal lamina of the epithelium.

3.4 Interstitial Tissues

The interstitial tissue between the gland tubules is relatively sparse; it is much more substantial around the extraglandular portion of the secretory duct. Between the tubules, the interstitium consists of sparse bundles of collagen fibers with occasional fibroblasts. It is highly vascular and also contains many lymphatics. In some species such as the mouse (Strum and Shear 1982) and the woodmouse *Apodemus sylvaticus* (Johnston et al. 1987) the capillaries are fenestrated; fenestrations may occur in the hamster, but they are not pronounced. The presence of sinusoids has recently been argued in the hamster (Menendez-Pelaez et al. 1990). In some species (e.g., the gerbil; Sakai and Yohro 1981) the gland is surrounded by a connective tissue capsule whose outer cellular layer is believed to act also as the wall of a retro-orbital venous sinus; the situation for the hamster is not known.

Unmyelinated nerves follow the course of the blood vessels; while some end on vessels, others make contact with myoepithelial cells and others are said to end directly on secretory cells (Bucana and Nadakavukaren 1972b; Vega-Alvarez et al. 1989). There are both cholinergic and adrenergic terminals (Norvell and Clabough 1972). Little is known of the functions of Harderian gland innervation, but cholinergic stimulation leads to the release of porphyrin-rich "blood tears" in rats (Tashiro et al. 1940). It is likely, therefore, that ductal transport is under neural control. The source of the nerve supply is not known in detail for the golden hamster; in the rat, cervical sympathectomy or destruction of the ophthalmic division of the trigeminal nerve both resulted in the loss of adrenergic terminals, while cutting the zygomatic nerve led to a loss of cholinergic fibers (Huhtala et al. 1977).

Mast cells are a prominent component of the interstitium in female hamsters; quantitative studies reveal that they are some 40 times more numerous in glands of females than in those of males (Payne et al. 1982). A sex difference in mast cell numbers has also been found in the Harderian glands of mice, with females possessing some four times more than males (Shirama et al. 1988). In neither species is there a clear sex difference in mast cell structure. In the hamster, castration of males leads to a rapid rise in mast cell numbers to female levels.

In some circumstances (after ovariectomy or in post-reproductive senesence), porphyrin occurs in the interstitium of the female hamster Harderian gland. It occurs in two locations, both readily seen by light microscopy and confirmed by TEM which reveals the crystalline structure (Payne et al. 1985; Spike et al. 1985, 1986).

1. Large interstitial deposits which appear exactly the same as intraluminal accretions. These probably arise as a result of the degeneration and disappearance of the surrounding section of tubule and are usually invested and sequestered by a foreign body giant cell.
2. Small interstitial deposits which appear within individual free macrophages. These deposits vary in form and density, perhaps suggesting different degrees of solubilization. The macrophages are often surrounded by groups of plasma cells which may suggest some form of immune response.

3.5 The Secretory Duct

There is no morphologically distinct duct system within the substance of the rodent Harderian gland, either in the golden hamster or in any other species which has been detailed. Despite the number of papers which have dealt with hamster Harderian gland morphology, the extraglandular duct has received little attention. There are three cell types in the hamster secretory duct (shown diagrammatically in Fig. 3).

1. Ciliated cells. These are less numerous than nonciliated cells, but are nevertheless readily found (see Fig. 4a, b). They are columnar cells with a centrally

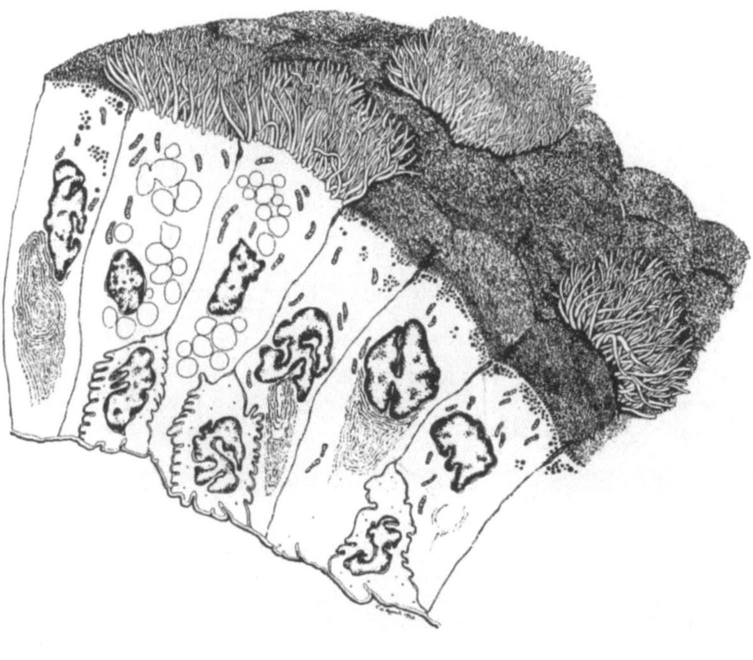

Fig. 3. A diagrammatic representation of cells of the duct of a hamster Harderian gland. Three cell types are present: (1) nonciliated cells with lamellar bodies and apical mucoid granules; (2) ciliated cells with large lipid vacuoles; (3) basal cells

Fig. 4a. A portion of the secretory duct of a hamster Harderian gland. In the center is a ciliated cell with large lipid vacuoles, while to the *left* is a nonciliated cell with dense apical mucoid granules. (× 4200). **b** SEM of the surface of a hamster Harderian gland duct. A single ciliated cell can be seen surrounded by nonciliated cells bearing stubby microvilli. (× 6000). **c** A high-powered view of the apical regions of a ciliated cell (*right*) showing lipid vacuoles and a nonciliated cell (*left*) showing apical mucoid granules. (× 12200). **d** A portion of the secretory duct of a hamster Harderian gland. This shows not only ciliated and nonciliated cells which abut the lumen, but also shows basal cells. (× 2500)

placed nucleus, which is generally small and regular. The cell surface bears not only true kinocilia of typical structure, but also thin elongated microvilli. A conspicuous feature of the cytoplasm of these cells is clusters of clear vacuoles which probably contain lipid; these can be both apical and basal to the nucleus.

2. Nonciliated cells. These form the majority of duct cells. They are also columnar epithelial cells with a centrally placed large irregularly shaped nucleus. The cytoplasm shows complex lammellar structures occupying the basal part of the cell. The cell surface bears numerous microvilli (Fig. 4b) and the apical region contains numerous small electron-dense granules (Fig. 4c). The ultrastructural appearance of these suggests they are mucoid granules, and this is borne out by the histochemical finding that the apical regions of these cells are PAS-positive.
3. Basal cells. Small squat pyramidal cells with irregularly shaped nuclei are basally placed within the duct epithelial wall (Fig. 4d). They have relatively few organelles and their lateral walls are greatly folded where they interdigitate with the ciliated or nonciliated cells. They are quite distinct from cells such as lymphocytes, which are observed migrating through the duct wall and from myoepithelial cells. Nevertheless, we cannot at present wholly rule out the possibility that their cytoplasm may make contact with the duct surface (i.e., that the epithelium is pseudo-stratified) though we believe this to be unlikely.

4 Hormones and the Harderian Gland

Since the hamster Harderian gland exhibits sex differences in a number of its structural characteristics (cell types, polytubular complexes, mast cell numbers, etc.) as well as in its biochemistry, it is pertinent to ask whether structure is genetically or hormonally controlled. All available information suggests that gonadal hormones are of major importance in controlling morphology and biochemistry. This is considered in detail in the chapters by Nadakavukaren and by Payne, Spike and Moore in this Volume, but the following points may be made:

1. Castration of male hamsters results in a complete change of gland structure to the female pattern. This is accompanied by an increase in porphyrin synthesis and accumulation and by changes in other biochemical parameters. These changes can be prevented by androgen administration.
2. Androgen administration to female hamsters results in masculinization of gland structure and a marked decrease in porphyrin synthesis and content. Ovariectomy per se does not masculinize the female gland, but is accompanied by degenerative changes in tubule structure and a relocation of porphyrin stores.

5 Development of the Hamster Harderian Gland

Few of the sexual dimorphisms present in the adult hamster Harderian gland are present at birth. The tubules are widely separated by connective tissue and the epithelial cells possess marked interdigitations (Bucana and Nadakavukaren 1973). Polytubular complexes in male glands appear during the second week after birth, and Bucana and Nadakavukaren (1973) have suggested that these arise from juxtanuclear structures and that their appearance coincides with the disappearance of porphyrin from the male gland. At birth, gland cells possess no lipid vacuoles (these appear at the end of the first week) and few microvilli. In glands of females, membranous formations appear during the second week after birth and lipid vacuole size (which is initially variable) conforms to the normal female pattern by one month after birth.

Mitotic figures are rare in the Harderian glands of adult hamsters, but they occur frequently in the glands of both sexes until 3 weeks postpartum, by which time sexual dimorphism of the gland is evident (Bucana and Nadakavukaren 1972c).

6 Comparisons with Other Species and Overview

The Harderian gland of the golden hamster resembles that of other species in its general characteristics, viz.,

1. that the gland tubules consist of a single layer of columnar epithelial cells with basal nuclei and large numbers of lipid vacuoles which are shed by exocytosis;
2. that these epithelial cells are surrounded by a layer of myoepithelial cells interposed between them and their basal lamina;
3. that there is no morphologically distinct duct system within the substance of the gland and the extraglandular secretory duct opens onto the surface of the nictitating membrane.

It differs markedly from other species so far studied in showing major sex differences in cell types, ultrastructure, and connective tissue elements, and in the presence of a virtually unique ultrastructural feature in male cells – the polytubular complexes. It differs in degree from other species in the sheer amount of porphyrin stores contained within the female (but not the male) gland.

It differs above all from other species in the extraordinary lability of gland structure to a variety of manipulations, and in that structure and gland activity co-vary in a robust and satisfying manner. The most obvious example is afforded by porphyrin biosynthesis. Here, every manipulation which alters gland synthesis also appears to alter gland morphology. This mixture of lability and co-variance of structure and activity makes the hamster Harderian gland a remarkable experimental model on which research can only intensify in the decade to come.

References

Brownscheidle CM (1974) The morphology and histochemistry of the Harderian gland of the Mongolian gerbil *Meriones unguiculatus*. PhD Thesis, State University of New York at Buffalo, Buffalo, NY

Brownscheidle CM, Niewenhuis RJ (1978) Ultrastructure of the Harderian gland in male albino rats. Anat Rec 190:735–754

Bucana CD, Nadakavukaren MJ (1972a) Fine structure of the hamster Harderian gland. Z ZellforschMikrosk Anat 129:178–187

Bucana CD, Nadakavukaren MJ (1972b) Innervation of the hamster Harderian gland. Science 175:205–206

Bucana CD, Nadakavukaren MJ (1972c) Ultrastructural investigation of the postnatal development of the hamster Harderian gland. I. Cell division. Z ZellforschMikrosk Anat 135:149–153

Bucana CD, Nadakavukaren MJ (1973) Ultrastructural investigation of the postnatal development of the hamster Harderian gland II. Male and female. Z Zellforsch Mikrosk Anat 142:1–12

Carriere R (1985) Ultrastructural visualization of intracellular porphyrin in the rat Harderian gland. Anat Rec 213:496–504

Chiquoine AD (1958) The identification and electron microscopy of myoepithelial cells in the Harderian glands. Anat Rec 132:569–583

Christensen F, Dam H (1953) A sexual dimorphism of the Harderian glands in hamsters. Acta Physiol Scand 27:333–336

Clabough JW, Norvell JE (1973) Effects of castration, blinding and the pineal gland on the Harderian glands of the male golden hamster. Neuroendocrinology 12:344–353

Cohn SA (1955) Histochemical observations on the Harderian gland of the albino mouse. J. Histochem Cytochem 3:342–353

Corbet GB, Hill JE (1986) A world list of mammalian species. British Museum (Natural History), London

Fulton GP (1968) The golden hamster in biomedical research. In: Hoffman RA, Robinson PF, Magalhaes (eds) The golden hamster: its biology and use in medical research. Iowa State University Press, Ames, IA

Gorgas K (1984) Peroxisomes in sebaceous glands. V. Complex peroxisomes in the mouse preputial gland; serial sectioning and three-dimensional reconstruction studies. Anat Embryol 169:261–270

Gorgas K, Zaar K (1984) Peroxisomes in sebaceous glands. III. Morphological similarities of peroxisomes with smooth endoplasmic reticulum and Golgi stacks in the circumanal gland of the dog. Anat Embryol 169:9–20

Grafflin AL (1942) Histological observations upon the porphyrin-excreting Harderian gland of the albino rat. Am J Anat 71:43–64

Hoffman RA, (1971) Influence of some endocrine glands, hormones and blinding on the histology of the Harderian glands of golden hamsters. Am J Anat 132:463–478

Hoffman RA, Jones JW (1981) Concentrations of metals in the Harderian glands of male and female hamsters. Comp Biochem Physiol 69A:153–156

Hoffman RA, Johnson LB, Reiter RJ (1985) Harderian glands of golden hamsters: temporal and sexual differences in immunoreactive melatonin. J Pineal Res 2:161–168

Hoh JH, Lin W-L, Nadakavukaren MJ (1984) Sexual dimorphism in the Harderian gland proteins of the golden hamster. Comp Biochem Physiol 77B:729–731

Huhtala A, Huikuri KT, Palkama A, Tervo T (1977) Innervation of the rat Harderian gland by adrenergic and cholinergic nerve fibres. Anat Rec 188:263–272

Johnston HS, McGadey J, Thompson GG, Moore MR, Payne AP (1983) The Harderian gland, its secretory duct and porphyrin content in the Mongolian gerbil, (*Meriones unguiculatus*). J Anat 137:615–630

Johnston HS, McGadey J, Thompson GG, Moore MR, Breed WG, Payne AP (1985) The Harderian gland, its secretory duct and porphyrin content in the Plains mouse (*Pseudomys australis*). J Anat 140:337–350

Johnston HS, McGadey J, Payne AP, Thompson GG, Moore MR, (1987) The Harderian gland, its secretory duct and porphyrin content in the woodmouse (*Apodemus sylvaticus*). J Anat 153:17–30

Jones CW, Hoffman RA (1976) Porphyrin concentration of the hamster (*Mesocricetus auratus*) Harderian gland: effects of incubation with delta-aminolevulinic acid and various hormones. Int J Biochem 7:135–140

Kittel R (1962) Vergleichend-anatomische Untersuchung über die Orbitaldrüsen der Rodentia. Wiss Z Univ Halle Math/Nat XI: 401–428

Leeson CR (1960) The histochemical identification of myoepithelium, with particular reference to the Harderian and exorbital lacrimal glands. Acta Anat 40:87–94

Lin W-L, Nadakavukaren MJ (1979) The androgenic effect on the fine structure of the Harderian gland in the male hamster. Cell Tissue Res 198:119–127

Lin W-L, Nadakavukaren MJ (1981) Harderian gland lipids of male and female golden hamsters, *Mesocricetus auratus*. Comp Biochem Physiol 70:627–630

Menendez-Pelaez A, Reiter RJ, Guerrero JM, Puig-Domingo M, Howes KA (1988) Sexual dimorphism in N-acetyltransferase activity, hydroxyindole-O-methyltransferase activity, and melatonin content in the Harderian gland of Syrian hamsters: Changes following gonadectomy. Proc Soc Exp Biol Med 187:287–291

Menendez-Pelaez A, Martinez-Telleria A, Vega JA, Coalla C, Reiter RJ (1989) Histochemical detection of monoamine oxidase and alcohol dehydrogenase activities in the Syrian hamster Harderian glands: existence of a sexual dimorphism. Histochem J 21:125–130

Menendez-Pelaez A, Tolivia D, Rodriguez-Colunga MJ (1990) Ultrastructure of the blood vessels in the Syrian hamster Harderian gland: Existence of sinusoids. J Morphol 204:257–263

Murphy MR (1971) Natural history of the Syrian golden hamster – a reconnaissance expedition. Am Zool 11:632

Norvell JE, Clabough JW (1972) Adrenergic and cholinergic innervation of the hamster Harderian gland. Science 178:1102–1103

Payne AP (1979) The attractiveness of Harderian gland smears to sexually naive and experienced male golden hamsters. Anim Behav 27:897–904

Payne AP (1990) The Harderian gland: its biology and role in porphyrin synthesis. Mol Aspects Med 11:145–152

Payne AP, McGadey J, Moore MR, Thompson GG (1977) Androgenic control of the Harderian gland in the male golden hamster. J Endocrinol 75:73–82

Payne AP, McGadey J, Johnston HS, Moore MR, Thompson GG (1982) Mast cells in the hamster Harderian gland: sex differences, hormonal control and relationship to porphyrin. J Anat 135:451–461

Payne AP, McGadey J, Johnston HS (1985) Interstitial porphyrin and tubule degeneration in the hamster Harderian gland. J Anat 140:25–36

Puig-Domingo M, Guerrero JM, Reiter RJ, Peinado MA, Menendez-Pelaez A, Santana C, Webb SM (1988) Androgenic control of immunoreactive somatostatin in the Harderian gland of the Syrian hamster. J Reprod Fertil 82:753–759

Sakai T, Yohro T (1981) A histological study of the Harderian gland of Mongolian gerbils, *Meriones meridianus*. Anat Rec 200:259–270

Shirama K, Furuya T, Takeo Y, Shimizu K, Maekawa K (1981) Influences of some endocrine glands and of hormone replacement on the porphyrins of the Harderian glands of mice. J Endocrinol 91:305–311

Shirama K, Kohda M, Hokano M (1988) Effects of endocrine glands and hormone replacement on the mast cell count of the Harderian gland of mice. Acta Anat 131:327–331

Spike RC, Johnston HS, McGadey J, Moore MR, Thompson GG, Payne AP (1985) Quantitative studies on the effects of hormones on structure and porphyrin biosynthesis in the Harderian gland of the female golden hamster. I. The effects of ovariectomy and androgen administration. J Anat 142:59–72

Spike RC, Johnston HS, McGadey J, Moore MR, Thompson GG, Payne AP (1986) Quantitative studies on the effects of hormones on structure and porphyrin biosynthesis in the Harderian gland of the female golden hamster. II. The time course of changes after ovariectomy. J Anat 145:67–77

Spike RC, Payne AP, Moore MR (1988) The effects of age on the structure and porphyrin synthesis of the Harderian gland of the female golden hamster. J Anat 160:157–166

Strum JM, Shear CR (1982) Harderian glands in mice: fluorescence, peroxidase activity and fine structure. Tissue Cell Res 14:135–148

Sun CY, Nadakavukaren MJ (1980) Effect of testosterone on the female hamster Harderian gland pigmentation and ultrastructure. Cell Tissue Res 207:511–517

Tashiro S, Smith CC, Badger E, Kezur E (1940) Chromadacryorrhea, a new criterion for biological assay of acetylcholine. Proc Soc Exp Biol Med 44:658–661

Vega-Alvarez JA, Menendez-Pelaez A, Coalla-Perez C, Martinez-Telleria A (1989) Sexual dimorphism in the cholinergic innervation of the hamster (*Mesocricetus auratus*) Harderian glands. Z Mikrosk Anat Forsch (Leipz) 103:166–172

Woodhouse MA, Rhodin JAG (1963) The ultrastructure of the Harderian gland of the mouse with particular reference to the formation of its secretory product. J Ultrastruct Res 9:76–98

Woolley GW, Worley J (1954) Sexual dimorphism in the Harderian gland of the hamster (*Cricetus auratus*). Anat Rec 118:416–417 (Abstr)

The Mammalian Harderian Gland: Ultrastructure, Sexual Dimorphism, and Its Regulation by Light and Steroids

MATHEW J. NADAKAVUKAREN

Department of Biological Sciences, Illinois State University, Normal, IL 61761, USA

1 Introduction

One of the earliest ultrastructural studies of the Harderian gland was by Chiquoine (1958), in which he examined the myoepithelial cells of the Harderian glands in mice, rabbits, and rats. Since then, there have been a number of reports of Harderian gland ultrastructure from a variety of animals. However, it is clear from a survey of Harderian gland literature that two groups of animals, the rodents and the birds, have received the most attention. General ultrastructure of the Harderian gland has been described in the armadillo (Weaker 1981), fowl (Rothwell et al. 1972; Kittner et al. 1978; Niedorf and Wolters 1978; Maxwell and Burns 1979), frog (DiMatteo et al. 1989), gerbil (Sakai and Yohro 1981; Johnston et al. 1983), golden hamster (Bucana and Nadakavukaren 1972a, b, c, 1973), lizard (Chieffi Baccari et al. 1990), mouse (Woodhouse and Rhodin 1963; Watanabe 1980; Strum and Shear 1982a; Johnston et al. 1985, 1987), rabbit (Björkman et al. 1960; Kühnel 1971; Wooding 1980), rat (Kelényi and Orbán 1965; Tsutsumi et al. 1966; Feria-Velasco and Tapia-Arizmendi 1974; Brownscheidle and Niewenhuis 1978), toad (Minucci et al. 1989), and turkey (Maxwell and Burns 1979; Maxwell et al. 1986). Although Harderian glands are known to be present in a number of other species (see review by Sakai 1981), ultrastructural studies of these are currently lacking.

The most common histological feature of all Harderian glands reported thus far is the secretory epithelium. In general, two types of secretory cells are found in the Harderian glands of rats, mice, armadillos, and male hamsters and only one type in gerbils and female hamsters. In the armadillo, the difference in cell types is based on secretory products rather than on vacuole size. In all mammals investigated, except the rabbit, there are no color or cellular differences between the lobes of a Harderian gland. Although the basic difference between the two cell types at the light microscope level is the size and number of secretory vacuoles, at the ultrastructural level there are several additional features that can be recognized, some of which are highly species-specific.

Webb et al. (Eds).
Harderian Glands

2 Secretory Epithelium

There are many common ultrastructural features present in the Harderian glands of mammalian species. The secretory epithelium has a single layer of columnar cells, each with a basally located nucleus. These cells usually have microvilli along their free surfaces and are connected at the apical region by typical junctional complexes. Interdigitations and desmosomes are often present along the lateral sides. Along the basal region, they are in contact with either the basal lamina or with the cytoplasmic extensions of the myoepithelial cells. Interdigitations along the basal lamina or with the myoepithelial cells are not uncommon, although desmosomes are more common between the secretory cells and myoepithelical cells than between the basal membrane and the secretory cells. In rodents, the secretory cells have many membrane-bounded lipid vacuoles, often concentrated at the apical region. The presence of a well-developed smooth endoplasmic reticulum and Golgi complex, sparse rough endoplasmic reticulum, numerous mitochondria, and free ribosomes are also common features of these cells in mammals. Porphyrin, when present, is usually found in the alveolar lumen and is seldom recognized in the cytoplasm, even though it is known to be synthesized by secretory cells of certain mammalian species.

2.1 Rabbit

The ultrastructure of the secretory epithelial cells of the rabbit Harderian gland was reported by Björkman et al. (1960) and Kühnel (1971). The rabbit Harderian gland is unique in being divided into a small white lobe and a larger red lobe. In general, all secretory cells of a given lobe are of the same kind, although occasional differences have been noticed (Björkman et al. 1960). The secretory epithelium of the white lobe consists of cells containing lipid vacuoles that are half the size of those in the red lobe. While the vacuoles in both lobes are surrounded by a membrane, an irregular black precipitate was seen at the inner surface of the membranes of those in the red lobe. Björkman et al. (1960) reported degenerating cells, especially in the white lobe. Since these authors had assumed a merocrine mode of secretion, this degenerative phenomenon may have been a natural part of the aging process rather than a holocrine mode of secretion. Björkman et al. (1960) also reported slits in the cytoplasm between vacuoles in the white lobe. The only other inclusion that is similar to the slits is the "slashes" that were reported in the Harderian gland of the gerbil (Johnston et al. 1983) and plains mouse (Johnston et al. 1985).

2.2 Armadillo

The Harderian gland of the nine-banded armadillo (*Dasypus novemcinctus*) was investigated ultrastructurally by Weaker (1981). The gland synthesized different secretory products in two different parts, the minor proximal region containing mucus-secreting alveoli and the major distal portion made up of lipid-secreting alveoli. There was only one type of secretory epithelial cell present in each alveolus. Two types of granules, amorphous dense-staining and granulated less dense-staining, were found along the periphery of the lipid-secreting cells, but in higher concentration near the luminal surface. Pleomorphic mitochondria with lamellar cristae were common in these cells. The mucus-secreting cells had a well-developed Golgi complex and rough endoplasmic reticulum. The most characteristic feature was the large number of secretory granules of varied size and density concentrated at the apical portion.

2.3 Gerbil

The ultrastructure of the secretory cells in the Harderian gland of the Mongolian gerbil (*Meriones meridianus*) was originally described by Sakai and Yohro (1981). The alveoli contained one type of secretory cell. The lipid-containing secretory vacuoles appeared to be released into the alveolar lumen by exocytosis. The secretory cells were characterized by a well-developed, tubular, smooth endoplasmic reticulum occupying most of the cell and a few scattered cisternae of rough endoplasmic reticulm. Rod-shaped mitochondria were usually found along the lateral sides of the cells.

Subsequently, Johnston et al. (1983) described the Harderian gland ultrastructure of a closely related species of the Mongolian gerbil (*Meriones unguiculatus*). The most apparent difference between the secretory cells of the two species is the presence of cytoplasmic "slashes", which are totally absent from *M. meridianus*. In addition, the rough endoplasmic reticulum seems to be more developed and the smooth endoplasmic reticulum less developed in the secretory cells of *M. unguiculatus*, which is the opposite of the situation in *M. meridianus*. Johnston et al. (1983) considered the cytoplasmic "slashes" to be an intermediate stage in the formation of the lipid-filled secretory vacuole. These cytoplasmic "slashes" are the most unique feature of *M. unguiculatus* Harderian gland.

2.4 Rat

The ultrastructure of the rat Harderian gland has been described by different investigators. Kelényi and Orbán (1965) reported two cell types in the secretory epithelium of the infantile rats and four cell types in the adult rats based on the number and size of the vacuoles. They concluded that these cell types must represent a transition between the different developmental stages of the same cell type. They also suggested that the secretory vacuoles are formed from a well-developed Golgi system found in Type 2 cells. Tsutsumi et al. (1966) reported two cell types, clear and dark, in the secretory epithelium, based on the electron density of the cytoplasmic matrix. They noted a number of 7–30-nm granules in cytoplasm of both cell types, and suggested that they may be complex salts of porphyrin associated with a certain metal.

A more detailed ultrastructural study of the rat Harderian gland was published by Brownscheidle and Niewenhuis (1978) in which they described two distinctly recognizable secretory cell types, arbitrarily called A and B. There were more A cells than B cells in most of the alveoli. Although the B cells were darker than A cells when viewed under a light microscope, the electron densities of these cells were somewhat similar on their micrographs. The A cells were characterized by the large size and fibrogranular appearance of the secretory vacuoles, which lacked a limiting membrane. The vacuoles of the B cells were smaller than those of the A cells and contained a lamellated, electron opaque material around the periphery. The mitochondrial outlines were larger in cell type A than in B and profiles of smooth endoplasmic reticulum appeared to be more prominent in A than in B. However, the smooth endoplasmic reticulum in Type B cells was frequently arranged in a lattice-work pattern, giving the appearance of tubular clusters, which was never seen in Type A cells. The ultrastructural differences of the secretory vacuoles between the two cell types were thought to have resulted from the differences in the solubility properties of the various lipid components present in the vacuoles. In both cell types, the secretory vacuoles were believed to have originated from the smooth endoplasmic reticulum and to be released into the alveolar lumen by a merocrine process. However, it should be pointed out that there are conflicting suggestions as to the origin of the vacuoles between these authors and Kelényi and Orbán (1965), who reported the origin of the vacuoles from the Golgi. Brownscheidle and Niewenhuis (1978) have suggested that their A cells may correspond to cell Type 4 described by Kelényi and Orbán (1965) and B cells to Cell Type 3.

2.5 Mouse

The ultrastructure of the Harderian gland in the laboratory mouse has been examined by several investigators (Woodhouse and Rhodin 1963; McCafferty and Clark 1969; Watanabe 1980; Strum and Shear 1982a). All of them reported two ultrastructurally distinct cell types, A and B, in the secretory epithelium. Except for a few minor differences, all reports agree on the basic characteristics of these two cell types. Strum and Shear (1982a) observed prominent gap junctions only between Types A and B cells. Woodhouse and Rhodin (1963) noticed increased basal plasma membrane infoldings in cell Type B. Smooth endoplasmic reticulum was prominent in both cell types, but less abundant in Type A, according to Watanabe (1980). Rough endoplasmic reticulum was scarce in both, but Strum and Shear (1982a) reported more in Type B than in A. Watanabe (1980) noticed larger mitochondria in Type A than in B.

Type A cell was the most numerous component in the alveoli and appeared to have a dense cytoplasm. These cells had numerous large homogenous lipid-containing membrane-bounded vacuoles. On the other hand, all lipid vacuoles in Type B cells were surrounded by a densely stained fibrillar material and concentric layers of membranes, which Strum and Shear (1982a) assumed to be deposits of phospholipids. They also reported unusual dense cylindrical structures contained within a double-layered membrane in Type B cells. Although these appeared to contribute to porphyrin pigment formation in the alveolar lumen, fluorescence microscope data do not support this role. Only Type A cells appear to secrete porphyrins.

Woodhouse and Rhodin (1963) conducted an ultrastructural study of the mouse Harderian gland with special reference to the formation of its secretory product. In Type A cells, the definitive secretory product is formed by the fusion between dense bodies and lipid-containing secretory vacuoles during the regenerative stage of the cell. Although there is some doubt as to the origin of the dense bodies and the secretory vacuoles, it has been suggested that both structures are derived from the Golgi apparatus. Considerably more secretory vacuoles than dense bodies were present in a cell during the stage of regeneration. The fusion between these two structures seems to take place in a synchronized manner. Incorporation of materials from the cytoplasm into these structures prior to fusion cannot be ruled out. The formation of the dense granule, the final secretory product of the Type B cell, seems to be less clear. Woodhouse and Rhodin (1963) described the dark granule in its definitive form as a thick, dense, osmiophilic membrane surrounding a less dense homogeneous central area in which a particulate precipitate is usually present. They proposed that these secretory granules are formed by the interaction of membrane-bounded structures called R and S forms, which are thought to be derived from the endoplasmic reticulum. In both cell types, the process of secretion was considered to be merocrine.

2.6 Plains Mouse

More recently, Johnston et al. (1985) described the ultrastructure of the Harderian glands in the plains mouse (*Pseudomys australis*). The glandular alveoli consisted of a single layer of secretory epithelial cells. The authors reported three different forms of secretory epithelial cells. Type I made up 60–80% of the secretory epithelium. This cell type had many apically located lipid vacuoles and lacked a typical Golgi. However, it contained numerous small mitochondria. The major difference between the Type I and Type II cells, which represented only 3–9% of the cell population, was the size, shape and distribution of mitochondria. These mitochondria were several times larger than those of Type I and were usually found in a "stack" formation and occasionally in concentric formations. Their cristae also were very distinct from those of mitochondria in the other cell types. Type II cells had well-formed Golgi. The lipid vacuoles in the Type II cells were intermediate in number and size to those of Type I. In addition, cytoplasmic "slashes" were present in some of the Type II cells. Approximately 11–29% of the secretory epithelium consisted of Type III cells, which contained a few very small lipid vacuoles, small but numerous mitochondria, tiny densely packed vesicles and no typical Golgi.

2.7 Woodmouse

In an ultrastructural study of the Harderian gland in the woodmouse (*Apodemus sylvaticus*) Johnston et al. (1987) reported two types of secretory epithelial cells, I and II, in approximately equal numbers. The Type I cells contained many small, clear lipid vacuoles while those in the Type II cells were large and electron dense. Most of the vacuoles in the Type II cells possessed an elaborate cap of multi-layered dense membranes, possibly deposits of phospholipids but of unknown function. The caps on the vacuoles of the Type I cells, when present, were less elaborate. The most unique feature of the Type II cells was the large number of polytubular bundles found in continuity with the smooth endoplasmic reticulum. Each bundle consisted of a variable number of tubules, each approximately 40 nm in diameter. The functional significance of these tubules is not understood. It is interesting to note that the only other mention of tubular structures in the Harderian literature is the clusters of cylindrical tubules present in the male hamster Harderian glands, the functional role of which is also unknown (Bucana and Nadakavukaren 1972a). Although the individual tubules in both cases are of the same diameter, the tubular clusters in the hamster were often membrane bounded and of unknown origin.

2.8 Hamster

The ultrastructure of the hamster (*Mesocricetus auratus*) Harderian gland was originally described by Bucana and Nadakavukaren (1972a). Several additional studies have been reported since then on various aspects of the gland, particularly related to its sexual dimorphism. The two unequal lobes of the gland are histologically similar, containing a single layer of secretory epithelium with two types of cells, those with large lipid vacuoles and those with small lipid vacuoles. In the male, approximately one third of the cells had large vacuoles and two thirds had small vacuoles, while cells in the female usually had only small vacuoles. Although the authors did not refer to these two different kinds of cells as Types I or II, by the current convention of nomenclature, the cells with the small vacuoles will be Type I and those with large vacuoles Type II. In addition to the two types based on the size of the vacuoles, two additional types based on electron density and other features were recognized at the ultrastructural level in both male and female hamsters. These were called "light" and "dark" cells (Fig. 1). Although there were some differences that can be recognized at the ultrastructural level between light and dark cells within the same sex, it is interesting that there were more differences between the secretory cells of the male and female than between light and dark cells within a given sex. The density difference was due to an abundance of ribosomes in the dark cell cytoplasm.

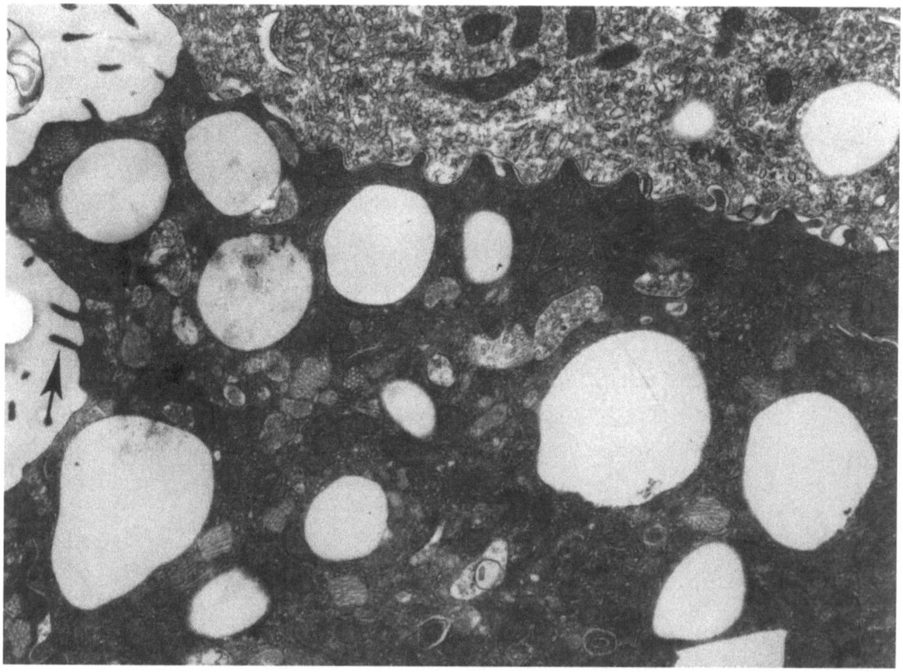

Fig. 1. Electron micrograph of parts of a light cell and a dark cell of a male golden hamster Harderian gland. Note interdigitations of the cell membranes. *Arrow* indicates a microvillus projecting into the alveolar lumen. × 15000. (Micrograph by Dr. Corazon Bucana)

Prominent interdigitations of the cell membranes of two adjacent light and dark cells were commonly seen (Fig. 1). Usually the dark cells had more mitochondria and larger vacuoles than the light cells. The vacuoles in all cells were bounded by a single membrane.

The light cells from the female were characterized by the presence of an extensive membrane system arranged in concentric lamellar formations or in the form of Golgi with proliferated cisternae (Fig. 2). The cytoplasm also contained many membranous whorls which were often associated with vacuoles. Although the origin of the secretory vacuoles could not be determined, there is ultrastructural evidence to suggest possible incorporation of Golgi products into the vacuoles. A characteristic feature of the dark cells in the female was the extensive distribution of dilated cisternae of smooth endoplasmic reticulum in the cytoplasm. A unique feature of the secretory cells from the Harderian glands of male hamsters was the presence of clusters of membrane-bounded cylindrical tubules that were randomly distributed in the cytoplasm of both light and dark cells (Fig. 3). Isolated tubules were also commonly seen. There was considerable variation in the number of tubules in a cluster. Each tubule was approximately 40 nm in diameter and appeared to be made up of a number of tubular subunits,

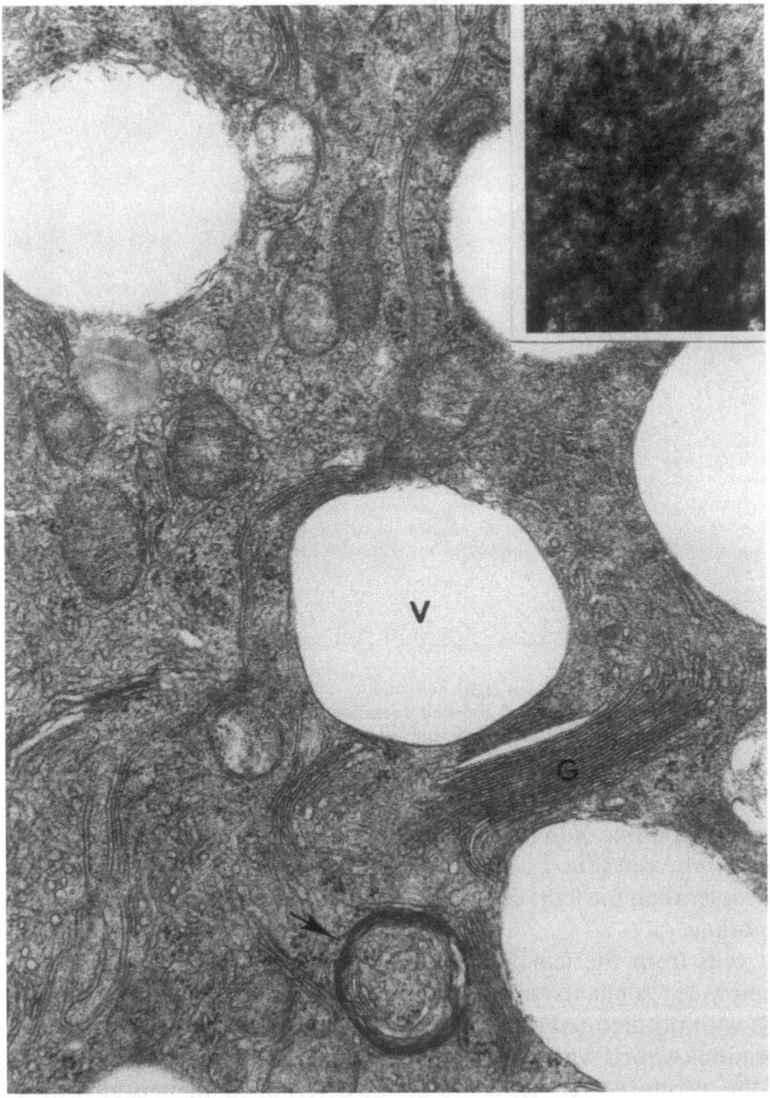

Fig. 2. Section through portion of a light cell from the Harderian gland of a female golden hamster showing extensive membrane formations. Note concentrically arranged lamellae (*arrow*) as well as Golgi with proliferated cisternae (*G*). Vacuoles (*V*) are often closely associated with these formations. ×45000. *Inset* Section of a pigment accretion in the alveolar lumen showing a fibro-granular needle-like substructure. ×62000. (Micrographs by Dr. Corazon Bucana)

the exact number of which was difficult to determine (Fig. 3, inset). These tubules were often seen in close proximity to vacuoles, nuclei, and mitochondria. There is ultrastructural evidence that indicates the incorporation of tubules into the

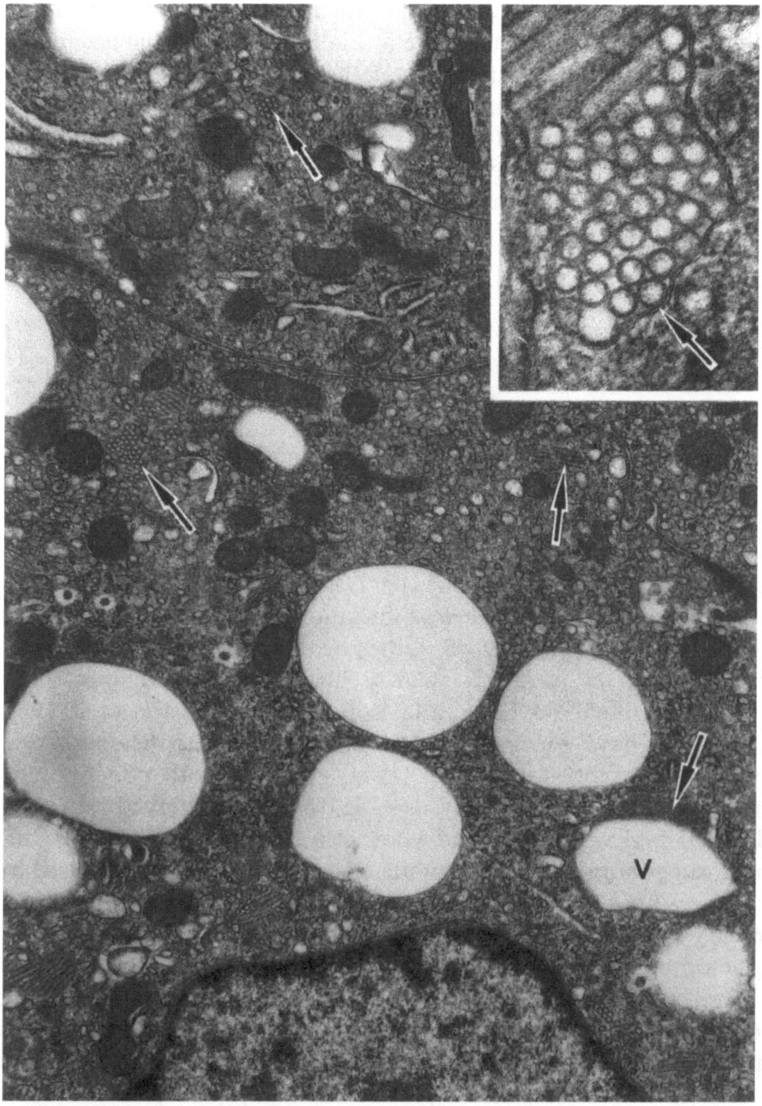

Fig. 3. Electron micrograph of secretory cells of the Harderian gland of a male golden hamster showing randomly distributed clusters of cylindrical tubules (*arrow*). These tubules are often associated with vacuoles (*V*). × 17000. *Inset* High magnification of a tubular cluster, which shows the substructure of the tubules as well as the membrane surrounding the cluster (*arrow*). × 120000. (Micrographs by Dr. Corazon Bucana)

vacuoles, the significance of which is not understood. The tubules were never seen in the Harderian glands of normal female hamsters. The functional significance of these tubular structures is yet to be determined. Nonetheless, based on the many reports that these tubules and porphyrin are mutually exclusive, it has been suggested that these tubules may have a role in preventing the accumulation of porphyrin in Harderian glands of male hamsters (Lin 1981). The dark speckling visible macroscopically in the Harderian glands of female hamsters was shown to be porphyrin pigment deposits in the alveolar lumina. These were manifested as electron dense fibrillar material when examined under the electron microscope (Fig.2, inset). Porphyrin accretions were not seen in the Harderian glands of normal male hamsters.

3 Secretory Products

3.1 Lipids

In general, the major secretory material in the mammalian Harderian gland is lipid. However, in the armadillo, the distal portion of the Harderian gland secretes lipids, while the proximal part secretes mucus (Weaker 1981). There are conflicting reports in the literature regarding the mechanism of secretion. Cohn (1955), Tsutsumi et al. (1966), and Kühnel (1971) have reported that the process by which the secretory vesicles are released into the alveolar lumen is apocrine, considered to be a form of pinching off. Other investigators (Woodhouse and Rhodin 1963; Rothwell et al. 1972; Brownscheidle and Niewenhuis 1978; Watanabe 1980; Wooding 1980; Sakai and Yohro 1981; Johnston et al. 1983; Maxwell et al. 1986) have reported the secretory process to be merocrine, otherwise known an exocytosis. Kühnel (1971) reported both apocrine and merocrine processes in the rabbit. Rarely, investigators have observed organelles (Bucana and Nadakavukaren 1972a) and whole cells (Johnston et al. 1983) in the alveolar lumina, suggesting a holocrine mode of secretion. Droplets released by the apocrine process are known to have a surrounding unit membrane, as in the case of mammary lipid droplets. Nonetheless, a detailed investigation of lipid secretion in the rabbit Harderian gland by Wooding (1980) showed that there are no unit membranes around the droplets in the lumen, suggesting that the secretory process is merocrine. According to Wooding (1980), each lipid droplet has a border that looks like an electron-dense line and is enclosed within a unit membrane, which in turn fuses with the plasmalemma, lipid droplet then being released into the lumen (Fig. 4). Winterhager and Kühnel (1983), using freeze-fracture techniques, have come to the same conclusion that the mode of secretion in the rabbit Harderian gland is merocrine. Currently, considerable ultrastructural evidence strongly suggests that the basic secretory mechanism in the Harderian glands of rodents, lagomorphs, and birds is merocrine. It is of interest to note, as Sakai and Yohro (1981) have stated, that the Harderian gland is the only example where lipid secretory vacuoles are released into the lumen by a merocrine process, whereas the mechanism of lipid vacuole secretion in the mammary gland is apocrine and that of the sebaceous gland is holocrine.

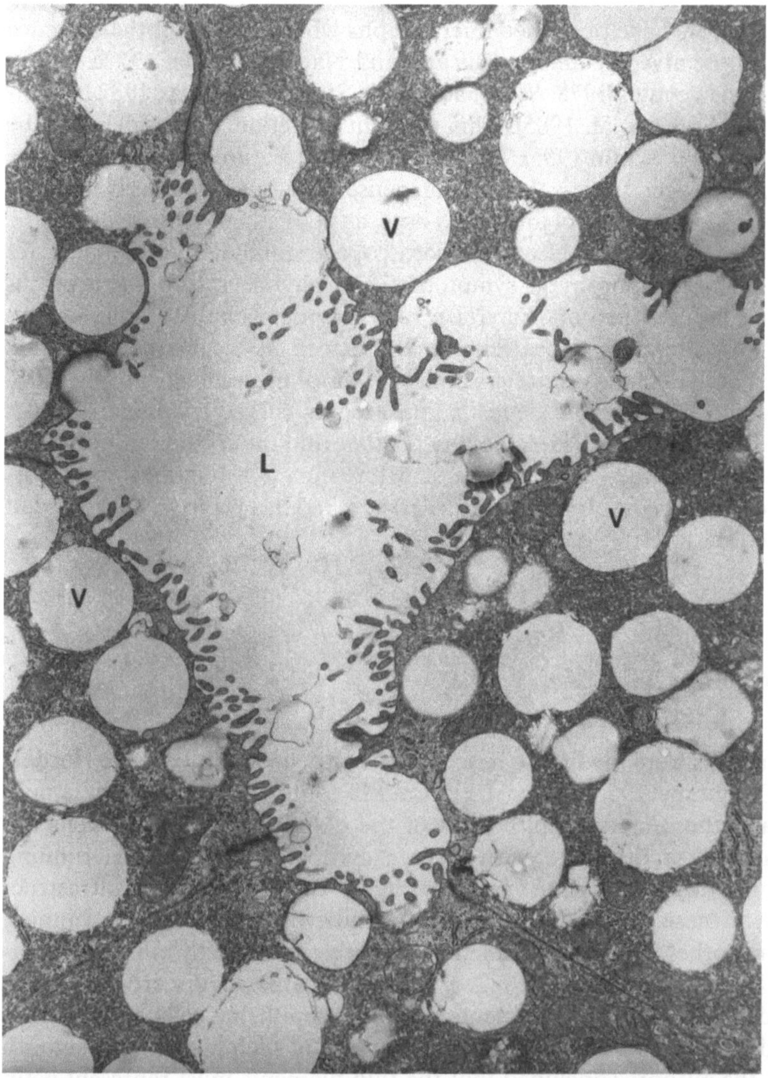

Fig. 4. Electron micrograph of the apical region of secretory cells from the Harderian gland of Mongolian gerbil showing merocrine mode of secretion. *V* vacuole; *L* lumen × 14 000. (Micrograph provided by Dr. Tatsuo Sakai)

3.2 Porphyrin

Porphyrin deposits have been reported in the Harderian glands of some rodents, such as female hamsters, gerbils, mice, and rats. The ultrastructural characteristics of porphyrin deposits in the Harderian gland are not well known.

Several investigators have published micrographs which document the presence of porphyrin in the alveolar lumina (Bucana and Nadakavukaren 1972a; Brownscheidle and Niewenhuis 1978; Watanabe 1980; Strum and Shear 1982a; Johnston et al. 1985; Spike et al. 1985). This pigment accretion is considered to be a lipoprotein complex (Cohn 1955). The ultrastructural feature that is common to all of the porphyrin accretions is the electron dense fibro-membranous needle-like appearance (Fig. 2, inset). It is of interest to note, as pointed out by Brownscheidle and Niewenhuis (1978), that electron microscope examination of hepatocytes from biopsies of human porphyria, symptomatic porphyria, erythropoietic protoporphyria, and induced protoporphyria revealed cytoplasmic needle-like structures or crystals similar in appearance to those seen in the Harderian gland. Recently, Carriere (1985) has reported visualization of intracellular porphyrin in the rat Harderian gland. These deposits appeared as curved trilaminar profiles and resembled the structures identified as protoporphyrin crystals in porphyric human and mice hepatocytes. A precise structure–function relationship of porphyrin synthesis and deposition in the Harderian gland has yet to be elucidated at the ultrastructural level.

4 Myoepithelial Cells

Myoepithelial cells were the first to be examined ultrastructurally in the Harderian gland (Chiquoine 1958). They are normally found between the basement membrane and the secretory epithelium of the gland (Fig. 5). These cells are usually elongated or branched with stellate cytoplasmic extensions running between the basal portions of the secretory cells. The characteristic ultrastructural feature of these cells is the presence of numerous myofilaments running parallel to each other. Only thin filaments have been reported by most investigators, except Strum and Shear (1982a), who were able to discern both thin (6–7 nm) and thick (12–17 nm) filaments in the myoepithelial cells of the mouse Harderian gland. Based on the size and appearance of such filaments in smooth muscle cells, the authors concluded that the thin filaments must represent actin and the thick filaments myosin.

Watanabe et al. (1981) examined the mouse Harderian gland, using both the transmission and scanning electron microscopes. Scanning electron microscopy showed the myoepithelial cell to have a central body and many long arms which enclosed the alveolus. In a similar study, Abe et al. (1981a) found that the flattened myoepithelial cells had an elongated central perikaryon from which radiated six to nine broad primary processes, which in turn bifurcated two or three times to produce a total of 15 to 28 terminals per cell. Weaker (1981) reported that, in the armadillo, the myoepithelial cells are attached to the lipid-secreting epithelial cells by desmosomes. Watanabe et al. (1981) found by

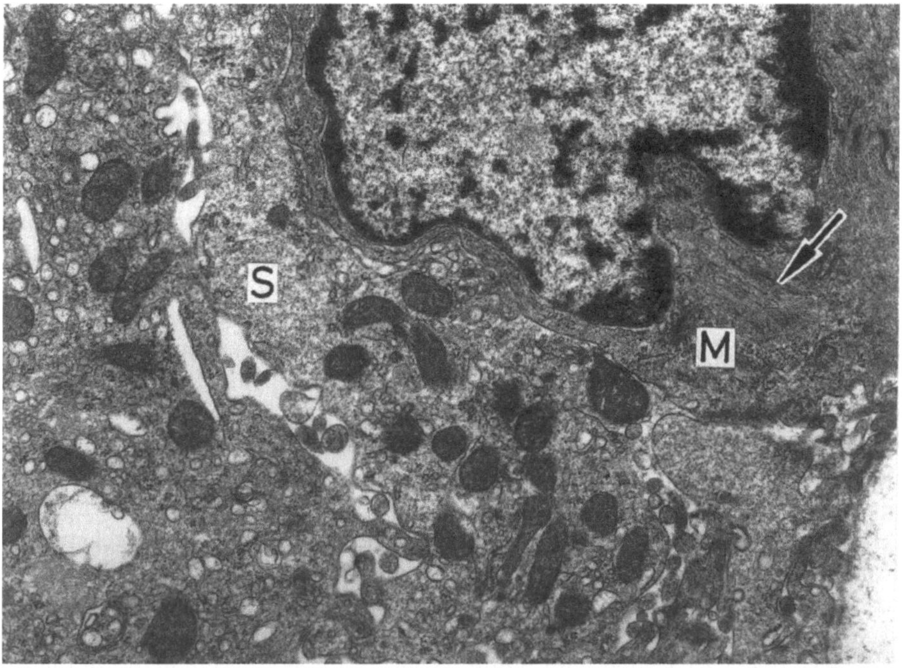

Fig. 5. Section from a Harderian gland of a male golden hamster showing the close association between a secretory cell (*S*) and a myoepithelial cell (*M*) with its characteristic myofilaments (*arrow*). × 25000. (Micrograph by Dr. Corazon Bucana)

serial sectioning that junctional complexes were present between myoepithelial cells and secretory cells, as well as between adjacent myoepithelial cells. They also reported nerve endings associated with the myoepithelial cells. Several other investigators have demonstrated similar associations between nerve endings and myoepithelial cells. However, only Bucana and Nadakavukaren (1972b) have shown a myoneural junction associated with the nerve terminals in the Harderian gland. Myoepithelial cells are known to be contractile and thus it is reasonable to assume from these reports that the secretory process in the Harderian gland is affected by neural stimulation of the myoepithelial cells. Some morphological support for this assumption has been provided by Abe et al. (1981b) in a scanning electron microscope study of normal and contracting myoepithelial cells in the rat Harderian gland. They demonstrated that when rats were injected with acetylcholine, there was a reduction in the volume of the alveoli and a concurrent appearance of irregular folds on the myoepithelial cells, suggesting contraction. Their observations were similar to the contracting states of the mammary gland alveoli induced by oxytocin, as reported by Abe (1979).

5 Innervation

A number of investigators have reported both adrenergic and cholinergic nerve terminals in the Harderian glands of mammals and nonmammalian species. Bucana and Nadakavukaren (1972b) were the first to report the presence of nerve endings in an ultrastructural study of any Harderian gland. They observed unmyelinated nerve fibers closely associated with blood vessels, cytoplasmic extensions of myoepithelial cells, and secretory epithelial cells of the hamster Harderian gland. The fibers associated with the blood vessels contained the characteristic dense-cored vesicles, suggesting their adrenergic nature, while the fibers associated with the epithelial cells were found to be cholinergic nerves. They also reported membrane evaginations of myoneural junctions, which suggest that the neurotransmitter passes a myoneural junction instead of diffusing through the connective tissue. Similar autonomic nerve fibers have been demonstrated ultrastructurally within the Harderian glands of the gerbil (Sakai and Yohro 1981), mouse (Watanabe 1980), rabbit (Kühnel 1971), and rat (Huhtala et al. 1977; Brownscheidle and Niewenhuis 1978). It can be generalized from these reports that in the Harderian gland adrenergic fibers are usually associated with the blood vessels, while cholinergic fibers are associated with the secretory and myoepithelial cells. Based on the large number of AChE-containing nerve fibers associated with blood vessels in the Harderian gland of the rat after sympathetic and trigeminal denervations, Huhtala et al. (1977) concluded that the blood vessels must also have a parasympathetic cholinergic innervation. Specific pre- and post-synaptic membranes were never reported by any of these investigators. Only in the gerbil (Sakai and Yohro 1981) have both myelinated and unmyelinated nerve fibers been seen. All other reports are of unmyelinated fibers. It appears that the Harderian glands are under the dual influence of adrenergic and cholinergic fibers. The fact that injecting rats with acetylecholine can produce "bloody tears", as reported by Tashiro et al. (1940), strongly suggests that innervation of the Harderian gland must play a role in the secretory process of the gland.

6 Secretory Duct

The secretory duct of the Harderian gland has not received much attention. Sakai and Yohro (1981) were the first to report an ultrastructural investigation of the secretory duct in the mammalian Harderian gland. They found the single duct in the gerbil (*Meriones meridianus*) to be surrounded by layers of connective tissue with their circularly oriented collagen fibers. Three types of cells were found in the epithelium of the excretory duct: surface cells, serous cells, and goblet cells. The serous cells had a large number of electron-dense and a few electron-lucent secretory granules. These granules were rarely seen in the surface cells, which had poorly developed microvilli on their luminal surface. The functional role of these cells was not discussed by the authors.

In a similar study, but of a related species of the gerbil (*M. unguiculatus*), Johnston et al. (1983) noticed within the gland long straight portions of the duct lined by a single layer of normal secretory cells. Outside the gland, the epithelial lining had three types of cells. Along the crests of the folds were cuboidal cells similar to the secretory cells but with variable number and size of lipid vacuoles and many small dense apical granules. A second type of cuboidal cell lining the clefts of the folds had dense granules and stubby microvilli but no lipid vacuoles. The crypts were lined by low columnar cells filled with irregularly shaped large dense granules. These cells also had an extensive rough endoplasmic reticulum, well-developed Golgi complex, and slender filamentous microvilli. Unusual aggregations of geometrically shaped crystalloid bodies of varying sizes and unknown function were present within the lumen of the cleft–crypt junction. These authors have suggested the possibility that the normal glandular secretions of lipid and porphyrin may be supplemented with serous and mucus secretions produced by the duct cells.

Johnston et al. (1985) found no intraglandular ducts in the plains mouse (*Pseudomys australis*) and noted that the extraglandular portion of the duct was lined with three types of cells. A surface epithelium of columnar cells with apical microvilli was found along the sinuous and straight parts. The cells along the straight part had a large vacuole containing small vesicles in the apical portion, while the cells along the sinuous portion had several vacuoles distributed randomly. Neither of these cell types showed any secretory activity. Pits that were present at the surface opening of the duct were lined with cells filled with mucus granules that were released into the lumen. Stratified squamous epithelial cells with microvilli covered the outer surface of the duct opening. These cells had a single apical vacuole and showed a high frequency of interdigitations. The secretory duct ultrastructure of the woodmouse (*Apodemus sylvaticus*), reported by Johnston et al. (1987), appeared to be similar to the plains mouse except for the melanocytes containing dense cytoplasmic granules found among the mucus cells.

In general, the secretory duct system of the armadillo described by Weaker (1981) appeared to be the most highly developed of all the reported species. The intralobular duct system of the mucus-secreting acini was better developed than that of the lipid-secreting acini. Cells adjoining the duct were cuboidal or squamous and were connected to the secretory cells and to each other by junctional complexes and desmosomes. A Golgi complex, cisternae of rough endoplasmic reticulum, and varying numbers of secretory granules were found in the duct cells. These cells were characterized by the presence of stacked microfilaments and numerous granules with varying electron densities.

With the limited information available, it is difficult to make any generalizations regarding the ultrastructure of the secretory ducts of the Harderian gland.

7 Sexual Dimorphism

Sexual dimorphism of the Harderian gland, particularly at the ultrastructural level, is not commonly encountered and is reported only in rodents. Porphyrin content in the Harderian glands was one of the early differences to be recognized. In such cases, the females always had higher amounts of porphyrin in their Harderian glands. The golden hamster is the only species in which a true sexual dimorphism of the Harderian gland has been established. The significance of this sexual dimorphism is not understood, although there are a number of reports related to the basis for this difference. Christensen and Dam (1953) were the first to report a marked macroscopic difference between the Harderian glands of the male and female hamsters (*Mesocricetus auratus*). The Harderian glands of the females had black granules that were absent in the males. They also showed that these pigment spots were localized in the alveolar lumina. In addition to the difference in pigmentation, Woolley and Worley (1954) reported that the alveolar cells of the Harderian glands of the female hamster contained only small vacuoles, while those in the male had large vacuoles.

Bucana and Nadakavukaren (1972a) were the first to report sexual dimorphism of the Harderian gland at the ultrastructural level. They recognized several structural differences between the Harderian glands of male and female golden hamsters, the most characteristic being the presence of tubular clusters in the secretory cells of the male and the absence of these in the female. These tubular clusters were often membrane-bounded and widely distributed in the cytoplasm (Fig. 3). The secretory cells of the Harderian glands of the female were characterized by the presence of an extensive smooth endoplasmic reticulum, often with dilated cisternae, numerous membranous structures arranged in concentric lamellar formations, and Golgi with highly proliferated cisternae (Fig. 2). The functional role of the tubular structures is uncertain. Bucana and Nadakavukaren (1973) reported an ultrastructural investigation of the postnatal development of the Harderian glands in male and female hamsters. They found ultrastructural evidence for sex difference during the 2nd week in the form of membrane-bounded juxtanuclear structures, which in the male were the foci for the development of cylindrical tubules, and in the female, the site of formation of membranous structures. Both male and female hamsters had pigment deposits at 2 weeks of age. The sex differences became fully established by the 4th week. The most recent structural difference noted between the Harderian glands of male and female hamsters was the number of mast cells reported by Payne et al. (1982), who found that the gland in the female contained approximately 40 times the number of mast cells present in the male.

Lin and Nadakavukaren (1982) reported significantly higher levels of delta-aminolevulinate (ALA) synthase activity in the Harderian glands of the female than in the male. However, the absence of porphyrin accretions in the Harderian glands of male hamsters may not be related only to the difference in activity of ALA synthase. It is possible that this enzyme or its product may be sequestered in the tubular clusters of the male, thus preventing the synthesis of porphyrin. Hoh et al. (1984) found a clearcut sexual difference in the Harderian gland proteins of the golden hamster. At least three major proteins present in the male were absent in the female. It is interesting that two of these proteins, HGTP1 and HGTP2, were found to be associated with the tubular clusters that are seen only in the male. Sexual differences with respect to protein composition were not found in the rat or mouse Harderian glands (Hoh and Nadakavukaren, unpubl. data). Lin and Nadakavukaren (1981) found that the predominant fatty acids in the Harderian glands of male hamsters ranged from C10 to C20, while those of females ranged from C16 to C20, with a prevalence of C16:0 and C18:1 fatty acids. This reflects a possible difference in the composition and function of lipids in the Harderian glands of male and female hamsters.

7.1 Influence of Light

Although there are many reports of the effects of light on the mammalian Harderian gland (see references in Olcese and Wesche 1989), only a few investigators have focused on the ultrastructural changes associated with these effects. Feria-Velasco et al. (1978) found ultrastructural evidence in newborn rats for increased glandular secretions when the animals were kept in continuous light, and decreased secretions when they were in the dark. The structural differences were limited to the apical portions of the secretory cells. They suggested that the increased secretory activity is directly related to melatonin production by the pineal gland. Strum and Shear (1982b) noted deleterious effects on the ultrastructure of the secretory cells and a related decrease in porphyrin content in the Harderian glands of female mice exposed to constant light. Nadakavukaren and Lin (1983) observed that exposure of hamsters of both sexes to a short photoperiod of 1L:23D induced conversion of the Harderian glands to the opposite type with the characteristic ultrastructural features, whereas the control group kept at 14L:10D maintained the normal type glands. The changes in the male included reduction of tubular clusters and appearance of membrane formations, while in the female a reduction of membrane formations coincided with the appearance of tubular clusters. These changes were similar to those induced by castration of the male or administration of testosterone to the female. DiIorio and Nadakavukaren (1984) demonstrated that pinealectomy can prevent the short-photoperiod-induced sexual and ultrastructural changes in the Harderian glands of hamsters. They found atrophy of the testes and uteri in the sham-operated hamsters maintained under short photoperiods, but not in the pinealectomized animals. They also noted that the ultrastructural changes coincided with the time of atrophy of the reproductive organs, indicating that the changes may have been induced by the altered hormone levels accompanying testicular and uterine atrophy. Animals which were either sham-operated or pinealectomized and then kept in the long photoperiod did not have any changes in their glands during the 16 weeks of the experiment.

Based on the influence of photoperiod, blinding, and gonadectomy on the mammalian Harderian gland, it has been suggested that the Harderian gland may be a component of the retinal-pineal-gonadal axis. The reports that female hamsters can synthesize melatonin in their Harderian glands (Menendez-Pelaez et al. 1987a) and that photoperiod can affect the activity of N-acetyltransferase and hydroxyindole-O-methyltransferase, two enzymes involved in the synthesis of melatonin (Menendez-Pelaez et al.1988), add support to this idea. However, the mechanism of interaction between the Harderian gland, pineal gland, and the gonads remains unknown. It is possible that the interaction may be mediated via a feed-back mechanism involving melatonin, which seems to be the common denominator for a number of light-regulated processes.

7.2 Regulation by Steroids

Woolley and Worley (1954) demonstrated that treatment with sex hormones or gonadectomy can reverse the sex difference in pigmentation, as well as vacuole size, of the hamster Harderian gland. Hoffman (1971) reported that testosterone propionate prevented the castration-induced changes in the male, while Clabough and Norvell (1973) found that blinding can induce a male-type gland in the female. These authors also suggested that gonadotropins may be responsible for these changes. However, more recent reports support the hypothesis that androgens are responsible for the sexual dimorphism of the hamster Harderian gland. Payne et al. (1977) showed that several naturally occurring androgens were able to prevent the transformation of the Harderian gland to the female type in castrated males. McMasters and Hoffman (1984) reported that ovariectomy in combination with blinding prevented the conversion of the female type gland to the male type that normally follows blinding, indicating that this conversion may require some ovarian substance, which they suggested to be androgens. Spike et al. (1985, 1986) found that the Harderian glands of ovariectomized females did not convert to the male type and had similar porphyrin content to that of the controls. When the ovariectomized hamsters were maintained on testosterone, there was a significant reduction in porphyrin content and delta-aminolevulinic acid synthetase activity, in addition to an increase in Type II cells and mast cells and the appearance of tubular clusters. Thus, their observations supported the concept of a link between gonadal hormones and gland structure–function. In a study of the Harderian glands of male hamsters after prolonged gonadectomy, Menendez-Pelaez et al. (1987b) observed ultrastructural changes that indicated enhanced transport in the secretory cells.

Investigators in our laboratory demonstrated that the changes in pigmentation and morphology of the hamster Harderian gland have an ultrastructural basis. Lin and Nadakavukaren (1979) reported that the clusters of cylindrical tubules, typical of the male type gland, decreased and disappeared almost completely 2 weeks after castration, while membranous structures, typical of the female type gland, became prominent in the secretory cells. These changes were prevented in the castrated male by daily administration of 2 mg testosterone propionate. The ultrastructural changes persisted even 20 days after the termination of the androgen treatment. These authors noted that the appearance of pigmentation coincided with the disappearance of the tubular clusters and suggested that there must be a possible relationship between the tubules and pigment production.

Sun and Nadakavukaren (1980) presented data that showed a distinct influence of testosterone on both pigmentation and ultrastructure of the female hamster Harderian gland. Daily injections of 2 mg testosterone propionate resulted in the appearance, starting on the 4th day, of tubular clusters in the secretory cells of the females and a reduction in pigmentation, starting on the 6th day. The number of tubular clusters increased with time during the course of the experiment, with a corresponding decrease in the number of membranous structures. They reported an apparent increase in rough endoplasmic reticulum and Golgi vesicles 2 days after testosterone injections. Pigmentation also decreased throughout the experiment and increased slowly between 21 and 49 days after the termination of testosterone injections. Similarly, the number of tubular clusters decreased with a corresponding increase in the number of membranous structures between the 21st and 49th day after termination of injections. This investigation demonstrated that there is a relationship between pigmentation and testosterone levels. It also confirmed that the disappearance of pigment from the Harderian glands of female hamsters maintained on testosterone coincided with the appearance of tubular clusters in the alveolar cells and that these two features are closely related in the hamster Harderian gland. The role of these tubular clusters in the expression of the male-type gland is an area of investigation that has not received proper attention. Initial attempts to isolate and characterize these tubules by Lin (1981) have not been fully successful. Determination of the chemical composition of the tubules has to be a primary objective in order to understand their role in the male type gland.

The above investigations clearly demonstrate a gonadal influence on the Harderian gland and it has been suggested that the male characteristics of the gland in hamsters are expressed when significant levels of androgens are present. Additional evidence that there are receptors for androgens in the Harderian glands of both sexes (Gustafsson and Pousette 1975; Vilchis et al. 1987; Vilchis and Perez-Palacios 1989) and that the gland is able to metabolize testosterone to its active form, 5-dihydrotestosterone (Menendez-Pelaez et al. 1990), strongly supports the hypothesis that androgens, and not gonadotropins, are responsible for the structural and functional maintenance of sexual dimorphism. However, the role of gonadal steroids as the sole regulators of sexual types has been questioned recently by Hoffman et al. (1989). Therefore, the mechanism by which sexual dimorphism in the Harderian gland is regulated remains unresolved.

Acknowledgements: I want to express my appreciation to my former students Corazon Bucana, Wen-lang Lin, Dennis DiIorio, Jan Hoh, Nancy Sun and James Shen for their contributions to our understanding of the hamster Harderian gland ultrastructure and sexual dimorphism. I owe many thanks to Dr. Bucana for initiating the "Harderian gland project" in our laboratory. I am also grateful to Dr. Lin for sending me his reprint collection, which greatly facilitated the timely completion of this chapter.

References

Abe J (1979) Scanning electron microscopic observations of the mammary gland myoepithelial cells of the rat under normal and experimental conditions. Kurume Med J 26:303–310

Abe J, Nakamura K, Murakami M (1981a) Scanning electron microscopy observations of the myoepithelial cells in rat Harderian gland. J Electron Microsc 30:100 (Abstr)

Abe J, Sugita A, Hamasaki M, Nakamura K, Iwanaga S, Nagae K, Atsuji K, Tsunawaki A, Abe T, Matsumoto T, Yo S, Murakami M (1981b) Scanning electron microscopic observations of the myoepithelial cells of normal and contracting status in the rat Harderian gland. Kurume Med J 28:103–112

Björkman N, Nicander L, Schantz B (1960) On the histology and ultrastructure of the Harderian gland in rabbit. Z Zellforsch 52:93–104

Brownscheidle CM, Niewenhuis RJ (1978) Ultrastructure of the Harderian gland in male albino rats. Anat Rec 190:735–753

Bucana CD, Nadakavukaren MJ (1972a) Fine structure of the hamster Harderian gland. Z Zellforsch 129:178–187

Bucana CD, Nadakavukaren MJ (1972b) Innervation of the hamster Harderian gland. Science 175:205–206

Bucana CD, Nadakavukaren MJ (1972c) Ultrastructural investigation of the postnatal development of the hamster Harderian gland. I. Cell division. Z Zellforsch 135:149–153

Bucana CD, Nadakavukaren MJ (1973) Ultrastructural investigation of the postnatal development of the hamster Harderian gland. II. Male and female. Z Zellforsch 142:1–12

Carriere R (1985) Ultrastructural visualization of intracellular porphyrin in the rat Harderian gland. Anat Rec 213:496–504

Chieffi Baccari G, Minucci S, DiMatteo L, Chieffi G (1990) Harderian gland and the lacrimal gland of the lizard *Podarcis S sicula*: histology, histochemistry and ultrastructure. Anat Rec 226:269–278

Chiquoine AD (1958) The identification and electron microscopy of myoepithelial cells in the Harderian gland. Anat Rec 132:569–583

Christensen F, Dam H (1953) A sexual dimorphism of the Harderian gland in hamsters. Acta Physiol Scand 27:333–336

Clabough JW, Norvell JE (1973) Effects of castration, blinding and the pineal gland on the Harderian glands of the male golden hamster. Neuroendocrinology 12:344–353

Cohn SA (1955) Histochemical observations on the Harderian gland of the albino mouse. J Histochem Cytochem 3:342–353

DiIorio DP, Nadakavukaren MJ (1984) Prevention by pinealectomy of short-photoperiod-induced ultrastructural changes in the hamster Harderian gland. Anat Rec 210:449–452

DiMatteo L, Minucci S, Chieffi Baccari G, Pellicciari C, d'Istria M, Chieffi G (1989) The Harderian gland of the frog, *Rana esculenta* during the annual cycle: histology, histochemistry and ultrastructure. Bas Appl Histochem 33:93–112

Feria-Velasco A, Tapia-Arizmendi G (1974) Further studies on the ultrastructure of Harderian gland in adult rats. Proc Elec Microsc Soc Am 32:288–289

Feria-Velasco A, Tapia-Arizmendi G, Malpica I (1978) Effects of the environmental light conditions on the ultrastructure of the Harderian gland in the newborn rat: biochemical correlation with the pineal gland. Arch Invest Med (Mexico) 9:67–84

Gustafson JA, Pousette A (1975) Demonstration and partial characterization of cytosol receptors for testosterone. Biochemistry 14:3094–3101

Hoffman RA (1971) Influences of some endocrine glands, hormones and blinding on the histology and porphyrins of the Harderian glands of golden hamsters. Am J Anat 132:463–478

Hoffman RA, Vaughan MK, Reiter RJ (1989) Photic input and pineal-Harderian-thyroid gland interactions in golden hamsters. In: Reiter RJ, Pang SF (eds) Advances in pineal research vol 3. John Libbey, London, pp 225–230

Hoh JH, Lin WL, Nadakavukaren MJ (1984) Sexual dimorphism in the Harderian gland proteins of the golden hamster. Comp Biochem Physiol 77B:729–731

Huhtala A, Huikuri KT, Palkama A, Tervo T (1977) Innervation of the rat Harderian gland by adrenergic and cholinergic nerve fibers. Anat Rec 188:263–271

Johnston HS, McGadey J, Thompson GG, Moore MR, Payne AP (1983) The Harderian gland, its secretory duct and porphyrin content in the Mongolian gerbil (*Meriones unguiculatus*). J Anat 137:615–630

Johnston HS, McGadey J, Thompson GG, Moore MR, Breed WG, Payne AP (1985) The Harderian gland, its secretory duct and porphyrin content in the plains mouse (*Pseudomys australis*). J Anat 140:337–350

Johnston HS, McGadey J, Payne AP, Thompson GG, Moore MR (1987) The Harderian gland, its secretory duct and porphyrin content in the wood mouse (*Apodemus sylvaticus*). J Anat 153:17–30

Kelényi G, Orban S (1965) Electron microscopy of the Harderian gland of the rat: maturation of the acinar cells and genesis of the secretory droplets. Acta Morphol 13:155–165

Kittner Z, Oláh I, Törö I (1978) Histology and ultrastructure of the Harderian glands – accessory lacrimal gland of the chicken. Acta Biol Acad Hung 29:29–41

Kühnel W (1971) Über den Feinbau und die stoffwechsel-möglichkeiten der Harderschen Drüse des Kaninchens. Verh Anat Ges 66:79–85

Lin WL (1981) An ultrastructural and biochemical investigation of sexual dimorphism in the hamster Harderian gland. PhD Diss, Illinois State University, Normal, Ill

Lin WL, Nadakavukaren MJ (1979) The androgenic effect on the fine structure of the Harderian gland in the male hamster. Cell Tissue Res 198:119–127

Lin WL, Nadakavukaren MJ (1981) Harderian gland lipids of male and female golden hamsters. Comp Biochem Physiol 70B:627–630

Lin WL, Nadakavukaren MJ (1982) Delta-aminolevulinate synthase activity in the Harderian glands of male and female golden hamsters. Comp Biochem Physiol 71B:539–540

Maxwell MH, Burns RB (1979) The ultrastructure of the epithelium of the ducts of the Harderian and lacrimal glands of the turkey, fowl and duck. J Anat 128:445-459

Maxwell MH, Rothwell B, Burns RB (1986) A fine structural study of the turkey Harderian gland. J Anat 148:147–157

McCafferty RE, Clark TS (1969) Harderian gland ultrastructure of mouse followed porphyrin administration. Anat Rec 163:315 (Abstr)

McMasters KM, Hoffman RA (1984) Harderian gland: regulation of sexual "type" by gonads and pineal gland. Biol Reprod 31:579– 585

Menendez-Pelaez A, Howes KA, Gonzalez-Brito A, Reiter RJ (1987a) N-acetyltransferase activity, hydroxyindole-O-methyltransferase activity and melatonin levels in the Harderian gland of the female Syrian hamster: changes during the light:dark cycle and the effect of 6-parachlorophenylalanine administration. Biochem Biophys Res Commun 145:1231–1238

Menendez-Pelaez A, Howes KA, Reiter RJ (1987b) Ultrastructural changes in the secretory cells of the male Syrian hamster's Harderian gland after prolonged gonadectomy. Anat Rec 218:90A (Abstr)

Menendez-Pelaez A, Santana C, Howes KA, Sabry I, Reiter RJ (1988) Effects of photoperiod or exogenous melatonin administration on the activity of N-acetyltransferase and hydoxyindole-O-methyltransferase and the melatonin content of the Harderian gland of two strains of female Syrian hamsters. J Pineal Res 5:293–300

Menendez-Pelaez A, Buzzell GR, Gonzalez-Brito A, Reiter RJ (1990) Androgenic control of N-acetyltransferase activity in the Harderian glands of the female Syrian hamster is mediated by 5-alpha-dihydrotestosterone. J Cell Biochem 42:95–100

Minucci S, Chieffi Baccari G, DiMatteo L, Chieffi G (1989) A sexual dimorphism of the Harderian gland of the toad, *Bufo viridis*. Bas Appl Histochem 33:299–310
Nadakavukaren MJ, Lin WL (1983) Influence of short photoperiods on the ultrastructure of hamster Harderian glands. Cell Tissue Res 233:209–213
Niedorf HR, Wolters B (1978) Development of the Harderian gland in the chicken: light and electron microscopic investigations. Invest Cell Pathol 1:205–215
Olcese J, Wesche A (1989) The Harderian gland. Comp Biochem Physiol 93A:655–665
Payne AP, McGadey J, Moore MR, Thompson GG (1977) Androgenic control of the Harderian gland in the male golden hamster. J Endocrinol 75:73–82
Payne AP, McGadey J, Johnston HS, Moore MR, Thompson GG (1982) Mast cells in the hamster Harderian gland: sex differences, hormonal control and relationship to porphyrin. J Anat 135:451–461
Rothwell B, Wight PAL, Burns RB, MacKenzie GM (1972) The Harderian glands of the domestic fowl. 3. Ultrastructure. J Anat 112:233–250
Sakai T (1981) The mammalian Harderian gland: morphology, biochemistry, function and phylogeny. Arch Histol Jpn 44:299–333
Sakai T, Yohro T (1981) A histological study of the Harderian gland of Mongolian gerbils, *Meriones meridianus*. Anat Rec 200:259–270
Spike RC, Johnston HS, McGadey J, Moore MR, Thompson GG, Payne AP (1985) Quantitative studies on the effects of hormones on structure and porphyrin biosynthesis in the Harderian gland of the female golden hamster. I. The effects of ovariectomy and androgen administration. J Anat 142:59–72
Spike RC, Johnston HS, McGadey J, Moore MR, Thompson GG, Payne AP (1986) Quantitative studies on the effects of hormones on structure and porphyrin biosynthesis in the Harderian gland of the female golden hamster. II. The time course of changes after ovariectomy. J Anat 145:67–77
Strum JM, Shear CR (1982a) Harderian glands in mice: fluorescence, peroxidase activity and fine structure. Tissue Cell 14:135–148
Strum JM, Shear CR (1982b) Constant light exposure induces damage and squamous metaplasia in Harderian glands of albino mice. Tissue Cell 14:149–161
Sun CY, Nadakavukaren MJ (1980) Effect of testosterone on the female hamster Harderian gland pigmentation and ultrastructure. Cell Tissue Res 207:511–517
Tashiro S, Smith CC, Badger E, Kezur E (1940) Chromadacryorrhea, a new criterion for biological assay of acetylcholine. Proc Soc Exp Biol Med 44:658–661
Tsutsumi A, Iwata K, Ogawa K, Matsuura K (1966) Histochemical and electron miscroscopic observations on the Harderian gland of the albino rat. Arch Histol Jpn 27:553–567 (in Japanese)
Vilchis F, Perez-Palacios G (1989) Steroid hormone receptors and the sexual phenotype of the Harderian gland in hamsters. J Endocrinol 121:149–156
Vilchis F, Hernandez A, Perez AE, Perez-Palacios G (1987) Hormone regulation of the rodent Harderian gland: binding properties of the androgen receptors in the male golden hamster. J Endocrinol 112:3–8
Watanabe M (1980) An autoradiographic, biochemical, and morphological study of the Harderian gland of the mouse. J Morphol 163:349–365
Watanabe M, Kihara T, Shimono R (1981) Electron microscopy of myoepithelial cells of the mouse Harderian gland. Bull Osaka Med Sch 27:1–7
Weaker FJ (1981) Light microscopic and ultrastructural features of the Harderian gland of the nine-banded armadillo. J Anat 133:49–65
Winterhager E, Kühnel W (1983) Membrane specializations of the cells of the Harderian gland of the rabbit with particular reference to the mechanism of exocytosis. Cell Tissue Res 231:623–636
Woodhouse MA, Rhodin JAG (1963) The ultrastructure of the Harderian gland of the mouse with particular reference to the formation of its secretory product. J Ultrastruct Res 9:76–78
Wooding FBP (1980) Lipid droplet secretion by the rabbit Harderian gland. J Ultrastruct Res 71:68–78
Woolley W, Worley J (1954) Sexual dimorphism in the Harderian gland of the hamster (*Cricetus auratus*). Anat Rec 118:416–417(Abstr)

The Harderian Gland of Amphibians and Reptiles

[1]G. Chieffi, [2]G. Chieffi Baccari, [1]L. Di Matteo, [1]M. d'Istria,
[1]C. Marmorino, [1]S. Minucci, and [1]B. Varriale

[1]Dipartimento di Fisiologia Umana e Funzioni Biologiche Integrate «Filippo Bottazzi».
[2]Istituto di Anatomia Umana Normale, I Facoltà di Medicina e Chirurgia,
Università di Napoli, and Stazione Zoologica di Napoli, Italy

1 Introduction

The Harderian glands (HG) of mammals have so far received the lion's share of attention. Amphibians, although being the most primitive tetrapods possessing HG, have been relatively neglected, as have other nonmammalian tetrapods, particularly reptiles.

The study of the HG in nonmammalian tetrapods gains significance not only for comparative and evolutionary implications, but also for better understanding of its still debated functions. The differences either in anatomical structure of the gland or in biochemical composition of the secretory products may reflect physiological adaptations to different environments. Another compelling reason for studying the HG in amphibians and reptiles is to discover new targets for sex hormones. Sexual dimorphism of the HG has been shown in some mammalian species and, therefore, the demonstration that the HG is influenced by sex hormones in amphibians and reptiles as well would be of great significance because of the scarcity of secondary sex characters in these vertebrates.

In this chapter, we review current knowledge on the HG of amphibians and reptiles from both morphological and physiological viewpoints.

2 Embryology

2.1 Amphibians

As in mammals, the HG originates from the conjunctival epithelium in amphibians. In *Rana pipiens*, the primordium of the HG appears at late metamorphosis when external and internal larval corneal layers fuse and a small nictitating membrane develops (Kollros and Kaltenbach 1952). The primordium of the HG in *Rana japonica* first appears in the orbit at late prometamorphic stage and becomes prominent at the metamorphic climax. In *Bufo b. japonicus*, the gland develops at the climax stage, while in *Xenopus laevis* the gland appears when the

Webb et al. (Eds).
Harderian Glands

tentacles reabsorb (Shirama et al. 1982). In *Rana esculenta*, cells budding from the conjunctival epithelium enter the mesenchyme at incipient metamorphosis, while in *Bufo viridis* tadpoles, the HG primordium develops later at climatic metamorphosis (Chieffi Baccari et al. 1990b). The appearance of this primordium is concomitant with the development of the nictitating membrane in both species. In *R. esculenta*, the HG is organized into acini at the preclimax stage while in *B. viridis*, the HG acquires an acinar structure later at climatic metamorphosis. At the completion of climax, the HG is fully organized in both species. In both *Rana* and *Bufo*, 40 days after metamorphosis, the HG is similar to that of the adult.

Therefore, in all the anuran species so far examined, the HG must be considered an "adult" organ. This conclusion is strengthened by the failure of the gland to develop following either hypophysectomy or thyroidectomy carried out at tailbud stage, and by its induction either by thyroxine or TSH in hypophysectomized *B. b. japonicus* larvae (Shirama et al. 1982).

2.2 Reptiles

No information is available on the organogenesis of the HG in reptiles. However, it is most likely that the HG originates from the conjunctival epithelium also in this group of tetrapods.

3 Gross Anatomy

3.1 Amphibians

The HG is the only orbital gland in anurans. It lies at the medial corner of the orbit associated with the nictitating membrane (Fig. 1). Urodeles have a row of glands along the lower eyelid, while no homologous gland develops in Apoda.

The HG of *Rana esculenta* (Di Matteo et al. 1989) and *Bufo viridis* (Minucci et al. 1989a) is an oval acinous gland, pinkish in color due to abundant vascularization, particularly on the surface. It is about 2.5 mm in length in both species. The gland is not lobed. The single excretory duct begins at the hilus and opens ventrally in the conjunctival fornix.

3.2 Reptiles

In reptiles, the HG is consistently present in all species so far examined and represents the only orbital gland in *Sphenodon punctatus* (Sphenodontidae), Geckonidae, *Brookesia spectrum* (Chamaeleontidae) and in the snakes where the lacrimal glands are absent (for review, see Saint-Girons 1985, 1988, 1989).

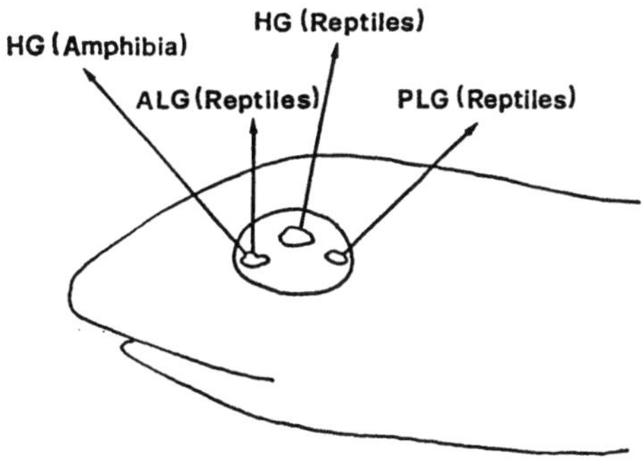

Fig. 1. Schematic localization of the orbital glands in amphibians and reptiles. *ALG* anterior lacrimal gland; *HG*, Harderian gland; *PLG* posterior lacrimal gland

The HG is located on the medial or medio-ventral aspect of the eyeball and its ducts run laterally and ventrally to open into the conjunctival sac in proximity to the cornea, or between this and the nictitating membrane if present (Fig. 1).

In chelonians, lacking the lacrimal canal, the secretory product remains in the conjunctival space. In Crocodilia, *Sphenodon*, and in the majority of lizards, the secretory product is also excreted into the conjuctival sac. From here, it can reach, via the lacrimal canal and a long naso-pharyngeal duct, a separate cavity from the palate in Crocodilia, and the lateral margin of the coane or coanal solcus and hence the palate in the other species. In Pygopodidae and Chamaeleontidae, the secretory product, via the lacrimal canal, can reach the coanal solcus directly or rise into the conjunctival sac (for review, see Saint-Girons 1985). In snakes, the conjunctival sac communicates with the nose by a single naso-lacrimal duct opening inside the vomeronasal organ of Jacobson. This is a drain through which the secretion flows into the mouth (Born 1883).

The volume of the HG may exhibit species-specific variation, but it is usually the largest orbital gland. It extends to the anterior portion of the orbit and may often reach the posterior orbital wall. In Anguidae, Amphisbaenidae, and snakes the fully developed HG has a large posterior enlargement. Only in the sea snake, the poorly developed gland hardly reaches the posterior wall of the eyeball (for review, see Saint-Girons 1988).

4 Histology and Ultrastructure

4.1 Amphibians

The most complete description of the HG in amphibians is by Di Matteo et al. (1989) in *Rana esculenta* and Minucci et al. (1989a) in *Bufo viridis*. In both species, the gland is composed of acini and is surrounded by a thin connective tissue capsule from which thin septa penetrate between acini. The gland is richly vascularized, particularly at the periphery. A single excretory duct begins at the hilus and opens ventrally. Melanocytes may be found either in the connective tissue or between acini. Numerous mast cells and plasma cells can be seen in the acinar spaces.

The acini, surrounded by myoepithelial cells, contain only one type of columnar cell which shows considerable variation in height, depending upon its functional state. Secretory activity is highest in the hottest months (July and August); it then declines drastically and recrudesces slowly from October onwards. The resumption of secretory activity is marked in some glandular cells by the appearance of nuclei which show a high affinity for aniline blue when stained with Mallory trichrome. This affinity is due to an increased rate of RNA synthesis (Chieffi Baccari et al. 1989). In *R. esculenta*, the number of "blue nuclei" increases also with the enhancement of secretion in May, while they completely disappear in August when nuclei are shrunken and apparently pycnotic (Fig. 2; Di Matteo et al. 1989).

At the ultrastructural level, the acinar cells are pyramidal or columnar with secretory granules usually concentrated at the apex. The glandular cells are characterized by the presence of well-developed lamellar rough endoplasmic reticulum (RER). The secretory granules, osmiophilic, pleiomorphic and

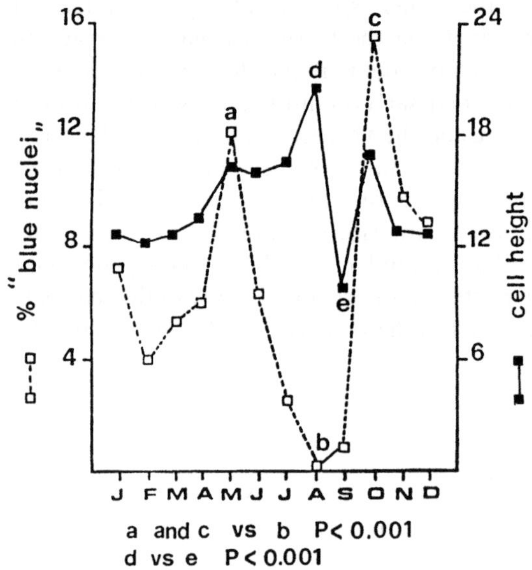

Fig. 2. Cell height and percentage of "blue nuclei" of the HG glandular cell of the frog, *Rana esculenta*, during the annual cycle

nonhomogeneous, are surrounded by a unit membrane (Fig. 3). In *B. viridis*, some granules show one or two crescent-like cups of lower density (Fig. 4).

The glandular cells of the female toad contain, at their base, numerous lipid droplets dispersed in a smooth endoplasmic reticulum (Fig. 5). Since these granules are absent in the male, this observation represents the only case of sexual dimorphism so far described in the HG of a nonmammalian tetrapod.

Secretion is mainly merocrine or apocrine. A single central duct lined by a single layer of cuboidal cells discharges the secretory product into the conjunctival fornix deep in the nictitating membrane.

Worthy of note are the circannual variations in the number of mastocytes in the HG of *R. esculenta*. They show a significant increase during winter (Chieffi Baccari et al 1991b).

4.2 Reptiles

In reptiles, the HG appears to be histologically similar in all the species studied. It is polistomatic and acinous, a compound tubular or tubulo-acinous gland, and is surrounded by a capsule of connective tissue more or less abundant in the various species. From this capsule, thin strands containing many blood vessels and nerve fibers penetrate into the glandular tissue but do not divide it into lobules. Only in those species in which the HG is highly developed, as in the snakes and Crocodilia, are incomplete lobules, bound together by a loose connective tissue stroma, observed (for review, see Saint-Girons 1985). In addition to the usual connective tissue cells, melanocytes and mast cells are scattered throughout the tissue. These cells are rare in snakes and lizards but numerous in chelonians. The height of the cells varies with the species, but also depends upon their functional state.

Under the light microscope, the secretory cells show similar cytological structure in all species. The secretory granules may occupy most of the cytoplasm or only its apical portion. They contain homogeneous materials of various density.

The HG of the lizard, *Podarcis s. sicula*, whose description by Chieffi Baccari et al. (1990b) seems to be the most complete among reptiles, can be divided into three contiguous but histologically recognizable zones: the more lateral tubulo-acinar of serous type, the intermediate acinar of seromucous type, and the medial of mucous type. The latter zone must be considered the anterior lacrimal gland, since its mucous secretory granules are ultrastructurally identical to those of the glandular cells of the posterior lacrimal gland. Ultrastructurally, the glandular cells of the intermediate zone show two types of secretory granules: one is of mucous type, the other is serous and occupies a different cell compartment (Fig. 6). The serous granule has been called "special secretory granule", owing to its strikingly unusual structure, consisting of three closely associated components, like atoms in a molecular model, which differ either in density or in structure (Fig. 9).

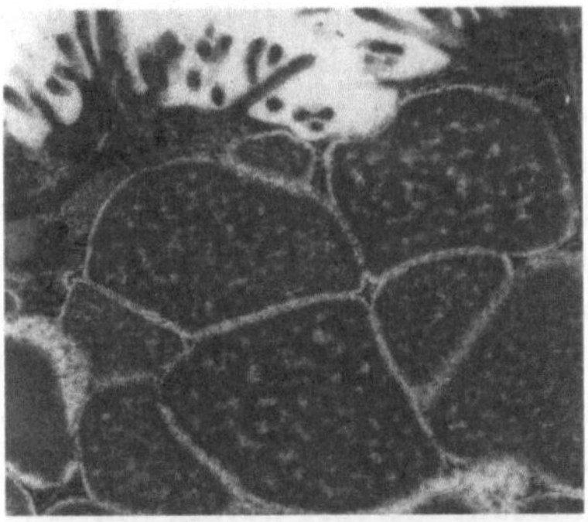
3

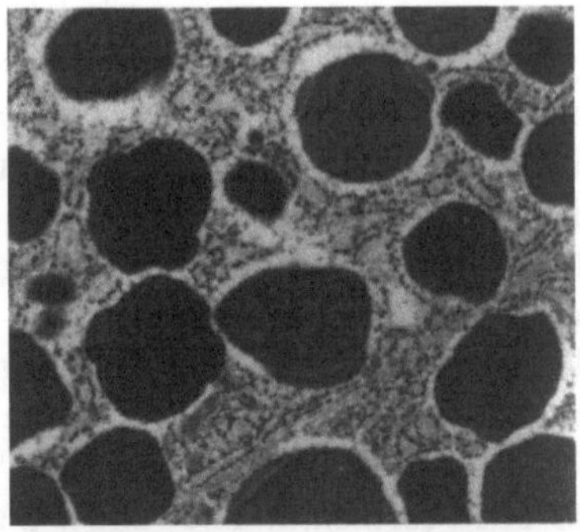
4

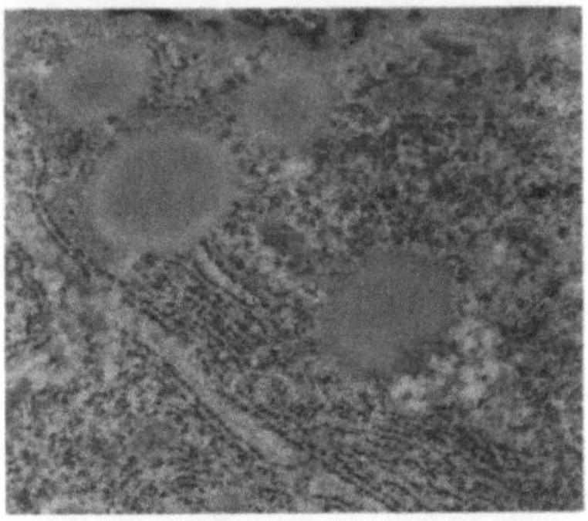
5

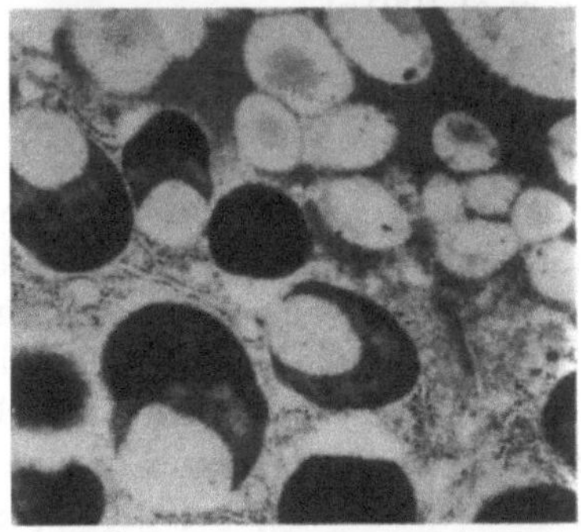
6

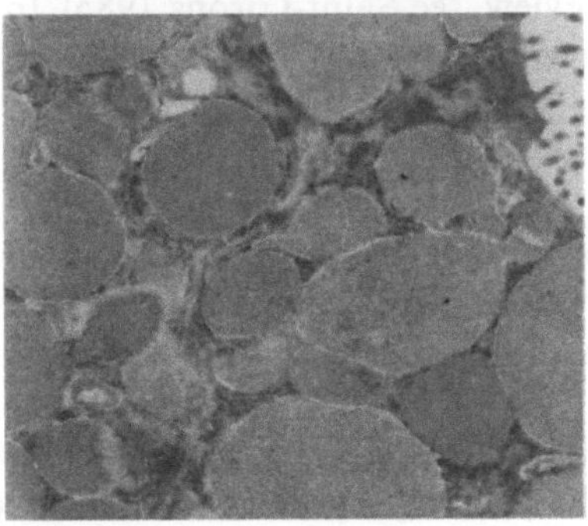
7

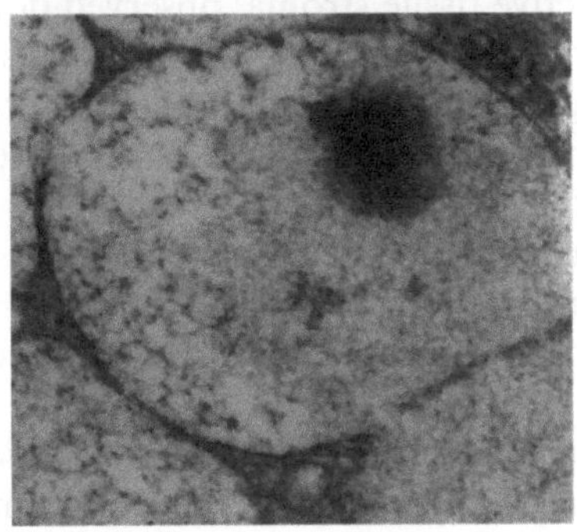
8

Fig. 3. Electron micrograph of a HG glandular cell of an August male frog, *Rana esculenta*. The secretory granules are dense, pleiomorphic and nonhomogeneous surrounded by a unit membrane. The free border of the cell shows numerous microvilli ×3600

Fig. 4. Electron micrograph of a HG glandular cell of a July male toad, *Bufo viridis*. The secretory granules are osmiophilic and show one or two crescent-like cups of lower density. ×3600

Fig. 5. Electron micrograph of a HG acinar cell of a July female toad, *Bufo viridis*, containing lipid droplets, as confirmed by the sudanophilia which is absent in the male glandular cells. ×4600

Fig. 6. Electron micrograph of a glandular cell of the HG intermediate zone of the lizard, *Podarcis s. sicula*. Note the presence of two types of secretory granules in different compartments of the same cell: one is mucous with a dense core surrounded by a zone of lower density; the second, serous, is described in Fig. 9. ×8000

In the HG of another squamate, the snake *Coluber viridiflavus*, the glandular epithelium is composed of two cell types (Minucci et al 1991). The more diffused type is characterized by the presence of an extraordinarily well developed RER. The secretory granules, bounded by a unit membrane, appear to be similar to the "special secretory granules" first described in the glandular cells of the lizard, *P. s. sicula* (Fig. 10). Only in the most anterior and posterior portions of the gland can a second cell type be seen. These cells contain tightly packed homogeneous secretory granules of low density (Fig. 7). Consequently, the anterior and posterior portions of the gland might be considered homologous to the anterior and posterior lacrimal glands of *P. s. sicula.*

The literature is rather confusing in regard to the HG of the chelonians, since it has been often mistaken for the posterior lacrimal gland (for review, see Saint-Girons 1985, 1988, 1989). In order to completely characterize the HG, both histochemical and ultrastructural characteristics have to be taken into consideration. In this paragraph, we refer to our circumstantial observations on the histology and ultrastructure of the HG of the freshwater terrapin, *Pseudemys scripta*, and the tortoise, *Testudo graeca*. These data appear to convincingly define the location of the gland.

The HG of the freshwater terrapin, *P. scripta*, consists of columnar or pyramidal glandular cells. At the electron microscopic (EM) level, four types are distinguishable (Fig. 11; Chieffi Baccari et al. 1991a). Between the secretory cells are numerous salt cells. These cells have their cytoplasm completely filled with tightly packed mitochondria and abundant glycogen granules; cysternae of the Golgi apparatus are observed between them. The free border of the cell shows short microvilli covered by a thick glycocalix. Also at its base, the cell develops numerous interdigitations characteristic of cells involved with active transport (Fig. 11).

The HG of *T. graeca* consists of tubuli with large lumina lined by a simple cuboidal epithelium. At the EM level, the glandular cells contain closely packed granules bounded by a unit membrane. They are usually formed by a moderately dense homogeneous material, but sometimes show a dense core (Fig. 8). Clumps of glycogen and vacuoles of various diameter are seen between the secretory granules. Among glandular cells, rare cells with the morphological characteristics of the "salt cells" are seen.

These observations, along with the histochemical details, justified our calling Harderian glands what previously had been called posterior lacrimal glands by Cowan (1969, 1971, 1973) and Saint-Girons (1985, 1988, 1989).

Fig. 7. Electron micrograph of the second type of glandular cell of the HG anterior zone of the snake, *Coluber viridiflavus*. This type of cell contains tightly packed homogeneous secretory granules of low density. ×4400

Fig. 8. Electron micrograph of the secretory granule of the HG glandular cell of the tortoise, *Testudo graeca*, which shows a dense core. ×22 000

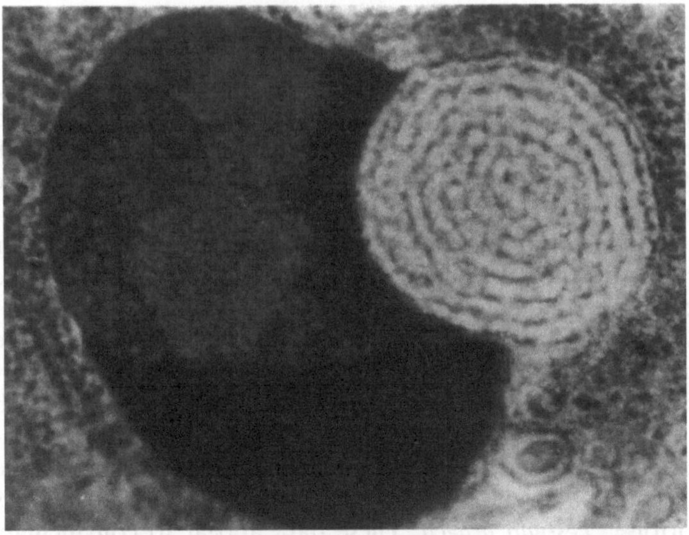

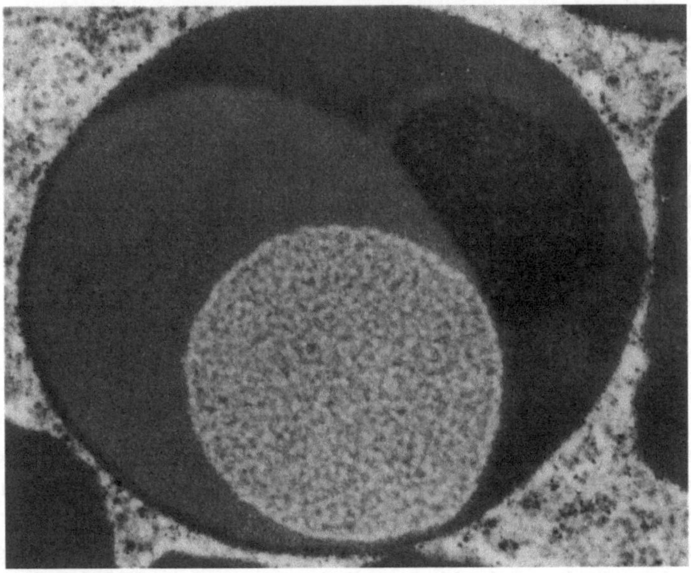

Fig. 9. The serous "special secretory granule" of the HG glandular cell of the lizard, *Podarcis s. sicula*, which consists of three closely associated components. × 30 000

Fig. 10. The "special secretory granule" of the HG glandular cells of the snake, *Coluber viridiflavus*, which is similar to that observed in the lizard. Note the presence of a crystalline structure in the peripherally more dense component of the granule. × 30 000

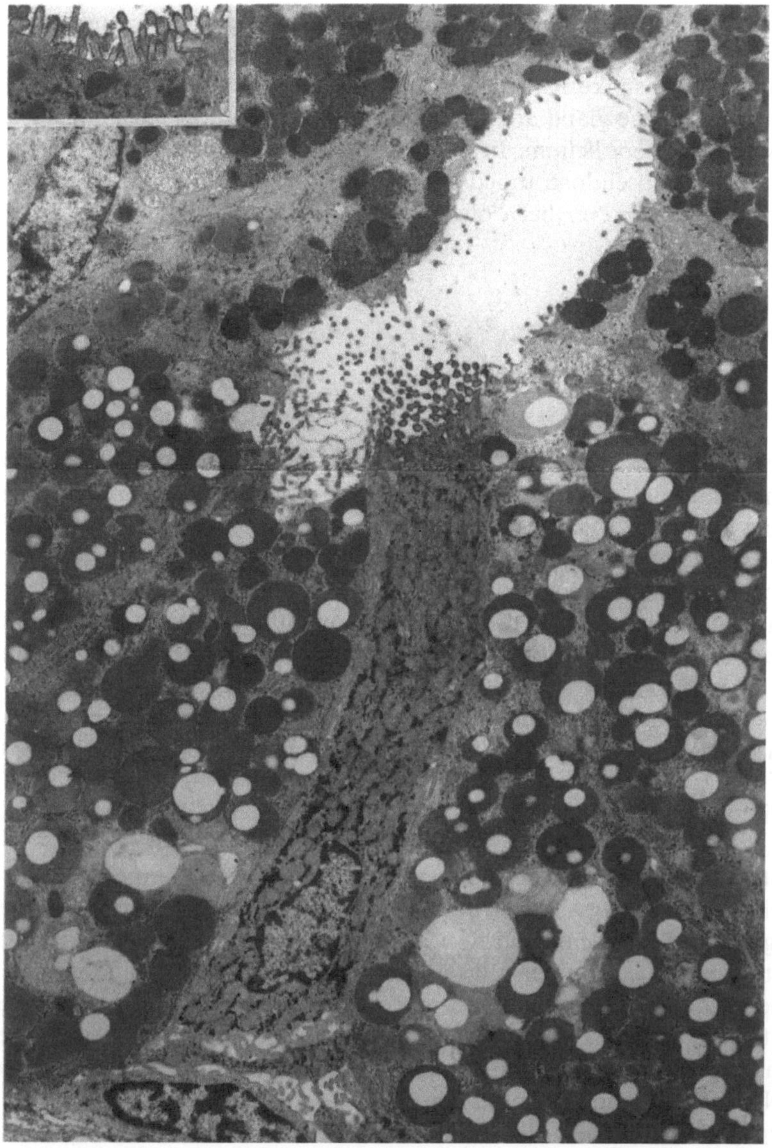

Fig. 11. Electron micrograph of a HG of the freshwater terrapin, *Pseudemys scripta*, showing different acinar cell types. Some secretory granules show a dense core while others have a light core surrounded by a zone of higher density. Among glandular cells, a salt cell can be identified by the presence of numerous mitochondria and glycogen accumulations. The free border of the cell shows short microvilli covered by a thick glycocalix (see *inset*). The salt cells also develop numerous interdigitations at their basal pole. × 3000; *inset* × 4400

In the reptilian HG, the mode of secretion is mainly merocrine and rarely apocrine. Since there is no evidence of mitotic activity, holocrine secretion seems unlikely. Only in the snake, *Coluber viridiflavus*, does a holocrine secretion possible. The ducts of the gland are not numerous and are often close together, lined by low cuboidal epithelium. In the larger ducts, the epithelium becomes columnar, sometimes pseudostratified columnar with centrally located nuclei. Sexual dimorphism has never been observed in reptiles.

5 Histochemistry

5.1 Amphibians

Periodic Acid Shiff (PAS)-positive material is present in the epithelium and lumina of the acini of all the species examined, e.g., *Rana catesbeiana* and *Rana pipiens* (Paule 1957). Di Matteo et al. (1989) and Minucci et al. (1989a) have extended the observations on *R. esculenta* and *B. viridis* throughout the year. The PAS-reaction is strongly positive in spring–summer. Alcian blue stains orthocromatic material in the epithelial cells and lumina. After the Alcian blue/PAS sequence, the majority of the granules in the acinar cytoplasm stain blue. Stronger reactions for sialic acid and hyaluronic acid are evident in the acinar epithelium than in the lumen content. In *R. esculenta*, no lipids were evident. On the contrary, in *B. viridis*, only glandular cells in females showed sudanophilic granules at their base. The sudanophilia increased in July. Sudanophilia was not detected in another toad, *Bufo bufo* (Minucci et al. unpubl.).

All positive reactions for mucosubstances, together with the mercury-bromophenol blue-positive test for protein, indicate that the secretion of the Harderian gland of *R. esculenta* and *B. viridis*, is seromucoid. These results agree with the rather high electron density shown by the secretory granules.

Histochemical tests for protein and mucosubstances confirm a progressive increase in the secretory activity of the acinar cells from May to August. These observations are supported by the determination, in the frog HG, of a high protein/DNA ratio in summer and a decline in September.

No porphyrin was found in the extract of the Harderian gland of either *R. esculenta* or *B. viridis*.

5.2 Reptiles

In the reptilian HG, the secretory cells of tubules and acini can be serous, seromucous and mucoserous. In all cases, the secretory product is rich in proteins and, with the exception of *Sphenodon*, contains little or no acid mucin. On the contrary, the reactivity of the glandular epithelium to Alcian blue-PAS varies greatly: it is usually negative or very weak, but is strong in Acrochordidae and somewhat less in Geckonidae, Scincidae, Varanidae, and *Laticauda colubrina* (Elapidae) (Saint-Girons 1988).

In the freshwater terrapin, *Pseudemys scripta* and the tortoise, *Testudo graeca*, strongly PAS-positive glandular cells can be found scattered between Alcian blue and Alcian blue-PAS positive cells. Since the PAS-positive reaction disappears after diastase digestion, these cells contain large amounts of glycogen (Chieffi Baccari et al. 1991a). These are the salt cells, whose characteristic ultrastructure has been described earlier.

In the HG of Anguidae, three types of cells can be distinguished in the glandular tubules: seromucous cells weakly positive to Alcian blue-PAS in the main body, seromucous cells, strongly Alcian blue-PAS positive in the central part, and mucoserous cells, rich in acidic mucin, but poor in proteins in the proximal portion (Saint-Girons 1988). This zonation is very similar to that described in the lizard, *Podarcis s. sicula*. In fact, the glandular cells of the HG lateral zone (main body) are weakly positive to Alcian blue-PAS reaction, while those of the intermediate part become more positive. The acinar cells are strongly positive in the anterior zone of the gland, whose secretory granules are ultrastructurally identical to those of the glandular cells of the posterior lacrimal gland (Chieffi Baccari et al. 1990).

Also in the snake, *Coluber viridiflavus*, two glandular cell types can be observed. The more numerous cells contain secretory granules in which it is possible to demonstrate two distinct parts of which only one, represented by a thin peripheral zone, is strongly positive to the Alcian blue-PAS reaction. The second cell type, concentrated in the most anterior and posterior zones of the gland, is Alcian blue-PAS positive and stains metachromatically with toluidine blue. This cell type contains homogeneous secretory granules of low electron density (Minucci et al. 1991).

Therefore, we suggest that the most anterior and posterior parts of the HG of these squamate be considered respectively the anterior and posterior lacrimal glands or a mixed Harderian-lacrimal gland.

Neither lipid droplets nor porphyrin were found in the reptilian HG. Thus, the secretory product of reptilian HG can be serous, seromucous, or mucoserous. In turtles and other water reptiles, the HG shares with the lacrimal gland the presence of salt cells.

6 Regulation of the Secretory Activity

Both exogenous and endogenous factors have been investigated as possible regulators of Harderian gland activity. Over many years, however, we have learned that these species display rather strong seasonal rhythms in responsiveness to experimental manipulations. In the green frog, for example, secretory activity is highest in July–August, declines to a low level in September and then gradually increases over the winter months (Di Matteo et al. 1989). Because of this circannual cyclicity, any investigation should be repeated during several seasons of the year.

6.1 Exogenous Factors

Since environmental factors like temperature and photoperiod are the extrinsic cues implicated in controlling seasonal cycles, experiments were performed during winter and summer in order to establish their respective role on the HG secretory activity of the green frog, *Rana esculenta.* The results clearly showed that seasonal temperature is more important than photoperiod. In fact, changes in photoperiod (8L:16D; 16L:8D) do not influence secretion at different temperatures selected (8 ± 1 °C; 24 ± 1 °C), while a marked effect of temperature is found both in the winter (Table 1) and in the summer (Table 2). Under experimental conditions, high temperature stimulates secretion, both at the histological and histochemical levels, while low temperature impairs secretory activity.

Although HG secretion in the green frog appears to be mainly controlled by environmental temperature, an influence of photoperiod (16L:8D) in terms of increased protein content was found when female winter frogs were kept at 24 °C (Minucci et al. 1990).

Table 1. Effects of several combinations of light and temperature (7-day treatment) on some morphological parameters of the HG of *Rana esculenta* in winter

Group	Height of glandular cells (μm)[a]	Area of acinar lumina (μm)[a]
Control (8 ± 1°C; L:D = 8:16)	14.98 ± 1.56*	2924 ± 428*
Group 1 (8 ± 1°C; L:D = 16:8)	14.97 ± 1.43*	2988 ± 395*
Group 2 (24 ± 1°C; L:D = 8:16)	18.43 ± 1.04**	21.2 ± 2.2**
Group 3 (24 ± 1°C; L:D = 16:8)	18.45 ± 0.65**	18.2 ± 3.4**

[a] Values are means ± SD of measurements of 300 glandular cells and 30 acinar lumina per animal (10 animals per group). *vs. ***P* < 0.001.

Table 2. Effects of several combinations of light and temperature (7-day treatment) on some morphological parameters of the HG of *Rana esculenta* in summer

Group	Height of glandular cells (μm)[a]	Area of acinal lumina (μm)[a]
Control (24 ± 1°C; L:D = 16:8)	20.38 ± 1.49*	12.8 ± 2.7*
Group 1 (24 ± 1°C; L:D = 8:16)	20.97 ± 1.61*	16.9 ± 2.7*
Group 2 (8 ± 1°C; L:D = 8:16)	10.98 ± 1.08**	3925 ± 687**
Group 3 (8 ± 1°C; L:D = 16:8)	10.87 ± 078**	4024 ± 454**

[a] Values are means ± SD of measurements of 300 glandular cells and 30 acinar lumina per animal (10 animals per group). *vs. ***P* < 0.001.

Temperature, whatever photoperiod used, affects the mast cell number. In fact exposure to higher temperatures casuses a decrease in mast cell number in the winter frogs, while exposure of the summer frogs to low temperatures provokes the opposite effect (Chieffi Baccari et al. 1991b).

6.2 Endogenous Factors

In addition to testing the influence of light and temperature on the secretion of the frog HG, studies were undertaken by us to ascertain the response of this gland to various endocrine manipulations.

7 Pinealectomy

The presence of melatonin in HG and pineal tissue extracts and in plasma of the frog, *R. esculenta*, has been radioimmunologically demonstrated (d'Istria et al. unpubl.). However, the melatonin content in HG and pineal was ten times lower than in plasma (80 pg/ml), suggesting that the major source of melatonin in this frog is very likely the retina as shown in *R. perezi* (Delgado and Vivien-Roels 1989) and *R. tigrina regulosa* (Pang et al. 1985).

A more detailed analysis carried out in *R. esculenta*, kept in continuous darkness for at least 2 weeks, showed a daily rhythm in plasma melatonin but not in pineal or HG melatonin (d'Istria et al. unpubl.). Whether HG melatonin is synthesized in situ is still unknown.

Two-month pinealectomized frogs (spring animals) were kept for 2 weeks under different photoperiods (24L:0D; 0L:24D; 12L:12D). No effect was observed in the HG morphology (d'Istria et al. unpublished). These data seem to exclude either a regulatory function of the pineal on the HG or a possible role of melatonin within the frog HG.

8 Hypophysectomy

The influence of hypophysis on secretory activity of frog (*R. esculenta*) HG was studied during representative periods of the annual cycle. Hypophysectomy, carried out during February, when the acinar cells start to elongate and accumulate secretory material, induced a further increase in the height of glandular cells which appear full of secretory granules (Minucci et al. 1989b). Hypophysectomy also affected the electrophoretic pattern of water soluble proteins preserving a protein fraction in the range of 480 kD, which disappeared in the captive sham-operated controls. Replacement therapy with crude homogenate of homologus pituitary restored the morphological and biochemical pattern of intact animals (Minucci et al. 1989b).

Among pituitary hormones tested in hypophysectomized frogs during winter, bTSH inhibited, while prolactin had no effect on the secretion of the glandular cells in either males or females. oACTH and oLH administration provoked a reduction of secretion in the female, while they had no effect in the male (Di Matteo et al. 1991).

We have not been able to carry out similar experiments in summer months owing to the high mortality (over 90%) following hypophysectomy.

Winter 1989–90 in Naples was remarkably mild and temperatures reached 22 °C. Consequently, the HG secretory activity was at its highest. Hypophysectomy performed in these animals was successful and provoked opposite results when compared to the previous study: acinar cell height decreased and the protein fraction of 480 kD disappeared. The replacement therapy restored the control protein pattern (Di Matteo et al. 1991).

It is pertinent that oACTH increased numbers of mastocytes, while bTSH had an effect only in the males and LH only in the females (Chieffi Baccari et al. 1991b).

The few reports available for mammals suggest an inhibitory effect of hypophysectomy in the rat (Boas and Bates 1954) and in the golden hamster (Hoffman 1971). However, in the frog, opposite results (inhibitory or stimulatory) can be obtained according to the physiological status of the HG and/or pituitary.

9 Sex Hormones

Oddly enough, only in a few mammalian species has the HG been shown to be a target for sex steroids (for review, see Olcese and Wesche 1989). That these steriods directly affect the gland is suggested by the occurrence of androgen receptors in the HG of golden hamster of both sexes (Vilchis et al. 1987; Vilchis and Perez-Palacios 1989). Only very small quantities of estrogen receptors were also reported (Vilchis and Perez-Palacios 1989).

In *R. esculenta*, we found not only a large number of cytosolic and nuclear androgen receptors, but also a parallel annual profile between plasma androgen concentration and receptor number sites (d'Istria et al. 1991a, b). Interestingly, we found a large number of androgen receptors in both sexes, which is not surprising since circulating androgens in the female are as high as in the male (d'Istria et al. 1974). No estrogen receptor was detected in the frog HG of either sex. The lack of a sexual dimorphism in the frog HG might be related to the similar pattern of circulating androgens in both sexes.

Notwithstanding the presence of androgen receptors in the frog HG, castration carried out in some periods of the year (February and November) had no effect on the morphology of the gland (Minucci et al. 1989b). However, the influence of testosterone on the HG was observed at the biochemical level. In fact, testosterone treatment induced a marked increase of some protein fractions which proved to be mostly glycoproteins; one of them had a MW of 480 kD, the same MW as the protein fraction which was influenced by hypophysectomy (see above). Cyproterone acetate treatment, given alone or in combination with testosterone, prevented these changes (Chieffi et al. 1991).

The role of testosterone on the secretory activity of the frog HG was further studied by in vitro incorporation of ^{3}H-uridine in the presence or absence of the steroid. Since the hormonal treatment resulted in a preferential uptake in the poly(A)$^+$ fraction, it is presumed that testosterone can trigger the specific biosynthesis of some mRNAs (Varriale et al. 1990.).

In *Bufo viridis*, despite the sexual dimorphism present at the ultrastructural and histochemical levels, neither androgen nor estrogen receptors were evident, although an androgen–estrogen binding protein has been found in the cytosolic fraction (d'Istria et al. 1991b). Minucci et al. (1989a) reported that ovariectomy caused the disappearance of sudanophilia in the glandular cells of the female toad, inducing a male type gland.

Using testosterone and estradiol as labeled ligands, no specific binding was detectable in *Podarcis s. sicula* (d'Istria et al. 1991b).

Overall, although sex hormones do not seem to influence the morphology of the frog HG, testosterone does modify, qualitatively, its secretion.

10 Putative Functions

There can be little doubt that among the several functions attributed to the HG, the basic one is the lubrication of the eye. Very likely this function is the most primitive, since the HG appeared during vertebrate invasion of the land and before the development of the lacrimal gland. However, it probably was not the only function, since the secretory product of amphibian HG is seromucous and not exclusively mucous as in lacrimal glands, which evolved later in reptiles.

Besides eye lubrication, other functions displayed by the HG may vary from species to species. For mammals, these include sexual dimorphism, hormone synthesis (melatonin, pheromones), thermoregulation, and photosensitivity, while in birds, the HG also plays an immunological role (for review, see Olcese and Wesche 1989).

In reptiles, Born (1883) suggested a possible role in digestion. Pratt (1948) and Bellairs and Boyd (1950) hypothesize a possible function of HG secretion as a vehicle of "sweet-smelling" substances to the Jacobson organ.

Unique among reptiles, chelonian HG participates as an extrarenal organ in the elimination of salt (Schmidt-Nielsen and Fänge 1958). The "salt gland" in the loggerhead turtle, *Caretta caretta*, corresponds most likely to the HG described by us in *Pseudemys scripta* and *Testudo graeca* (Chieffi Baccari et al. 1991a). In fact, we found a large number of salt cells intermingled with serous and mucoserous glandular cells. The salt cells contain the machinery relevant to the salt-secreting function, i.e., the vast absorptive surface at both poles of the cells, a mucopolysaccharide associated with a thick glycocalix at the apical pole, and the large number of mitochondria surrounded by numerous glycogen particles. The anionic charge of the surface coat of mucopolysaccharide is considered an effective means of attracting, trapping, and concentrating cations. The packed mitochondria, along with the extreme richness in glycogen, complete the machinery with their high energy support for this osmoregulatory device.

The amphibians were the first tetrapods to develop the HG that remained the only orbital gland. Therefore, its role is confined mainly to the lubrication of the cornea and the nictitating membrane. The increase of secretory activity in correlation with increasing temperature (Di Matteo et al. 1989; Minucci et al. 1990) is also suggestive of this hypothesis. However, the sexual dimorphism observed in *Bufo viridis* with lipids only in the female, is indicative of a pheromonal function which may influence the sexual behavior of the male (Minucci et al. 1989a).

11 Concluding Remarks

The comparative study of the HG is in statu nascenti. The variety of suspected functions in mammals is probably related to ecological adaptations. If this is true for mammals, it will be not surprising to find new and unexpected functions in nonmammalian tetrapods. The comparative study will surely throw more light on the significance of this still relatively unknown structure.

References

Bellairs A, Boyd JD (1950) The lacrymal apparatus in lizards and snakes. II The anterior part of the lacrymal duct and its relationship with palate and with the nasal and vomeronasal organs. Proc Zool Soc Lond 120:269–310

Boas NF, Bates RW (1954) Role of the thyroid and anterior pituitary glands in the maintenance of the Harderian glands of the rat. Endocrinology 55:601–612

Born G (1883) Die Nasenhöhlen und der Tränen-Nasengang der amnioten Wirbeltiere. Gegenbaur's Morphol Falrb 8:188–232

Chieffi G, Chieffi Baccari G, Di Matteo L, d'Istria M, Marmorino C, Minucci S, Varriale B (1991) Hormonal control of the Harderian gland of *Rana esculenta.* Belgian J Zool 121:130–132 (abstr)

Chieffi Baccari G, Marmorino C, Minucci S, Di Matteo L, d'Istria M, Chieffi G (1989) Mallory stain can unveil cell activity status through nuclear color change. Bas Appl Histochem 33/suppl: 30 (abstr)

Chieffi Baccari G, Minucci S, Di Matteo L, Chieffi G (1990a) Harderian gland and the lacrimal gland of the lizard *Podarcis s. sicula*: histology, histochemistry, and ultrastructure. Anat Rec 226:269–278

Chieffi Baccari G, Minucci S, Di Matteo L (1990b) Organogenesis of the Harderian gland in *Rana esculenta* and *Bufo viridis.* Boll Zool 57:221–224

Chieffi Baccari G, Di Matteo L, Minucci S (1991a) The orbital glands of the chelonians *Pseudemys scripta* and *Testudo graeca*: comparative histological, histochemical and ultrastructural investigations. J Anat (in press)

Chieffi Baccari G, Minucci S, Marmorino C, Izzo Vitiello I (1991b) Number of mast cells in the Harderian gland of the green frog, *Rana esculenta*: the annual cycle and its relation to environmental and hormonal factor. J Anat (in press)

Cowan FBM (1969) Gross and microscopic anatomy of the orbital glands of *Malaclemys* and other emydine turtles. Can J Zool 47:723–729

Cowan FBM (1971) The ultrastructure of the lachrymal "salt" gland and the Harderian gland in the euryhaline *Malaclemys* and some closely related stenohaline emydines. Can J Zool 49:691–698

Cowan FBM (1973) The homology of cranial glands in turtles with special reference to the nomenclature of "salt glands". J Morphol 141:157–170

Delgado MJ, Vivien-Roels B (1989) Effect of environmental temperature and photoperiod on the melatonin levels in the pineal, lateral eye, and plasma of the frog *Rana perezi*: importance of ocular melatonin. Gen Comp Endocrinol 75:46–53

Di Matteo L, Minucci S, Chieffi Baccari G, Pellicciari C, d'Istria M, Chieffi G (1989) The Harderian gland of the frog, *Rana esculenta*, during the annual cycle: histology, histochemistry and ultrastructure. Bas Appl Histochem 33:93–112

Di Matteo L, Minucci S, Chieffi Baccari G, Varriale B (1991) Effects of hypophysectomy and replacement therapy on the Harderian gland of *Rana esculenta*. Rend Fis Acc Lincei S 9, 2:415–419

d'Istria M, Delrio G, Botte V, Chieffi G (1974) Radioimmunoassay of testosterone, 17β-estradiol and estrone in the male and female plasma of *Rana esculenta* during sexual cycle. Steroids Lipids Res 5:42–48

d'Istria M, Chieffi Baccari G, Di Matteo L, Minucci S, Varriale B, Chieffi G (1991a) Androgen receptor in the Harderian gland of *Rana esculenta*. J Endocrinol 129(2):227–232

d'Istria M, Chieffi Baccari G, Di Matteo L, Minucci S, Varriale B (1991b) Sex steroid binding proteins in the Harderian gland of non-mammalian tetrapods. Rend Fis Acc Lincei S 19, 2:421–424

Hoffman RA (1971) Influence of some endocrine glands, hormones and blinding on the histology and porphyrins of the Harderian glands of golden hamster. Am J Anat 132:463–478

Kollros JJ, Kaltenbach JC (1952) Local metamorphosis of larval skin in *Rana pipiens*. Physiol Zool 25:163–170

Minucci S, Chieffi Baccari G, Di Matteo L, Chieffi G (1989a) A sexual dimorphism of the Harderian gland of the toad, *Bufo viridis*. Bas Appl Histochem 33:299–310

Minucci S, Di Matteo L, Chieffi Baccari G, Marmorino C, Chieffi G (1989b) On the effect of hypophysectomy and castration on the Harderian gland on the frog *Rana esculenta*. Gen Comp Endocrinol 74:263–264 (abstr)

Minucci S, Chieffi Baccari G, Di Matteo L, Marmorino C, d'Istria M, Chieffi G (1990) Influence of light and temperature on the secretory activity of the Harderian gland of the green frog, *Rana esculenta*. Comp Biochem Physiol 95A:249–252

Minucci S, Chieffi Baccari G, Di Matteo L (1991) Histology, histochemistry, and ultrastructure of the Harderian gland of the snake *Coluber viridiflavus*. J Morphol 210:1–6

Olcese J, Wesche A (1989) The Harderian gland. Comp Biochem Physiol 93A:655–665

Pang SF, Shiu SYW, Tse SF (1985) Effect of photic manipulation on the level of melatonin in the retina of frogs (*Rana tigrina regulosa*). Gen Comp Endocrinol 58:464–470

Paule WJ (1957) The comparative histochemistry of the Harderian gland. Doctoral Dissertation, Ohio State University, Columbus, Ohio

Pratt CWN (1948) The morphology of the ethmoidal region of *Sphenodon* and lizards. Proc Zool Soc Lond 119:171–201

Saint-Girons H (1985) Histologie des glandes orbitaires des crocodiles et des tortues, et comparaison avec lec lépidosauriens. Ann Sci Nat Zool, Paris 7:249–264

Saint-Girons H (1988) Les glandes céphaliques exocrines des reptiles. I. Données anatomiques et histologiques. Ann Sci Nat Zool, Paris 9:221–255

Saint-Girons H (1989) Les glandes céphaliques exocrines des reptiles. II. Considérations fonctionnelles et évolutives. Ann Sci Nat Zool, Paris 10:1–17

Schmidt-Nielsen K, Fange R (1958) Salt glands in marine reptiles. Nature 182:783–785

Shirama K, Kikuyama S, Takeo Y, Shimizu K, Maekawa K (1982) Development of Harderian gland during metamorphosis in anurans. Anat Rec 202:371–378

Varriale B, Chieffi Baccari G, Di Matteo L, Minucci S, d'Istria M (1990) Effect of testosterone on specific 3H-Uridine incorporation in the frog (*Rana esculenta*) Harderian gland. Gen Comp Endocrinol 82:218 (abstr)

Vilchis F, Hernandez A, Perez AE, Pérez-Palacios G (1987) Hormone regulation of the rodent Harderian gland: binding properties of the androgen receptor in the male golden hamster. J Endocrinol 112:3–8

Vilchis F, Pérez-Palacios G (1989) Steroid hormone receptors and the sexual phenotype of the Harderian gland in hamsters. J Endocrinol 121:149–159

Morphology of the Harderian Gland in the Rabbit. A Short Review

WOLFGANG KÜHNEL

Department of Anatomy, Medical University, Ratzeburger Allee 160,
D-2400 Lübeck 1, FRG

1 Introduction

The lacrimal systems are functionally related to the eye, in particular to the cornea, which depends on the normal secretion of tears for its transparency. The tear liquid cleans the conjunctival sac and also nourishes and moistens the cornea. The components contributing to its variable composition arise from the secretory activities of the various orbital glands.

In the rabbit, the glandula lacrimalis, the glandula infraorbitalis, the glandula orbitalis externa, the gland of the nictitating membrane, and the Harderian gland open into the conjunctival sac. Kittel (1962) presented comparative topographical and histological investigations concerning the orbital glands in rodents and Kühnel (1966, 1971a, 1972a, 1974, 1983), Kühnel and Beier (1973), Kühnel and Burger (1969) and Kühnel and Wrobel (1966a, b; 1968) reported the findings of light and electron microscopy and histochemistry with particular emphasis on the Harderian gland in rabbits (see also Peters 1890; Duboc 1926; Mukai 1926; Mycytowicz 1966; Brownscheidle and Niewenhuis 1978).

The Harderian gland is widespread throughout the animal kingdom, but differs in its structure among various species. Even though it was discovered as early as 1694 by Johann Jacob Harder, a noted physician born on September 17, 1656, in Basel, Switzerland, who first described it in the deer (*Dama vulgaris*), little is known about the function and metabolism of this orbital organ. Since that time, the Harderian gland has been observed in many terrestrial species, absent only in the orbit of primates. However, the Harderian gland is rudimentary in monkeys and may even be found in vestigial form in the human fetus from the 11th to the 30th week or occasionally in adults with abnormalities (Kennedy 1970; Olcese and Wesche 1989). The Harderian gland often is named Glandula palpebrae tertiae profunda. However, the gland is not present in all animals with a third eyelid (Loewenthal 1892; Lutz 1899).

Webb et al. (Eds).
Harderian Glands

2 Gross Anatomy

The rabbit Harderian gland is irregularly kidney-shaped. It is convex on the outer side and concave on the inner, to comply with the shape of the eyeball. It is about 20 mm long, 15 mm wide, and 5 mm thick. There is a deep notch on its inferior anterior border, in which the inferior oblique muscle rests (Davis 1929). The gland consists of two lobes: the upper is white while the lower is pink in color. The lower lobe is about twice the size of the upper, although the ratio is variable. The two lobes are loosely connected by thin fibers of connective tissue and may be easily separated by gross dissection so that they suggest two glands rather than one. Sometimes the white and the pink lobe may be found mixed. The secretion is discharged by a single duct opening on the concave surface of the palpebra tertia, well back and near its inferior border (Davis 1929; Loewenthal 1892).

3 Morphology

3.1 Light Microscopy

Examination of numerous sections of the rabbit Harderian gland demonstrates a tubuloalveolar structure, uniform in all areas, with each part divided into lobules separated from each other by a small amount of connective tissue containing fibroblasts, small blood vessels, and nerve fibers (Kühnel 1971a). Sometimes alveoli from both lobes are mingled and occasionally some alveoli contain pink and white cells side by side (Björkman et al. 1960; Kühnel 1971a). The light microscopic picture yields no apparent morphological differences compared with findings of Buschke (1933) and Kühnel (1971a) with respect to material embedded with paraffin and stained by hematoxylin-eosin, azan, iron-hematoxylin or chromalaun-hematoxylin-phloxin following formaline fixation. The pink lobe is characterized by large alveoli with a cuboidal epithelium. The basal nuclei are round with uneven or serrated outlines (Fig. 1). In the white part, the alveoli have a smaller diameter and the lumina are narrower. The epithelium is high columnar and the cells are more regular than in the pink lobe. The rounded nuclei are regularly placed in the basal compartment of the cell (Fig. 2).

Fig. 3. Epithelium of the pink lobe of the rabbit Harderian gland. The cytoplasm is filled with lipid vacuoles. Vacuoles lying in the apical part of the cells are often separated from the opening by a thin membrane only. The spindle-shaped nuclei on the bases belong to myoepithelial cells. There is a capillary (*C*) below the epithelial cells. Semi-thin section, Methylene blue-azure II stain. ×560

Fig. 4. Columnar epithelial cells of the white lobe. Vacuoles in the cytoplasm are much smaller than those in the pink lobe. The nuclei are regularly placed in the basal parts. There are two capillaries (*C*) between the alveoli. Semi-thin section, methylene blue-azure II stain. ×560

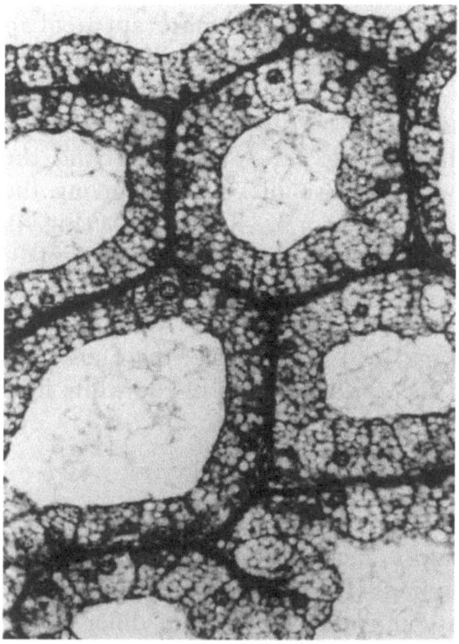

1

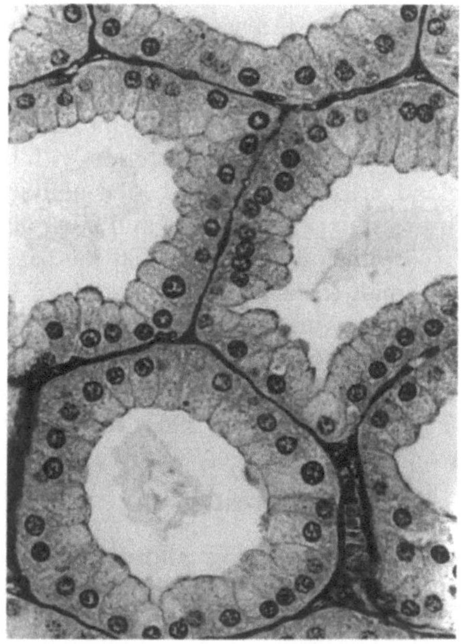

2

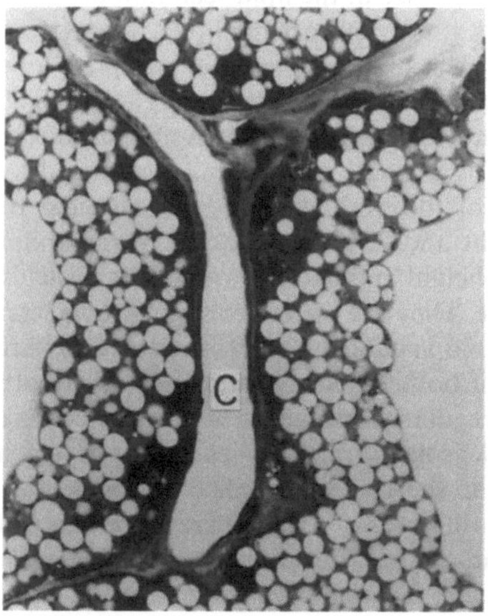

3

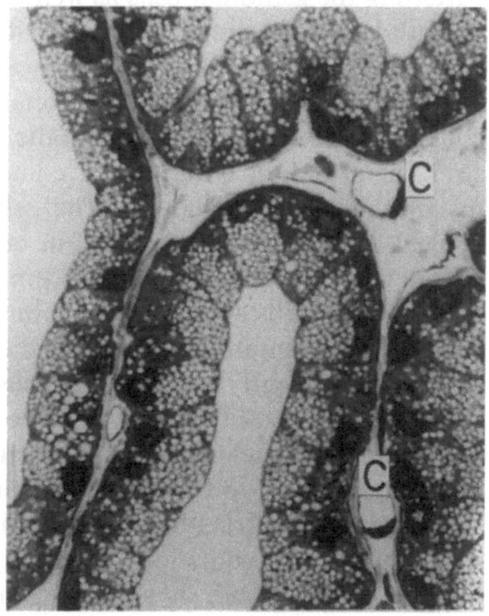

4

Fig. 1. Several alveoli of the pink lobe. Each alveolus is composed of a single layer of cuboidal secretory cells. The cytoplasm has poor stainability. It contains large vacuoles. Mayer's hemalaun-eosin stain. ×350

Fig. 2. Alveoli of the white lobe. Each alveolus is composed of a single layer of columnar secretory cells. The cells are more regular than in the pink part. The cytoplasm shows a diffuse basophilia. Gomori's hemalaun-phloxine stain. ×350

In contrast, a vacuolization of the cytoplasm becomes more apparent in semi-thin sections stained by methylene blue-azure II. The gland cells of the pink lobe contain large vacuoles which remain separated from one another by small parts of cytoplasm. Apically located vacuoles may be ruptured, thus indicating a direct contact of the secretion with the lumen (Fig. 3). On the other hand, the gland cells of the white lobe contain noticeably small vacuoles, giving the impression of foamy cells and most of the cells show a basal basophilia (Fig. 4). Myoepithelial cells appear at the base of the epithelium of both lobes.

Small to large lipid vacuoles in the pink lobe become visible in cryostat sections by using Scarlet Red and Sudan Black B, and these correspond to the rough vacuolization of the cytoplasm (Figs. 5, 6). In contrast, there are only finely dispersed sudanophilic particles in the epithelial cells of the white lobe (Fig. 6) (Kahán et al. 1967; Kühnel 1971a).

3.2 Electron Microscopy

Electron microscopic photographs indicate a vacuolization of the cytoplasm corresponding to the light microscopic findings (Björkman et al. 1960; Kühnel 1971a). The photographs also show that pink and white lobes differ in the pattern of the glandular cells as well as in the size of the lipid secretion droplets. In both lobes, the preparatory techniques totally extract the lipids, thus giving rise to microscopically empty vacuoles. The glandular cells of the pink lobe contain large lipid secretory droplets with a wide variation in size (Fig. 9). In contrast, the secretion products of the white lobe cells are more homogeneous in size. The alveoli of both lobes are lined by a single epithelium, surrounded by myoepithelial cells (Fig. 10). The cells of the pink lobe appear cuboidal or conical, measuring about 16 μm in height and 20 μm in width. The glandular cells of the white lobe are about 15 μm in height and 6 μm in width. The nuclei in both kinds of cells are about equal in size. They are round, or slightly oval, and often show indentations caused by the lipid vacuoles. Cells with two nuclei are seen very frequently in glandular cells of both lobes and in glands from both male and female rabbits. One or more irregularly outlined granular nucleoli are present in each nucleus. The double nuclear membrane has fenestrations which sometimes appear to be more numerous in the glandular cells of the pink lobe.

The mitochondria are numerous in both kinds of cells but are more frequent in the cells of the white lobe, especially in the basal parts of the cells. They are medium-sized, round, or slightly elongated. The lamelliform cristae are often short and stub-like, and are not very closely packed in the more or less dense matrix.

Apart from the nucleus and the Golgi apparatus the small amount of cytoplasm left between the lipid vacuoles is densely packed with anastomosing tubules of smooth endoplasmic reticulum, which is denser in the glandular cells of the white lobe (Fig. 10). The Golgi complex is found above the nucleus, often displaced laterally. It is relatively small and slightly elongated. In the white lobe

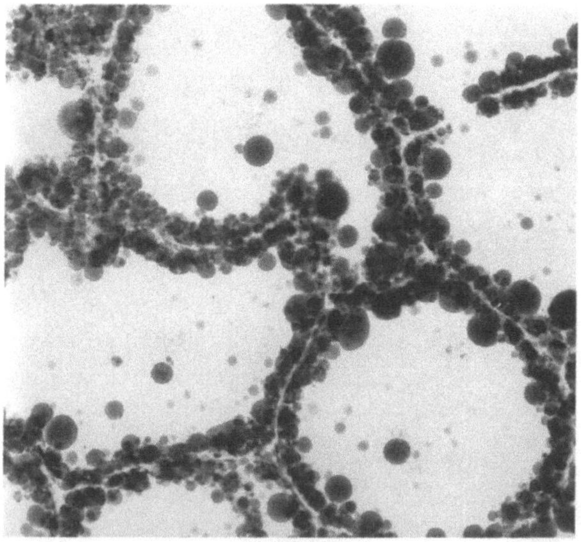

5

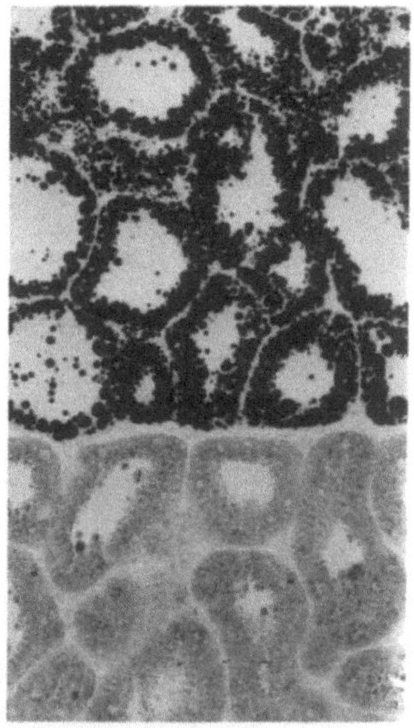

6

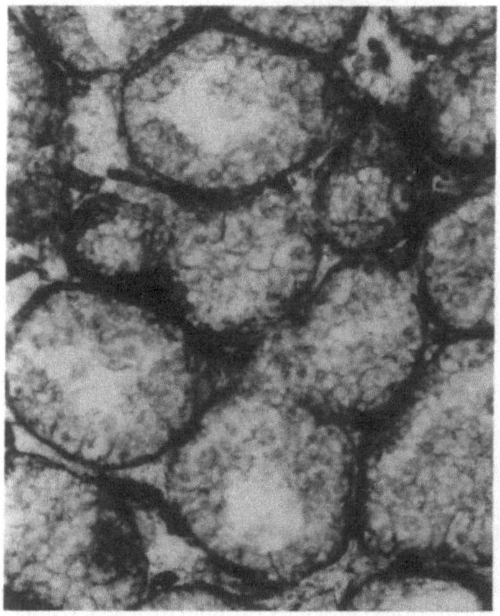

7

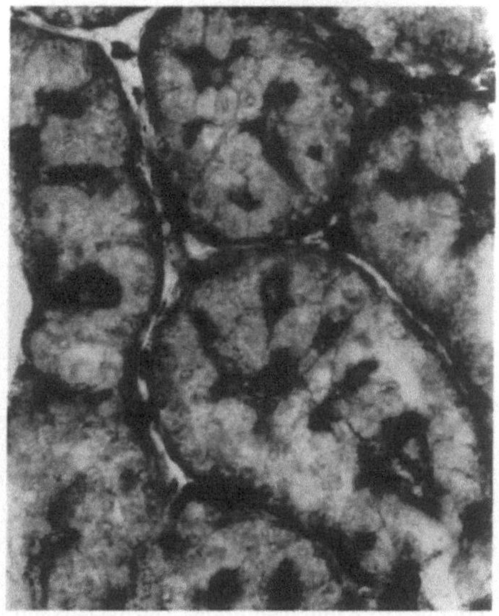

8

Fig. 5. Alveoli of the pink lobe. Large lipid droplets are seen in the cryostat section. Scarlet red stain. ×350

Fig. 6. Alveoli of the pink lobe are at the top of the figure, the white lobe below. Cryostat section, Sudan black B stain. The cells of the pink lobe show large lipid droplets. In the cells of the white lobe, only finely dispersed sudanophilic granules can be seen. ×80

Fig. 7. Pink lobe of the rabbit Harderian gland stained for alkaline phosphatase. Note the positive staining for ALK-Pase in the myoepithelial cells. ×140

Fig. 8. White lobe of the rabbit Harderian gland stained for ATPase. Note the positive staining at the apical and lateral membranes of the gland cells. ×140

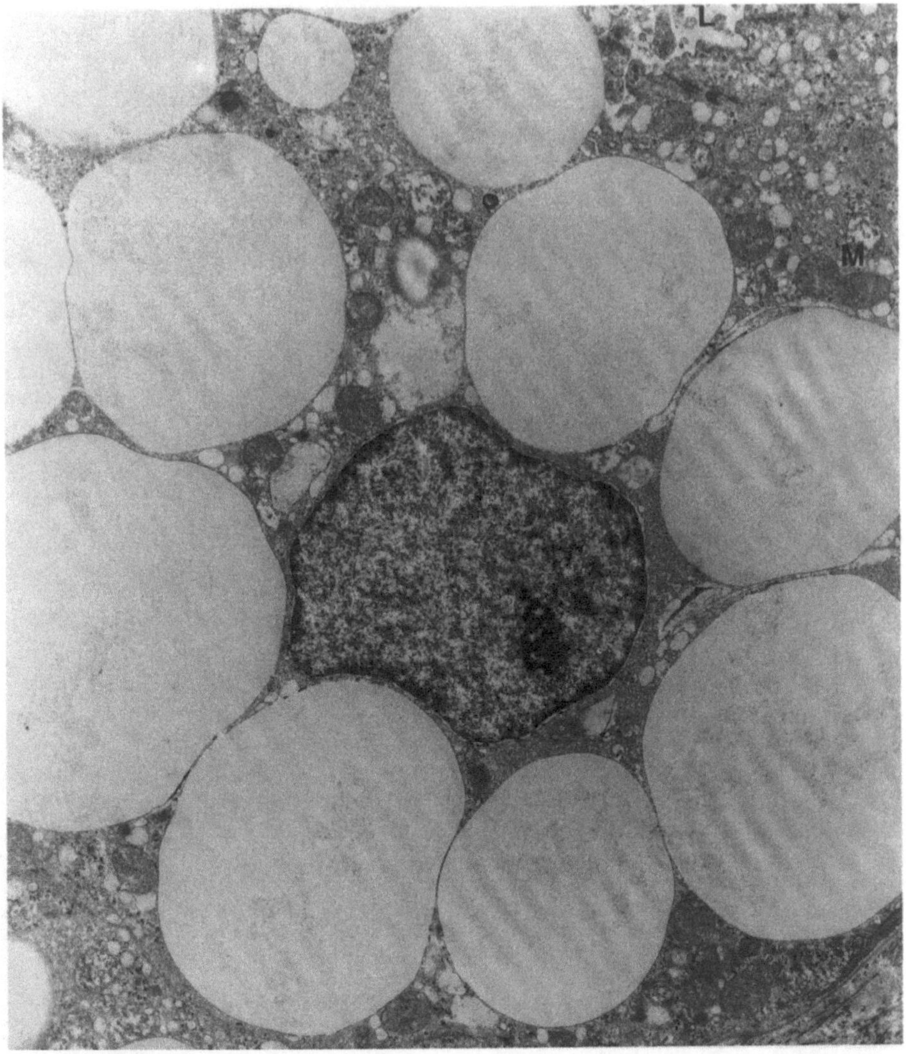

Fig. 9. Electron micrograph showing large lipid vacuoles in the cytoplasm of the pink lobe. They are bounded by a unit membrane. The nucleus shows indentations caused by the lipid droplets. *L* lumen of the alveolus; *M* mitochondria. × 10500

it is oriented along the longitudinal axis of the cells. The lamellae are rather few, but small opaque vesicles are numerous. The cytoplasm also contains numerous vesicles and vacuoles, the smallest of which resemble the cisternae of the Golgi zone, especially in the cells of the pink lobe. Some of the larger ones may be intermittently stippled by a few ribosomes; others contain a flocculent material.

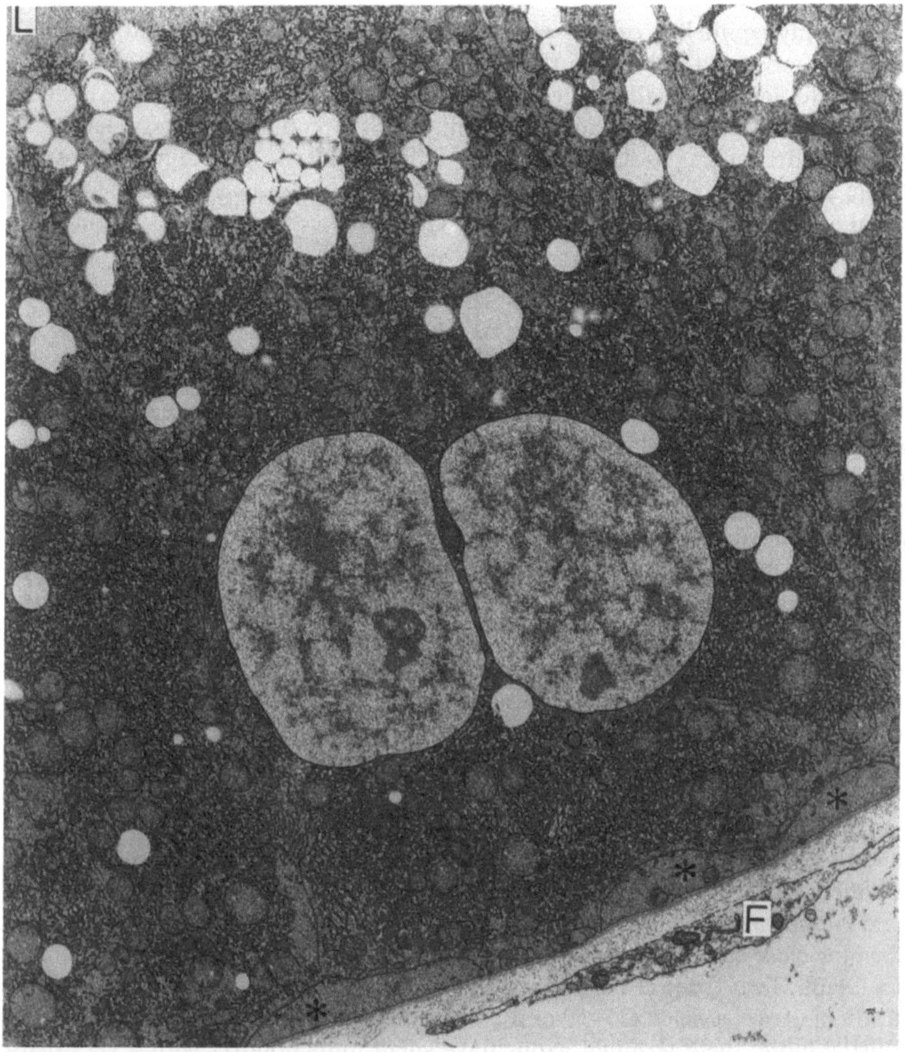

Fig. 10. Electron micrograph showing three gland cells of the white lobe. Note the cell with two nuclei, the high amount of smooth endoplasmic reticulum, the numerous mitochondria and the small lipid vacuoles. Compare with Fig. 9. *L* lumen; * = myoepithelial cells; *F* fibroblast. × 7000

The apical plasma membrane of the glandular cells forms a number of irregular cytoplasmic processes and short microvilli. The basal plasma membranes show areas of extensive irregular folding, alternating with smooth areas, irrespective of the presence or absence of myoepithelial processes. These foldings continue over the lateral plasma membranes forming interdigitations, only to be interrupted by desmosomes. A usually short tight junction is present at the luminal end of the lateral plasma membranes.

The myoepithelial cells are quite extensive, their cytoplasmic processes covering most of the basal surface of the glandular epithelium (Figs. 11, 12). Their cytoplasm is usually more electron-dense than that of the epithelial cells. It contains numerous thin (6 nm) and some thick (10 nm) filaments, a few small mitochondria, some isolated cisternae of endoplasmic reticulum, and small Golgi complexes. Free ribosomes are scarce. The plasma membrane facing the basal lamina shows numerous caveolae. Glandular cells and myoepithelial cells are often connected by desmosomes.

3.3 Innervation

Autonomic nerve fibers have been found in the white and pink lobe of the rabbit Harderian gland (Kühnel 1971b). It is commonly accepted that secretion by the mammalian exocrine glands is regulated by the autonomic nervous system (see review by Bargmann 1977). The nerve fibers are located around the acini or tubules, and some of them also penetrate between the acinus or duct cells (Kühnel 1971a,b, 1972a,b, 1974, 1983; Kühnel and Beier 1973; Kühnel and Scheele 1979; Schramm and Kühnel 1979; Nikkinen et al. 1984; Radke et al. 1985; Mendoza and Kühnel 1987; Tsukahara and Jacobowitz 1987). These axons form terminal swellings, which are embedded at the base of the secretory or duct cells with a separation of 20 nm. The contents of these terminals include many agranular vesicles, large granular vesicles, and mitochondria. Thickenings of pre- and postsynaptic membranes are not developed.

Electron microscope investigations on the rabbit Harderian gland confirm that it contains nerve fibers similar to those in other exocrine glands of various species (Figs. 11, 12). Blood vessels and nerve fascicles containing both myelinated and unmyelinated axons are present in the interstices between the alveoli. A few isolated axon varicosities are seen inside the basal lamina; they indent the basal and lateral surface gland cells or are found between gland and myoepthelial cells, forming close contacts with these cells. The diameter of the fibers ranges from 0.2 to 1.2 μm. Two types of nerve fibers are discernible: the first contains accumulations of clear vesicles (35–50 nm) and some dense-cored vesicles (50–65 nm). Furthermore, these terminal axons have small mitochondria with a very dense matrix and longitudinally arranged cristae, neurotubules, and glycogen particles. The second type of nerve fibers contains only a few clear vesicles, but many membrane-bound granules of different sizes and various electron densities, some small dark mitochondria, and occasional multilamellate bodies. The cleft between nerve fibers and gland cells is 10–20 nm wide. Postsynaptic membranes or specializations of the gland cell surface are not developed.

3.4 Lipid Droplet Secretion

The secretion of lipids is usually by release of lipid-filled whole cells. The mammary and the Harderian gland of the rabbit can secrete individual lipid

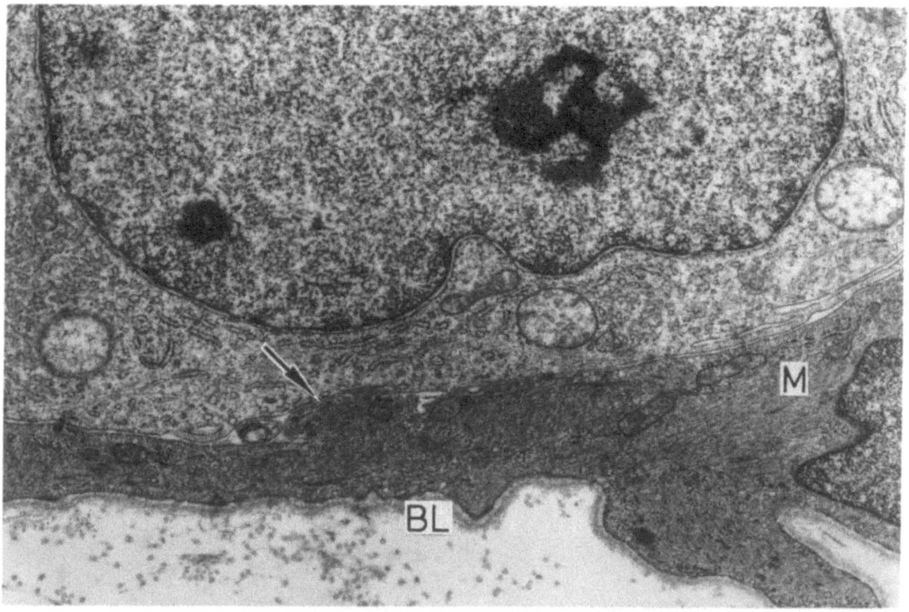

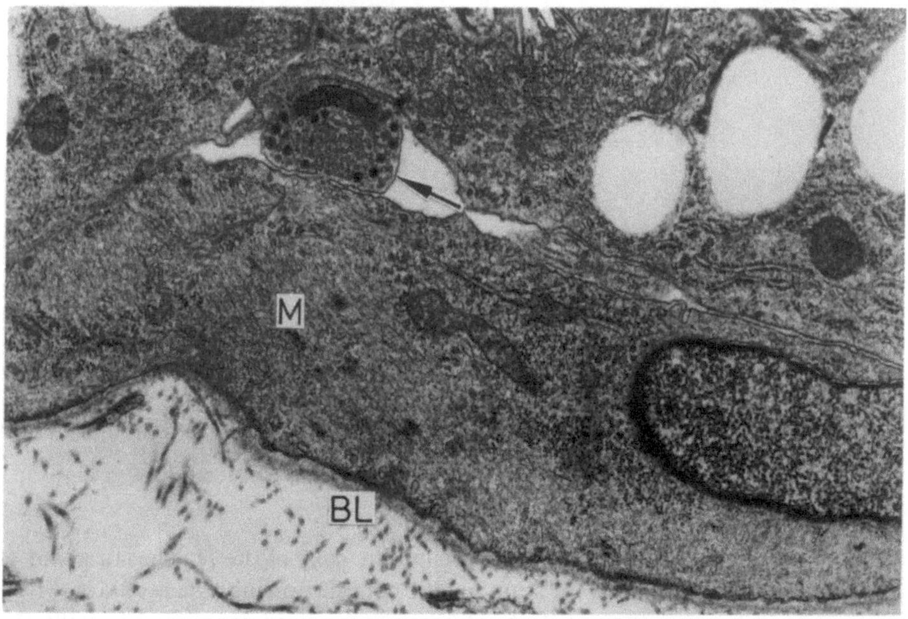

Fig. 11. Basal part of a white glandular cell and an extensive myoepithelial cell. Note the nerve terminal (↑) between the gland and the myoepithelial cell. *BL* basal lamina; *M* myoepithelial cell. ×23 000

Fig. 12. Basal part of a pink glandular cell and a myoepithelial cell. Note the nerve terminal (↑) with clear and dark vesicles. *M* myoepithelial cell; *BL* basal lamina. ×29 000

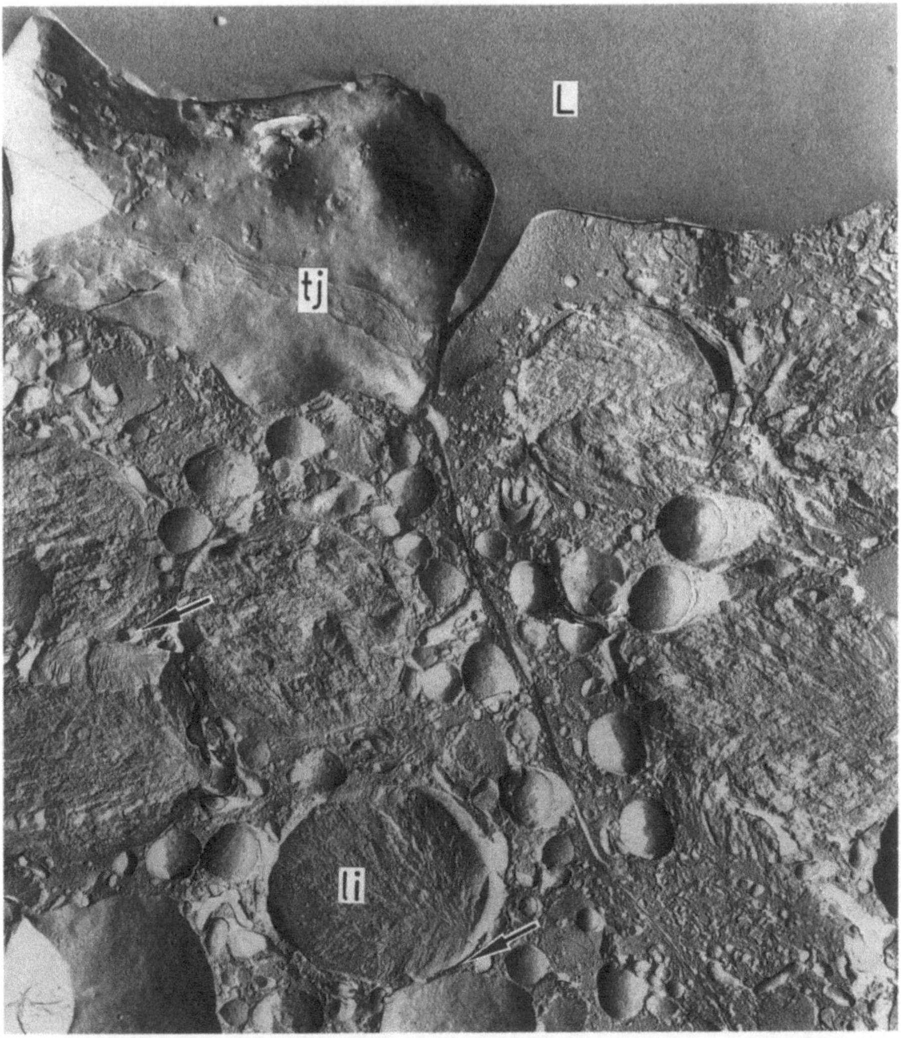

Fig. 13. Freeze-fracture replica of the pink lobe of the Harderian gland. The glandular cells contain numerous lipid droplets, differing in size and sometimes fusing with one another (*arrows*). The outline of the lipid (*li*) appears relatively smooth. *L* lumen; *tj* tight junction; *li* lipid. ×15400

droplets continuously from persistent cells. The cells of the Harderian gland of the rabbit are a good model for studying the complex membrane events involved in exocytosis because they contain extremely large lipid droplets, especially the cells of the pink lobe, that are discharged into the lumen via merocrine secretion (Wooding 1980).

Several years ago, we studied membrane specializations of the cells of the pink and white lobes of the Harderian gland. Membrane differentiations involved in exocytosis, as well as granule interactions were investigated by means of the freeze-fracture and thin-section techniques (Winterhager and Kühnel 1982, 1983). In the Harderian gland of the rabbit, the lipid droplets of both the pink and the white lobe are surrounded by a unit membrane (Figs. 13–15). Fracture of the lipid membrane of the droplets from cells of both lobes reveals a concave P-leaflet with about 870 IMP/μm^2. The leaflets often show a network-like aggregation of particles and smooth membrane patches. The texture of the E-face is, apart from the smooth zones, somewhat irregular. The particle background makes it difficult to determine the true IMP packing density, which is estimated to about 1200/μm^2. Larger lipid droplets of both glandular parts are formed by fusion and show particle-free areas at the fusion site of the granular membranes. During exocytosis, the large secretion droplets of the pink-lobe cells often cause the plasma membrane to bulge extensively into the lumen. The over-lying plasmalemma shows IMP-free patches. Secretion droplets of the white-lobe cells, smaller in volume, deform the plasma membrane

Fig. 14. Freeze-fracture replica of the white lobe of the Harderian gland. The numerous lipid droplets, when compared to those in the pink part, are smaller in size and the secretory product is formed in the manner of myelin whorls. *L* lumen; *tj* tight junction; *li* lipid. ×14700

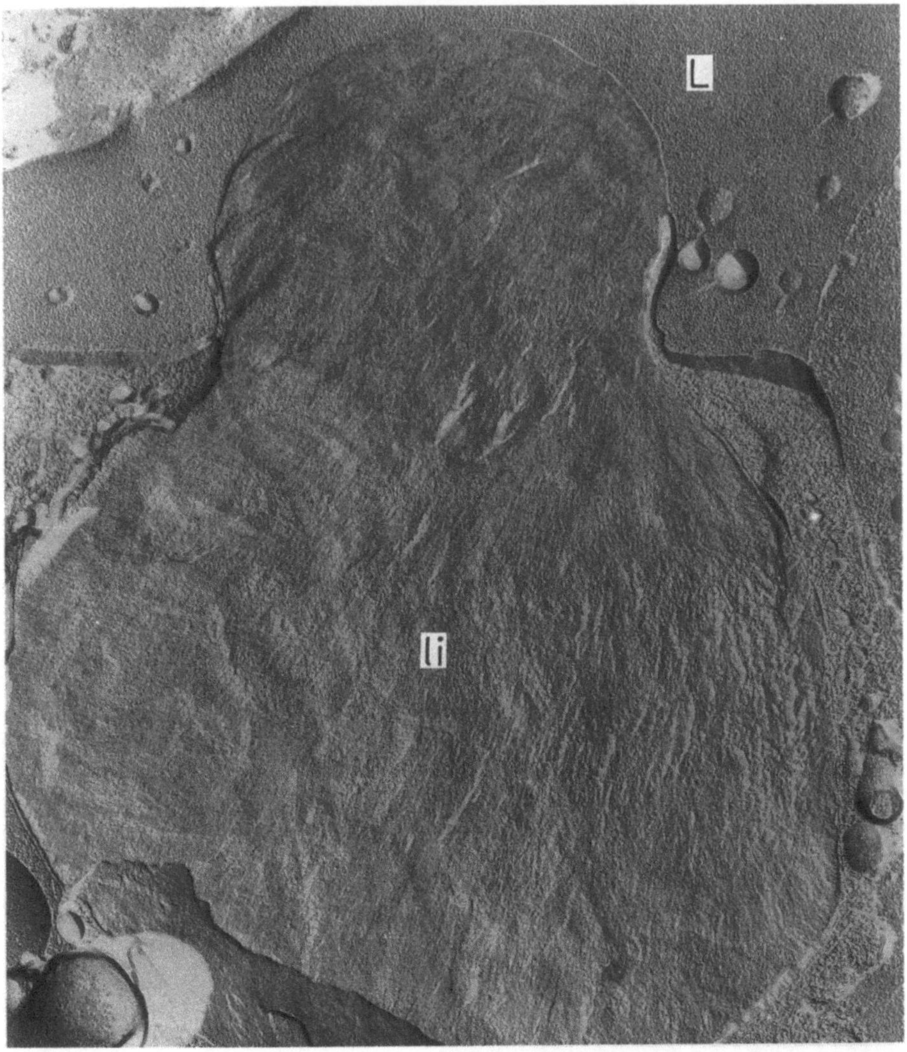

Fig. 15. Process of lipid discharge into the glandular lumen. Release of the secretory product starts with a knob protruding through a relatively narrow exocytotic opening. *L* lumen; *li* lipid. ×46000

only slightly when lying in close apposition. The distribution of IMP on the plasmalemma of the white-lobe cells remains random at the side of impending fusion. Thin sections sometimes reveal, in this stage of exocytosis, a single unit membrane but never a pentalaminar structure. Both cellular components of the Harderian gland, when pretreated identically, show different membrane reactions with respect to IMP-clearing during exocytosis. This leads to the conclusion that IMP-clearing may not be responsible for membrane fusion but for other complex events in the regulation of membrane-to-membrane interactions (Winterhager and Kühnel 1983).

4 Histochemistry – Biochemistry

A more expedient approach towards an unknown organ should include a histochemical and a biochemical analysis, with particular reference to the biochemical components of the organ and the prevalence of certain metabolic pathways. Since very little biochemical and histochemical information about the Harderian gland of rabbits could be found, Jost et al. (1974), Kühnel (1966), and Kühnel and Wrobel (1966a,b, 1968) reported investigations on the latter aspect. From the analysis of chemical composition, enzyme patterns, and respiratory activity of isolated mitochondria, the first metabolic profile of the rabbit Harderian gland was presented by Jost et al. (1974). In addition, the possible function of the gland was elucidated by a detailed analysis of its lipid components (Jost 1972, 1974; Jost et al. 1974; Table 1).

From the macroscopic division of the gland into two parts, it could be inferred that the white lobe is particularly rich in fat, but this is not so. The percentage of lipids in the white lobe amounts to 16.7%. The pink lobe, in contrast, contains 41.5% fat per gram of fresh weight. This represents a very high percentage of fat, for even pure adipose tissue contains only 65 to 80% fat per gram of fresh weight.

The metabolism of the gland was described by enzyme patterns and respiratory activity of the mitochondria by Jost et a. (1974). In both lobes, activities of enzymes involved in the Embden-Meyerhof-pathway and the citric acid cycle as well as the transaminase reactions are low. However, activities of certain enzymes, such as citrate synthase, hydroxy-acyl-dehydrogenase and, in particular, glucose-6-phosphate dehydrogenase, are very high, indicating the importance of these pathways and a high lipid turnover in the gland.

Mitochondria isolated from both lobes show controlled respiration and a nearly normal P/O ratio when tested with usual substrates. Respiratory activity expressed in μ atoms oxygen (per gram mitochondrial protein per second) shows exceptionally high activity for α-glycerol-3-phosphate (Jost et al. 1974). Furthermore, Rock and Snyder (1975) and Rock et al. (1976) found that

Table 1. Lipid, protein, and glycogen content of the rabbit Harderian gland in mg/g wet wt. (Jost et al. 1974)

White lobe:	167 lipid, 33.6 soluble protein, 0.3 glycogen
Pink lobe:	415 lipid, 24.8 soluble protein, 0.3 glycogen

the three-position of the alkyldiacylglycerols from the pink lobe of the rabbit Harderian gland is occupied exclusively by isovaleric acid. They described a microsomal 1-alkyl-2-acyl-su-glycerol acyltransferase from the Harderian gland which specifically incorporates short-chain acyl-CoAs into the three-position of alkylacyl-glycerols. The enzyme was most active in the presence of CoA esters with chain lengths similar to isovaleric acid and was inactive in the presence of acetyl CoA and long-chain acyl-CoAs. No evidence was found for an enzyme that would transfer long-chain acyl-CoAs to the same substrate. The specificity of this acyltransferase can account for the exclusion of long-chain acyl moieties from the three-position of the alkyldiacylglycerols in the Harderian gland of rabbits.

Radominska-Pyrek et al. (1979) investigated the synthesis and content of ether-linked glycero-phospholipids in the Harderian gland of rabbits. Although Harderian glands are rich in neutral glycerolipids with ether bonds, less than 20% of the choline glycerophospholipids have ether bonds in the white and pink portions of the adult rabbit Harderian gland. Only 6% of these are plasmalogens, while 94% are alkylacyl glycerophosphocholins. The ethanolamine glycerophospholipids include 37% with ether bonds in both white and pink portions. In the white portion, 96% are plasmalogens but only 19% are plasmalogens in the pink lobe. The incorporation of radioactivity from CDP(^{14}C) ethanolamine and CDP(^{14}C) choline into ethanolamine and choline plasmalogens was increased several-fold by addition of alkylacylglycerols, but was not increased substantially by addition of diacylglycerols.

Enzyme histochemical studies were made by Kühnel (1966), Kühnel and Wrobel (1966a,b, 1968). They found high activities of NADH-T-Red, NADPH-T-Red, SDH, β-HBDH, G-6-PDH, ATPase (Fig. 8) and ALK-Pase (Fig. 7) in both the white and the pink lobe, whereas the activities of the enzymes Cyt-OX, NAD-MDH, LDH and GPDH were low or very low. Further comparative histochemical studies on the Harderian gland were made by Eichhorn (1988), Paule and Hayes (1958) and Schneir and Hayes (1951).

5 Conclusion

Reviewing the list of microscopical, histochemical, and biochemical contributions dealing with the morphology of the Harderian gland of the rabbit, one might argue that the physiological role of this gland is well known; but this is not true. The function of the Harderian gland is still unknown, but it is a good model for studying cytological details of exocytosis. In contrast to the Harderian glands of other animals, the Harderian gland of the rabbit shows no sexual dimorphism, either biochemical (porphyrin content, melatonin synthesis etc.) or morphological. Also the enzymatic patterns of the two portions of the gland are not really very different. In contrast, chemical analysis shows a difference in lipid content of the white and the pink lobes. The most important compound of the pink lobe is diacyl glycerol ether, and in view of the attention paid to this class of lipids, the Harderian gland may be particularly interesting. The extremely high content of this relatively rare lipid demands special attention since such a high concentration of diacyl glycerol ether has only been found in some fish livers. The work of Emmerie and Engel (1962), who isolated similar compounds from the codfish liver, is very interesting in relation to the significance of these lipid components for the function of the Harderian gland. They were able to prove that such compounds inhibit the growth of some bacteria, and glycerol ethers are also thought to be effective in wound healing. Considering that the Harderian gland secrets its lipids into the conjunctival sac and thereby moistens the cornea, all these properties could be of great importance for the protection of the eye. But in spite of these speculations, it will be necessary, for example, to analyze the correlations between the Harderian gland and the gonadal function as well as the connection between the pineal and this fascinating orbital gland.

References

Bargmann W (1977) Histologie und mikroskopische Anatomie des Menschen. Thieme, Stuttgart

Björkman N, Nicander L, Schantz B (1960) On the histology and ultrastructure of the Harderian gland in rabbits. Z Zellforsch Mikrosk Anat 52:93–104

Brownscheidle CM, Niewenhuis RJ (1978) Ultrastructure of the Harderian gland in male albino rats. Anat Rec 190:735–754

Buschke W (1933) Die Hautdrüsenorgane (Hardersche Drüsen, Inguinaldrüsen, Analdrüsen, Kieferdrüsen) der Laboratoriumsnagetiere und die Frage ihrer Abhängigkeit von den Geschlechtsdrüsen. Z Zellforsch Mikrosk Anat 18:217–243

Davis FA (1929) The anatomy and histology of the eye and orbit of the rabbit. Trans Am Ophthalmol Soc 27:401–441

Duboc T (1926) Recherches histochimiques sur la glande de Harder du lapin. Anat Ber 6:361–362

Eichhorn M (1988) The presence of carbonic anhydrase in orbital glands. An enzyme-histochemical study in cynomolgous monkeys and rabbits. Histochemistry 88:475–479

Emmerie A, Engel C (1962) Die α-Glycerinäther des Lebertrans. Fette, Seifen, Anstrichmittel. Die Ernährungsindustrie 64:813–816

Harder JJ (1694) Glandula nova lachrymalis una cum ductu excretorio in cervis et damis. Acta Erudit Publ Lipsiae, pp 49–53

Jost U (1972) Die Hardersche Drüse des Kaninchens. Eine erste biochemische Analyse, insbesondere die Charakterisierung der darin enthaltenen Lipide und ihre Strukturaufklärung von Glycerinäther/Ester. Inaugural-Dissertation (Med Fak) Heidelberg

Jost U (1974) 1-Alkyl-2,3-diacyl-su-glycerol, the major lipid in the Harderian gland of rabbits. Hoppe-Seyler's Z Physiol Chem 355:422–426

Jost U, Kühnel W, Schimassek H (1974) A morphological and biochemical analysis of the Harderian gland in the rabbit. Cytobiologie 8:440–456

Kahán IL, Juhász K, Mindszenti S (1967) Lipids of the Harderian gland of rabbits. Acta Biol Hung 18:295–301

Kennedy GY (1970) Harderoporphyrin: a new porphyrin from the Harderian gland of the rat. Comp Biochem Physiol 36:21–36

Kittel R (1962) Vergleichend-anatomische Untersuchungen über die Orbitaldrüse der Rodentia. Wiss Z Univ Halle, Math-Nat XI/4:401–428

Kühnel W (1966) Enzymhistochemische Untersuchungen an der Harderschen Drüse des Kaninchens. Histochemie 7:230–244

Kühnel W (1971a) Struktur and Cytochemie der Harderschen Drüse von Kaninchen. Z Zellforsch Mikrosk Anat 119:384–404

Kühnel W (1971b) Synapsen an exokrinen Drüsen. Naturwissenschaften 58:369

Kühnel W (1972a) Über die Stoffwechselmöglichkeiten der Harderschen Drüse des Kaninchens. Verh Anat Ges 66 (Anat Anz Suppl 130):79–82

Kühnel W (1972b) On the innervation of the salt gland. Z Zellforsch Mikrosk Anat 134:435–438

Kühnel W (1974) Zur funktionellen Morphologie der Harderschen Drüse des Schweines (*Sus scropha* L.) mit Bemerkungen zur Frage ihrer Innervation. Berl Münch Tierärztl Wochenschr 87:9–15

Kühnel W (1983) Über die Glandula infraorbitalis des Goldhamsters (Mesocricetus auratus Waterhouse). Acta Anat 116:245–256

Kühnel W, Beier HM (1973) Morphologie und Cytochemie der Harderschen Drüse von Anatiden. Z Zellforsch Mikrosk Anat 141:255–283

Kühnel W, Burger G (1969) Cytochemische Lokalisation einiger Hydrolasen in drei Orbitaldrüsen des Kaninchens. Histochemie 20:87–98

Kühnel W, Scheele G (1979) Zur Feinstruktur der Glandula lacrimalis des Schweines (*Sus scropha* L.). Anat Anz 145:87–106

Kühnel W, Wrobel KH (1966a) Die Histotopik von Aldolase und Alkohol-Dehydrogenase in der Harderschen Drüse des Kaninchens. Histochemie 7:245–250

Kühnel W, Wrobel KH (1966b) Über die histochemisch faßbare Aktivität der β-D-Glucuronidase und der β-D-Galactosidase in der Harderschen Drüse des Kaninchens. Albrecht von Graefes Arch Klin Exp Ophthalmol 171:173–183

Kühnel W, Wrobel KH (1968) Histochemische Studien an der Harderschen Drüse des Kaninchens. Gegenbaurs Morphoe Jahrb 111:493–500

Loewenthal N (1892) Beitrag zur Kenntnis der Harderschen Drüse bei den Säugetieren. Anat Anz 7:546–556

Lutz A (1899) Beiträge zur Kenntnis der Drüsen des dritten Augenlides. Gustav Fischer, Jena

Mendoza A, Kühnel W (1987) Morphological evidence for a direct innervation of the mouse vomeronasal glands. Cell Tissue Res 247:457–459

Mukai H (1926) Über die feinere Struktur der Harderschen Drüse beim Kaninchen. Albrecht von Graefes Arch Ophathalmol 117:243–272

Mycytowicz R (1966) Observations in odiferous and other glands in the Australian wild rabbit and harc. CSIRO Wildlife Res 11:49–53

Nikkinen A, Lehtosalo JI, Wusitalo H, Palkama A, Panula P (1984) The lacrimal glands of the rat and the guinea pig are innervated by nerve fibers containing immunoreactivities for substance P and vasoactive intestinal polypeptide. Histochemistry 81:23–27

Olcese J, Wesche A (1989) The Harderian gland (mini review). Comp Biochem Physiol 93A:655–665

Paule WJ, Hayes ER (1958) Comparative histochemical studies on the Harderian gland (Abst). Anat Rec 130:436

Peters A (1890) Beitrag zur Kenntis der Harderschen Drüse. Arch Mikrosk Anat 36:192–203

Radke R, Radke M, Radke C (1985) Licht- und elektronenmikroskopische Untersuchungen zur Innervation des exokrinen Pankreas. Z Mikrosk Anat Forsch (Leipz) 99:735–751

Radominska-Pyrek A, Dabrowiecki Z, Horrocks LA (1979) Synthesis and content of ether-linked glycerophospholipids in the Harderian gland of rabbits. Biochim Biophys Acta 574:248–257

Rock CO, Snyder F (1975) A short-chain acyl-CoA:1-alkyl-2-acyl-su-glycerol acyltransferase from a microsomal fraction of the rabbit Harderian gland. Biochim Biophys Acta 388:226–230

Rock CO, Fitzgerald V, Rainey WT, Snyder F (1976) Mass spectral identification of 2-(o-acyl) hydroxy fatty acid esters in the white portion of the rabbit Harderian gland. Chem Phys Lipids 17:207–212

Schneir ES, Hayes ER (1951) The histochemistry of the Harderian gland of the rabbit. J Nat Cancer Inst 12:257–258 (Abst)

Schramm U. Kühnel W (1979) Morphologie der glandula lacrimalis der Schleichkatze (*Herpestes edwardsi, Viverridae*). Anat Anz 145:120–130

Tsukahara S, Jacobowitz DM (1987) Peptidergic innervation of the rat Harderian gland. Histochemistry 87:233–236

Winterhager E, Kühnel W (1982) Zum Mechanismus der Fettextrusion der Harderschen Drüse von Kaninchen. Verh Anat Ges 76 (Anat Anz Suppl 152):47–49
Winterhager E, Kühnel W (1983) Membrane specializations of the cells of the Harderian gland of the rabbit with particular reference to the mechanism of exocytosis. Cell Tissue Res 231:623–636
Wooding FBP (1980) Lipid droplet secretion by the rabbit Harderian gland. J. Ultrastruct Res 71:68–78

The Function of the Harderian Gland in the Mongolian Gerbil, *Meriones unguiculatus*

DELBERT THIESSEN

Department of Psychology, University of Texas, Austin, Texas 78712, USA

1 Introduction

This chapter summarizes 15 years of research on the Harderian gland of the Mongolian gerbil, *Meriones unguiculatus*. Partial reviews and discussions of research methodology are found in Thiessen (1977, 1983a, b, 1988) and Thiessen and Harriman (1983). Three aspects of the gland are discussed: (1) release of Harderian gland secretions to the outside of the body, (2) chemosignaling qualities of Harderian gland exudates, and (3) thermoregulatory and osmoregulatory functions of the Harderian gland. The divisions are arbitrary, in that secretion, chemocommunication, and physiological regulation are intimately related. The chapter ends with a general discussion of the importance of the Harderian gland for the regulation of behaviour and physiology.

The taxonomic distribution of the Harderian gland is only partially known (see Table 1). It has been described in a wide range of vertebrate species, but there is little apparent association between the presence of the gland and the phylogenetic or ecological status of the organism. It is found in species that are diurnal and nocturnal, that are genetically and phylogenetically dissimilar, that occupy radically different ecologies, and that differ widely in sociability. Animals with heavy body hair and nictitating membranes generally have the gland, and aquatic vertebrates generally do not, although those relationships are weak. Primates, including *Homo sapiens*, do not possess Harderian glands.

The Harderian gland is bilobular and Z-shaped in the gerbil. It is surprisingly large: approximately 182 mg in the adult male and 169 mg in the adult female. When these weights are corrected for sex differences in body weight, the gland weights in the two sexes are little different.

Payne and his colleagues (Payne et al. 1982) have described the anatomical characteristics of the gland in *Meriones unguiculatus*, and Sakai and Yohro (1981) have described the histology of the gland in a closely related species, *Meriones meridanus*. These advances should eventually reveal a great deal about the relationship between the structure of the gland and its functional significance.

Webb et al. (Eds).
Harderian Glands

Table 1. Taxonomic representation of mammalian Harderian glands (After Sakai 1981)

Marsupials	Opossums *Didelphis*, *Caluromys*, and *Marmosa*
	Hedgehog *Erinaceus*
	Big-clawed shrew *Sorex*
	Short-tailed shrew *Blarina*
Edentata	Nine-banded armadillo *Dasypus*
Lagomorphs	Rabbit *Oryctolagus* and *Sylvilagus*
	Pika *Ochotona*
Rodents	Mouse *Mus*
	Rat *Rattus* and *Norvegicus*
	Hamster *Mesocricetus* and *Cricetalus*
	Gerbil *Meriones* (sps.)
	Guinea pig *Cavia*
	Squirrels *Tamiasciurus*, *Sciurus* (sps.), and *Citellus*
	Chipmunk *Tamias*
	Deer mouse *Peromyscus*
Cetaceans	Whale, Dolphin Mysticetes B. *alaenoptera* (sps.), *Megaptera*, *Balaena*, and Odontocetes
	Delphinnus, *Phocaena*, *Delphinapterus*, and *Hyperodon Platanista* and *Pontoporia*
Pinnipeds	Walrus *Odobenus*
	Seal *Phoca*
Sirenians	Dugong *Dugong*
Artiodactyls	Pig *Sus*
	Deer *Cervus* and *Dama*

2 The Release of Harderian Gland Secretions

The Mongolian gerbil releases Harderian gland material to the outside of the body by way of the lacrimal canals (Thiessen et al. 1976). The point of exit is the outer nares of the nose, although evidence of secretion can be found in the anterior corner of the eye and in the mouth. Release into the mouth is presumably through the nasal palatine duct. The position of the gland in the orbital cavity is seen in Fig. 1.

The glands are easily removed by snipping the nictitating membrane and extracting the glands with forceps. Minor bleeding occurs, but that can be controlled by pressing the eyelids together for a few seconds after the removal of the gland. Electrodesiccation of the anterior nictitating membrane is sufficient to disrupt the discharge of Harderian material, and can therefore be used to prevent the occurrence of exudates, while leaving the gland intact in the ocular cavity of the eye. This latter procedure induces glandular atrophy within 2 weeks of the operation. Neither gland removal nor distruption of the ducts seems to interfere with the visual capabilities of the animal.

The secretory material has not been identified in detail. Lipids and pigments predominate. The primary fatty acids are derivatives of 2, 3-alkanediol diacyl esters (Otsuru et al. 1983; Otsuka et al. 1986) and the primary pigment is

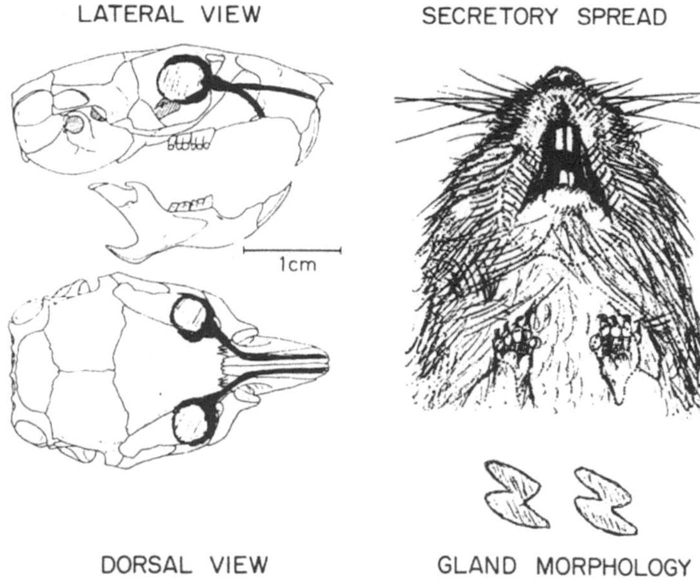

Fig. 1. Location of Harderian gland and ducts in the ocular cavity of the gerbil eye (*left*) and the typical areas of the body where Harderian material spread can be found immediately following an autogroom (*right*). The gross morphology of the gland is indicated

protoporphyrin (Kennedy 1970; Sakai 1981). Melatonin may also be produced, as it is in some rodents, suggesting a relationship between Harderian gland status, pineal activity and reproductive activities (Reiter et al. 1981). The major lipids in the gerbil are indicated in Table 2. The protoporphyrin exits from the gland along with the lipids and can be visualized under long-wave ultraviolet stimulation (UV 360 nm). Thus, protoporphyrin is a convenient index of the gland's activity.

The release of material to the outside of the nose is associated with increases in body temperature (Thiessen et al. 1977). It is released at the same time saliva is produced, and both materials are spread widely over the body during an autogroom (Grant and Thiessen 1989). Figure 2 shows the major components of the autogroom. The release can be indexed by noting the red fluorescence of the protoporphyrin immediately following an autogroom. The protoporphyrin, and possibly the biologically active lipids, have a half-life of approximately 2 min (Thiessen 1977) as do the behavioral effects. It is likely that components in the saliva degrade the protoporphyrin and lipid molecules when the two substances are mixed during a groom.

The pathways of release can be visualized by injecting small amounts of India ink into the conjunctival sac of the eye immediately prior to a groom. The ink traverses the lacrimal ducts and is spread during an autogroom. Upon examination, the ducts are visible.

Table 2. Major lipids (derived from 2,3-alkanediol diacyl esters) of the Harderian gland of *Meriones unguiculatus*. (Otsuru et al. 1983; Otsuka et al. 1986)

Capriç	3.0%
Lauriac	16.3%
Myristic	16.5%
Stearic	5.1%
Arachedic	16.4%
Behenic	9.8%

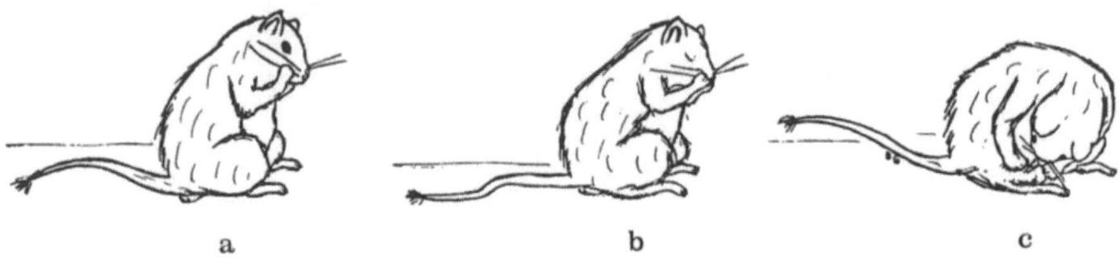

Fig. 2a–c. A typical sequence of movements for a gerbil autogroom. Note the contraction of the eye (**b**) which is the point in time that Harderian material is released to the outer nares of the nose

Because of the strong correlation between the release of the fluorescent pigment, protoporphyrin, and the lipids, the amount of exudates can be indexed by the intensity of UV reactivity. Ordinarily, the fluorescence is scored as absent, moderate, or intense. Recently, Dr. James Fox and I have been able to refine the measure by comparing the fluorescence under dark conditions with a fluorescence scale that varies from intense (0) to absent (15). The scale is pictured in Fig. 3. Thus, we are able to describe the wide variation in Harderian gland exudate and to associate it with variations in environmental and social conditions.

The intensity of Harderian gland discharge is associated with contraction of the eyelids during an autogroom. Usually three to five eyelid closures occur during a groom. Video analyses allow the investigator to measure eyelid changes and correlate these with secretion release at the nares of the nose. Frequency of eye blinks is correlated with the measured intensity of fluorescence ($r = 0.33$). A more robust measure of release is the average change in eye width during the autogroom. More specifically, the lower the ratio of the maximum eye width during grooming to the eyewidth immediately prior to grooming, the higher the quantity of secretion ($r = 0.52$). Thus, the overall muscle contraction of the eye is associated with release of the material. Our scale of measurement is seen in Fig. 4. This visual measure of Harderian gland secretion is especially useful where UV examination of secretion is not feasible during behavioral observations. The measurement of eyewidths can be taken directly from the video monitor during frozen frames.

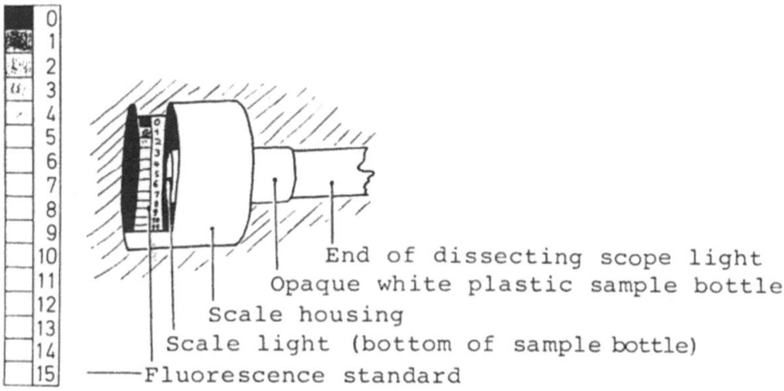

Fig. 3. The fluorescence scale is used to measure both the relative fluorescence of the nose area smeared with Harderian gland material (relative amount) and the rate of degradation (half-life) of the substance. The scale is a strip of 16 patches of red paint of diminishing shades matching the range of variation of the Harderian gland exudate seen under ultraviolet stimulation. Each patch is assigned an *index number*, such that a decrease in red intensity is halved every third patch down the scale. Thus, an initial secretion of index *1* is considered to be twice as much as an initial index of *4*, and a change from *1* to *4*, measured in seconds, represents the half-life of the secretion

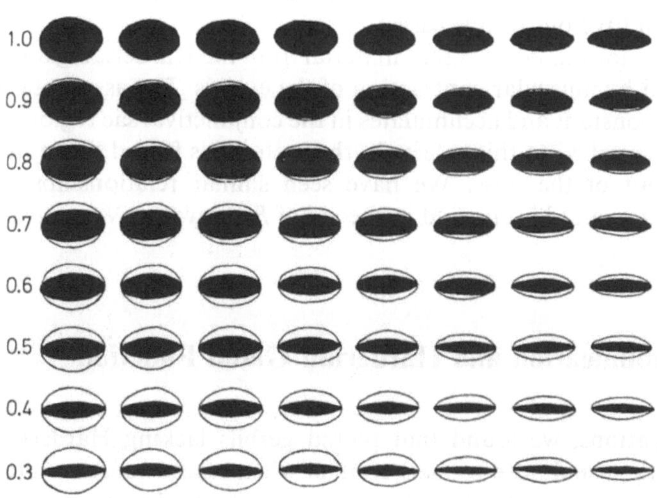

Fig. 4. The eyeblink intensity chart is designed to index the degree of muscle activity associated with an autogroom. The proportional change in eye width size from its initial value is assessed during a normal eyeblink. Each *vertical column* represents a series of proportional changes in eye width size beginning with its initial value pictured at the top of that column. Thus to derive a measure, the initial size is matched with a sample along the *top row*, and its proportional change is then determined from *changes depicted down that column*

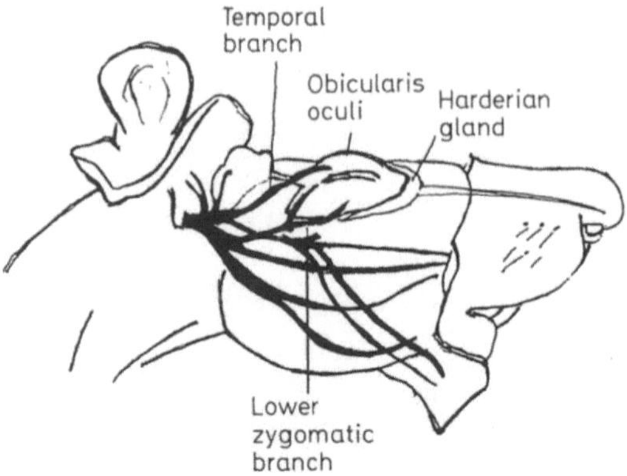

Fig. 5. Seventh cranial nerve controlling muscles of the face and jaw. Severing of the temporal branch of the nerve which controls the *obicularis oculi* muscles disrupts eyelid contraction and the release of Harderian gland material from the nose

The apparent relationship between muscle activity of the eye and Harderian gland releases was verified with partial sectioning of the 7th cranial nerve innervating the *obicularis oculi* muscles surrounding the eyeball and controlling the contraction of the eyelids (Fig. 5). Sectioning of the temporal branch of the nerve resulted in a reduced eyeblink intensity and diminished release (six out of six cases), whereas a control nerve section did not alter eyeblink intensity, or release of the secretion (five out of five cases).

What these observations indicate is that material from the Harderian glands is periodically released by muscular contraction of the eyelids. The assumption is that the secretion is constant and accumulates in the conjunctival sac region of the eye. When force is applied to this reservoir, the material is forced down the lacrimal canal and out of the nose. We have seen similar relationships in *Meriones tristrami, shawi,* and *libycus*; and in species of *Peromyscus, Norvegicus,* and *Mesocricetus.*

3 Chemical Communication and Harderian Gland Function

Early in our investigations, we found that paired gerbils lacking Harderian glands did not show the usual nose-to-nose contact when placed together after being separated for short periods. A lack of an attractant pheromone was suspected. This hypothesis was strengthened when we observed that individuals of small groups showed synchrony of Harderian gland spread (fluorescence of protoporphyrin). Either all members of a group showed similar intensities of release from the nose, or none showed the release. As we watched, we saw periodic bursts of autogrooming that would sweep through the population. It was immediately after the grooming that high levels of fluorescence were seen around the nose and the mouth areas. Apparently grooming was contagious and was related to the synchronous release of Harderian gland materials. These observations led to our descriptions of the relationships between autogrooming, Harderian gland release, and social attractiveness (Thiessen 1977).

Subsequent experiments strongly implicated olfactory and possibly gustatory components. In summary we found:

1. Animals learn to inhibit bar-pressing for food when Harderian gland extract in an airflow is followed by foot shock (Thiessen et al. 1976).
2. Animals learn to avoid food paired with Harderian gland material when made sick with lithium carbonate following the ingestion of food smeared with the gland extract. This effect was apparent even when the animals were made anosmic with zinc sulfate (Thiessen et al. 1976). Possibly the animals use gustatory cues, or vomeronasal stimulation.
3. Animals approach each other and engage in nose-to-nose sniffing more frequently following an autogroom. This occurs under red light conditions, as well as under white light conditions, suggesting that visual cues are not involved. Autogrooming in pairs often becomes synchronized. Mutual investigation declines significantly when the animals can smell each other through a mesh screen, but cannot make contact (Thiessen and Gerald, unpubl.). Therefore, it appears that tactile cues may be important.
4. Animals orient toward visually hidden Harderian gland extracts, and toward Harderianectomized animals whose faces are smeared with extracts (Thiessen 1977). Incidentally, the presence of the extract in an open field increases ventral scent marking.
5. Males direct their sandbathing to areas in which other males have sandbathed (Kittrell and Thiessen 1981). Harderianectomized males prefer sand marked by intact males, and the reverse, suggesting that Harderian gland material acquires meaning depending upon experiential factors.
6. Autogrooming does not generate low- or high-frequency auditory signals and so does not contribute to the mutual investigation that follows a groom. Ultrasounds can, however, evoke autogrooming, suggesting that various sensory systems are interrelated.
7. In a number of situations, we have found that Harderianectomy reduces male aggression, and that dominant intact males autogroom more frequently than subordinate intact males. These observations are similar to those found for the golden hamster (Payne 1979).
8. In unpublished studies, we have found that lactating females smear their pups with Harderian gland material (Pendergrass and Thiessen, unpubl.) and that offspring of Harderianectomized parents have difficulty in finding their way back to the nest (Clancy and Thiessen, unpublished).
9. Females apparently use olfactory cues from the males to guide their approach behaviors (proceptive behaviors). Females direct fewer proceptive behaviors toward males who are Harderianectomized (Harriman and

Thiessen 1983; Thiessen and Harriman 1986). In addition, estrus females are aggressively defensive toward males who do not have their Harderian glands, and prefer to spend more time in the vicinity of intact males.

We believe that the pheromonal consequences of Harderian gland release are more extensive than we have seen. It is a difficult system to study and there is virtually nothing in the literature to guide our work. The most evident features of chemocommunication are mutual attraction and synchronization of behaviors. The release of the material from the nose provokes investigation by conspecifics, and this material is apparently a cue used by pups in returning to the nest.

Harderian gland material may also synchronize social behaviors. Release of the material is associated with an autogroom, and that consequence, in turn, can provoke grooming in conspecifics. We have seen a relationship between social dominance and grooming, and we suspect that Harderian gland pheromones may serve in maintaining relationships within groups. Harderian gland signals by the male are also important for the female in her display of sexual interst. Many other social behaviors may be modified and synchronized by chemosignals from the Harderian glands. Certainly, there is more exciting research to do.

4 Thermoregulation and Osmoregulation

One of the most compelling relationships is between changes in body temperature and Harderian gland release. An increase in body temperature of about 0.5–1.0 °C evokes an autogroom and the release of material from the Harderian glands (Thiessen et al. 1977). Any environmental or physiological event that elevates body temperature will provoke grooming, including exposure to novelty, general activity, and ingestion of food. Also, increased metabolism that occurs under cold conditions will increase the frequency of grooming and the release of Harderian gland substances (Ie 1988). For example, gerbils groom on the average 7.9 times during 30 min at 7 °C, and an average of 6.4 times at 22 °C. Moreover, under low ambient temperatures, each groom averages about 167, but under high ambient temperatures the groom averages about 114 s. Under cold conditions, the gerbil spreads the Harderian gland material over more body parts, often grooming the body from head to tail. In other words, when cold, the gerbil releases more Harderian gland material and spreads the material more effectively over the pelage.

Contrary to this effect, saliva production is positively correlated with ambient temperature (Grant and Thiessen 1989). Under hot conditions, more saliva is produced, and it is spread more locally on hairless regions of the body (face and ears) where evaporative cooling is effective. The relationships between ambient temperature, Harderian gland secretion and saliva secretion are seen in Table 3.

Table 3. Effects of ambient temperature on saliva and Harderian gland secretion and latency to groom for *Meriones unguiculatus*. (Grant and Thiessen 1989)

Ambient temperature (°C)	Saliva secretion (mg)[a]	Harderian gland secretion (mg)[b]
7	52.4 ± 8.0	5.8 ± 0.47
22	49.2 ± 5.4	5.0 ± 0.66
34	100.2 ± 10.6	4.8 ± 0.51

[a] Saliva collected on cotton pledget.
[b] Harderian material blotted from nose.

These relationships make sense if it is assumed that Harderian material is used to insulate the hair under cold ambient conditions, and that saliva is produced in copious amounts when evaporative cooling is important under hot ambient conditions.

The importance of Harderian gland secretion for insulation has been demonstrated in a number of ways. Harderianectomy results in a reduced level pelage lipids within 8 to 16 days (Kittrell and Thiessen 1981). When Harderianectomized animals, or animals shampooed to remove hair oils, are exposed to an ice water dip for 1 s, their body temperature drops much faster than for control animals. Apparently the oils on the pelage help insulate the gerbil against cold and wetness.

The gerbil is very efficient at maintaining an optimal level of Harderian gland lipids on its pelage. Two mechanisms are employed, autogrooming and sandbathing. Autogrooming is used to apply lipids and pigments to the hair, and sandbathing to remove these materials. As expected, both mechanisms are temperature-dependent. Under cold ambient conditions, animals acquire hair lipids at a higher rate than under warm ambient conditions (Thiessen and Pendergrass 1985), and under warm ambient conditions, animals remove body lipids and pigments at an increased rate (Pendergrass and Thiessen 1983). Apparently the gerbil uses two antithetical behaviors to maintain uniform body temperature under changing ambient temperatures.

The lipids and pigments on the pelage also affect the degree of solar reflectance and body temperature. This is extremely important for *Meriones unguiculatus*, as this species is diurnal or nocturnal, depending upon seasonal and daily temperatures (Randall and Thiessen 1980). In general, the material from the Harderian glands spread over the pelage, darkens the hair, increases radiant absorption, and increases body temperature (Thiessen and Pendergrass 1985). Conversely, loss of the material by sandbathing, shampooing, or Harderianectomy lightens the hair color, increases radiant reflectance, and helps the animal maintain a lower body temperature under heat stress. These data are seen in Table 4.

Clearly, Harderian gland material spread over the pelage has a thermoregulatory function. The material acts as a lipid barrier to cold and wetness and also alters the reflectance quality of the pelage. Autogrooming is the primary mechanism for spreading the lipids and pigments, and by sandbathing the animal removes excess material. Both of these behaviors are temperature-sensitive and are closely related to each other.

The Harderian gland material on the pelage also has a profound effect on osmotic balance (Harriman and Thiessen 1983). Body weight loss during water deprivation is greater in intact animals than in Harderianectomized animals when sandbathing is not allowed (living on wire mesh), but is equivalent when sandbathing is allowed. The variation in relative humidity is associated with the magnitude of the difference in body weight loss during water deprivation between intact and Harderianectomized gerbils. This relationship is seen in Table 5.

Hence, when Harderian gland lipids are allowed to accumulate, body water is lost through insensible perspiration. The effect is exaggerated at lower relative humidities, and apparently can be attenuated by sandbathing. With the accumulation of lipids, the body hairs mat together, thus exposing more skin area to the

Table 4. Effects of variations in Harderian-pelage status on pelage reflectance and body temperature increases under radiant energy for *Meriones unguiculatus*. (Thiessen and Pendergrass 1985)

Condition	Solar reflectance (%)	Temperature increase (%)
Harderianectomized	22.12 ± 3.38	3.30 ± 0.40
Control	19.56 ± 3.10	5.84 ± 0.49
Shampooed	16.98 ± 2.95	1.34 ± 0.49
Control	14.63 ± 2.59	2.68 ± 0.31
Sand-exposed	24.74 ± 2.90	3.34 ± 0.56
Control	18.70 ± 2.12	4.84 ± 0.38

Table 5. Mean body weight losses (g) by 12 Harderianectomized (Hx) and 12 intact gerbils given 4-day water deprivation tests conducted under three conditions of relative humidity. (After Harriman and Thiessen 1983)

	Level of relative humidity					
	70%		50%		30%	
Operative state	M	SD	M	SD	M	SD
Hx	13.6	3.2	12.8	2.1	13.0	2.1
Intact	14.5	2.0	14.8	1.5	15.5	2.0
Mean difference (g)	0.9		2.0		2.5	

environment, and allowing more water evaporation. The animal normally avoids excessive water loss by removing excess lipids from the hair.

5 The Delicate Balance

Our most startling finding is that the Harderian complex is closely bound to behavior and other physiological processes. Behavior is the instrument by which Harderian gland material is used to control social interactions, body temperature, and osmotic balance. The secretion of the material to the outside of the body is dependent upon ambient conditions. If the gerbil is in a cold environment, more Harderian gland material than saliva is released and spread. Apparently the gerbil thus meets the need for the thermal insulation of the pelage. On the other hand, if the gerbil is in a hot environment where evaporative cooling is essential, greater amounts of saliva are spread, and less material from the Harderian gland.

The insulative capabilities of Harderian gland spread are enhanced by the radiant absorption characteristics of the lipids and pigments. Not only does the gerbil maximize its insulative capacity by spreading the material, but, because the lipids and pigments darken the pelage, an additional thermal advantage is gained through radiant absorption.

Whereas autogrooming controls the application of lipids and pigments, another behavior, sandbathing, controls the removal of excess Harderian material. Thus, under hot ambient conditions where hair insulation and solar radiant absorption would drive body temperature up, the gerbil removes the material by sandbathing. Lipid removal is important for another reason, namely to condition the pelage so that insensible evaporation is minimized. This removal becomes crucial at low relative humidities where evaporation is greatest.

Thus, one can see the crucial interrelationship of production, spread and removal of Harderian gland secretion in the regulation of body temperature, and osmotic balance. The delicate balance is maintained by autogrooming and sandbathing – two behaviors that seem to be critically sensitive to variations in body and ambient temperatures, and to relative humidity.

So far we have found no direct physiological consequences of Harderianectomy. The observed consequences seem indirect and mediated by the mechanical movement of lipids and pigments. There is no evidence from our work that gonadal hormones affect Harderian functioning, or the reverse, although further study might reveal such influences. We suspect, too, that circadian and circannual variations in light may affect the Harderian gland behavioral complex by way of the hypothalamus and pineal gland. We simply do not know.

Another facet of Harderian gland function, at least in the gerbil, is the use of the secretion as a chemosignal. Learned or not learned, the spread of Harderian gland material alters several behaviors by way of the olfactory system. It induces investigation by conspecifics – often social investigation – and it reduces social aggression. Females make sexual and social choices based upon the ability of males to release Harderian gland material. It appears that such signals coordinate social interactions and stabilize interactions among individual members. We have only begun to understand the chemosignaling properties of this system.

Our overall findings and assumptions are illustrated in Fig. 6. Obviously, the interactions are varied and complex. Within this framework we hope to be able to illustrate the behaviors, physiological and ecological significance of the Harderian gland, and to set the stage for comparative investigations.

Also, we should say something further about the release mechanisms of secretion. There is a feedback relationship between the animal's thermal and osmotic state and its effects on neural and sensory mechanisms that control Harderian gland secretion, autogrooming, and sandbathing. The timing among these events is superb and acts to stabilize the animal's physiology in the face of changing events. The animal detects its own homeostatic balance and provokes those behaviors which will keep thermal and osmotic levels within a restricted range.

In addition, depending upon the state of the organism and the ambient conditions, the animal is able to carefully adjust the hair levels of Harderian gland lipids and pigments. Our understanding of these sensory devices is

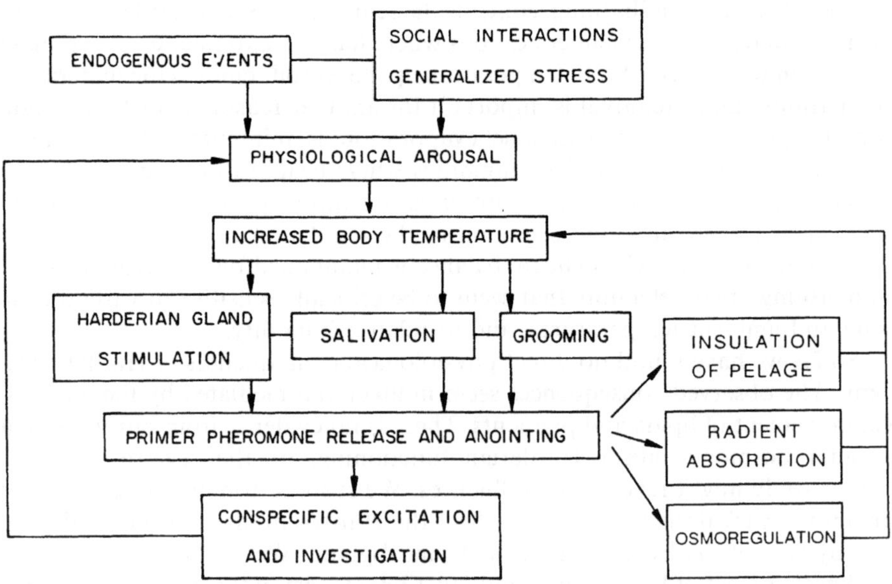

Fig. 6. A conceptual notion of the interaction between environmental events, physiological changes, and behavior

incomplete, to say the least. Viewing this complexity, one's appreciation of the animal's ability to sense its own physiological and environmental situation and make behavioral corrections is certainly enhanced.

Acknowledgements. I am indebted to my many colleagues and students who have participated in the work described in this chapter. Their contributions are illustrated in part by the publications listed in the References. In addition, they have contributed strongly to the overall orientation of the research and to the development of useful methodology. I am also grateful to the National Institute of Mental Health for nearly 22 years of continuous research support. The writing of this chapter was supported by NIMH grant MH 14076-22. Finally, I would like to thank the originators of this volume, and the participants who have made it possible. For those of us who have worked on this little-known and often overlooked organ of the body, the Harderian gland, this volume is a welcome acknowledgement of the significance of this unique and fascinating organ.

References

Grant M, Thiessen DD (1989) The possible interaction of Harderian material and saliva for thermoregulation in the Mongolian gerbil, *Meriones unguiculatus*. Percept Mot Skills 68:3–10

Harriman AE, Thiessen DD (1983) Removal of Harderian exudates by sandbathing contributes to osmotic balance in Mongolian gerbils. Physiol Behav 31:317–323

Ie Y (1988) The effects of ambient temperature on autogrooming in the Mongolian gerbil: *Meriones unguiculatus*. MA Thesis University of Texas, Austin, TX

Kennedy GY (1970) Harderoporphyrin: a new porphyrin from the Harderian glands of the rat. Comp Biochem Physiol 36:21–36

Kittrell EMW, Thiessen DD (1981) Does removal of the Harderian gland affect the physiology of the Mongolian gerbil, *Meriones unguiculatus?* Physiol Psychol 9:299–304

Otsuka H, Otsuru O, Kasama T, Kawaguchi A, Yamasaki K, Seyama Y (1986) Stereochemistry of 2,3-alkanediols obtained from the Harderian gland of Mongolian gerbil (*Meriones unguiculatus*). J Biochem 99:1339–1344

Otsuru O, Otsuka H, Kasama T, Seyama Y, Sakai T, Yohro T (1983) The characterization of 2,3-alkanediol diacyl esters obtained from the Harderian glands of Mongolian gerbil (*Meriones unguiculatus*). J Biochem 94:2049–2054

Payne AP (1979) The attractiveness of Harderian gland smears to sexually naive and experienced male golden hamsters. Anim Behav 27:897–904

Payne AP, Johnston HS, McGadey J (1982) The Harderian gland and its duct in the Mongolian gerbil (*Meriones unguiculatus*). J Anat 135:850–851

Pendergrass M, Thiessen D (1983) Sandbathing is thermoregulatory in the Mongolian gerbil, *Meriones unguiculatus*. Behav Neural Biol 37:125–133

Randall JA, Thiessen D (1980) Seasonal activity and thermoregulation in *Meriones unguiculatus*: A gerbil's choice. Behav Ecol Sociobiol 7:267–272

Reiter RJ, Richardson BA, Hurlbut EC (1981) Pineal, retinal and Harderian gland melatonin in a diurnal species, the Richardson's ground squirrel *Spermophilus richardsonii*. Neurosci Lett 22:285–288

Sakai T (1981) The mammalian Harderian gland: morphology, biochemistry, function and phylogeny. Arch Histol Jn 44:299–333

Sakai T, Yohro T (1981) A histological study of the Harderian gland of Mongolian gerbils, *Meriones meridianus*. Anat Rec 200:259–270

Thiessen DD (1977) Thermoenergetics and the evolution of pheromone communication. In: Sprague JM, Epstein AN (eds) Progress in psychobiology and physiological psychology, vol 7. Academic Press, New York, pp 91–191

Thiessen DD (1983a) Thermal constraints and influences on communication. In: Muller-Schwarze D, Silverstein RM (eds) Chemical signals in vertebrates 3. Plenum, New York, pp 147–189

Thiessen DD (1983b) The thermoenergetics of communication and social interactions among Mongolian gerbils. In: Rosemblum L, Moltz H (eds) Symbiosis in parent–young interactions. Plenum, New York, pp 113–144

Thiessen DD (1988) Body temperature and grooming in the Mongolian gerbil. Ann NY Acad Sci 525:27–39

Thiessen DD, Harriman AE (1983) Thermal and osmolarity properties of pheromonal communication. In: Muller-Schwarze D, Silverstein RM (eds) Chemical signals in vertebrates 3. Plenum, New York

Thiessen DD, Harriman AE (1986) Harderian gland exudates in the male *Meriones unguiculatus* regulate female proceptive behavior, aggression, and investigation. J Comp Psychol 100:85–87

Thiessen DD, Pendergrass M (1985) Changes of pelage lipids in the Mongolian gerbil, *Meriones unguiculatus*, as the result of autogrooming and sandliving. J Mammal 66:469–475

Thiessen DD, Clancy A, Goodwin M (1976) Harderian gland pheromone in the Mongolian gerbil. J Chem Ecol 2:231–238

Thiessen DD, Graham M, Perkins J, Marcks S (1977) Temperature regulation and social grooming in the Mongolian gerbil (*Meriones unguiculatus*). Behav Biol 19:279–288

Thiessen D, Pendergrass M, Harriman AE (1982) The thermoenergetics of coat colour maintenance by the Mongolian gerbil (*Meriones unguiculatus*) J Therm Biol 7:51–56

Gonadal Hormones, Structure, and Porphyrin Synthesis in the Rodent Harderian Gland

A.P. Payne[1], R.C. Spike[1], and M.R. Moore[2]

[1] Department of Anatomy, Glasgow University, Glasgow G12 8QQ, UK
[2] Department of Medicine and Therapeutics, The Gardiner Institute, Western Infirmary, Glasgow G11 6NT, UK

To establish whether there are links between gonadal hormones, haem biosynthetic capacity and structure within the rodent Harderian gland, it is necessary to pose the following simple questions:

1. Are there sex differences in porphyrin synthesis or gland structure?
2. Are there changes in synthesis or structure during natural periods of hormonal change?
3. Do experimental manipulations of hormone levels result in changes of either structure or synthesis?

If the answers to these are positive, supplementary concerns include whether the phenomena are specific to the Harderian gland or occur in other porphyrin-synthesising tissues; by what mechanisms do gonadal hormones act; are morphology and biosynthesis affected independently or do they co-vary in some meaningful fashion; what might the functional significance of this particular control mechanism be? Furthermore, these concerns spring merely from the question of whether hormones affect the Harderian gland. A wholly separate problem is whether the Harderian gland affects gonadal hormones – as might well be expected if it is indeed part of a retinal-pineal axis.

1 Sex Differences in Porphyrin Synthesis and Gland Structure

Porphyrins seem to occur only in the Harderian glands of rodents where they are both synthesised and stored. Although it was initially supposed that the gland was a passive store for porphyrins synthesised elsewhere (Derrien and Turchini 1924), it is now clear that the gland shows activity of all the major enzymes necessary for porphyrin synthesis (Thompson et al. 1984) and can synthesise porphyrins from appropriate precursors (Davidheiser and Figge 1955; Tomio and Grinstein 1968; Jones and Hoffman 1976). Furthermore, in some rodent species such as the golden hamster, enzymes such as 5-aminolaevulinate synthase (ALA-S), the rate-limiting enzyme for haem biosynthesis (Sinclair and Granick 1976; Moore and Disler 1985), show up to 100 times more

Webb et al. (Eds).
Harderian Glands

activity in the Harderian gland than in organs such as liver or kidney (Payne 1990; Payne and Moore 1987). Porphyrins are normally stored in the tubule lumina of the gland as solid accretions which are large enough to be seen with the naked eye. These accretions exhibit concentric layers, suggesting successive phases of deposition. Also the porphyrin may well be complexed with lipids (the major gland secretory product) or proteins. A minor proportion of gland porphyrin occurs in the interstitial tissue (Payne et al. 1982; Spike et al. 1986). Biochemical assays of porphyrin content cannot distinguish between newly formed intracellular porphyrin, stored intra-luminal porphyrin, and interstitial porphyrin; thus, they can give only limited information about current gland synthesis unless taken in conjunction with analyses of enzyme activity. Bearing in mind these limitations, porphyrin has been assayed in the Harderian glands of several rodent species (see Table 1). The data suggest that:

1. The glands of all rodents assayed to date contain porphyrin,
2. the amount varies greatly between different rodent species and between individuals, but
3. there is a general pattern for female glands to contain more than male glands. However, there are exceptions. In the rat, there appears to be no sex difference, while in the mouse there are sex differences in some strains but not in others (Strong 1942; Margolis 1971; Shirama et al. 1981).

Of the species shown in Table 1, enzyme activities are available only for the golden hamster. Sex differences in ALA-S activity in the hamster Harderian gland have been reported by Mindegaard (1976) and Lin and Nadakavukaren (1982). The most complete study is by Thompson et al. (1984) in which the activities of seven enzymes were measured. Here, marked sex differences in activity were found for five enzymes, including ALA-S, porphobilinogen deaminase, uroporphyrin decarboxylase, coproporphyrinogen oxidase and

Table 1. Levels of porphyrin in the Harderian glands of several species of rodent (nmol/g tissue; means ± SEM)

	Male	Female
Plains mouse (1) (*Pseudomys australis*)	237 ± 83	1932 ± 772
Mongolian gerbil (2) (*Meriones unguiculatus*)	89 ± 35	213 ± 69
Woodmouse (3) (*Apodemus sylvaticus*)	61 ± 22	125 ± 64
Golden hamster (4) (*Mesocricetus auratus*)	42 ± 9	3300 ± 302

[a] (1) Johnston et al. (1985); (2) Johnston et al. (1983); (3) Johnston et al. (1987); (4) Payne and Moore (1987).

protoporphyrinogen oxygenase. In addition, high-performance liquid chromatography suggests there are sex differences in the proportion of different porphyrins in the hamster Harderian gland (Spike et al. 1990).

Sex differences have been found in other aspects of hamster Harderian gland chemistry. The major Harderian gland product in mammals is lipid, and the female hamster gland has significantly higher proportions of C16 and C18 fatty acids than the male gland (Lin and Nadakavukaren 1981). There are reported sex differences in the proportions of different lipids in the mouse and rat Harderian glands, but these appear to be of a more minor nature (Bareggi et al. 1979). There are also sex differences in major proteins within the hamster Harderian gland, which were not found in the glands of rats or mice (Hoh et al. 1984). Female hamster glands have higher concentrations of sodium, manganese and calcium than those of males, and lower concentrations of iron and molybdenum (Hoffman and Jones 1981); they also have higher levels of immunoreactive melatonin and hydroxyindole-O-methyltransferase, lower levels of N-acetyltransferase and monoamine oxidase (Hoffman et al. 1985; Menendez-Pelaez et al. 1988, 1989), and higher levels of immunoreactive somatostatin (Puig-Domingo et al. 1988).

Morphological sex differences also occur in the Harderian gland of the golden hamster, the species where the sex difference in porphyrin content is highest. Several are well documented (see Bucana and Nadakavukaren 1972b, 1973; Payne et al. 1977):

1. *Cell Types.* Epithelial cell types have been distinguished on the basis of cytoplasmic lipid vacuole size. In the male hamster, the Harderian gland possesses two cell types; one has small vacuoles (Type I), the other has very large ones (Type II). The female gland has only one cell type resembling the male Type I.

2. *Polytubular Complexes.* These ultrastructural features occur only in male hamster Harderian glands and are found in both epithelial cell types; individual cells contain great numbers. They may be derived from vacuolated smooth endoplasmic reticulum. Conversely, female epithelial cells contain membranous structures arranged in concentric lamellae (Bucana and Nadakavukaren 1972b) which are not conspicuous in the male.

3. *Mast Cells.* These occur frequently in the interstitium of the Harderian gland and show a remarkable sex difference in number, being some 40 times more numerous in glands of females than in those of males (Payne et al. 1982). C3H/HE mice also show a sex difference in mast cell numbers, with females having some four times as many as males (Shirama et al. 1988).

It is likely (though this requires quantitative verification) that a fourth morphological sex difference may be the number of ciliated cells lining the gland duct, which appear to be greater in females. Similarly, a fifth difference may be the density of cholinergic nerve terminals which are said to be more numerous in glands of males (Vega-Alvarez et al. 1989; but see contrary report by Bucana and Nadakavukaren 1972a).

To date, no other species has been found with such obvious sex differences in Harderian gland morphology as the golden hamster.

2 Hormone Changes, Gland Structure and Porphyrin Synthesis

Since there are sex differences in porphyrin content in the Harderian glands of several rodent species, and since this coincides with sex differences in enzyme activity and structure in at least one species, the golden hamster, it is natural to consider whether this has a hormonal basis. Again, most information comes from studies using the hamster and has permitted investigators to pose simple questions of the form: "why does the male gland make so little porphyrin?" and "why does the female gland make so much?". We can consider the two sexes separately.

2.1 The Male Hamster Harderian Gland

The very restricted synthesis of porphyrins which occurs in the male hamster Harderian gland suggests that androgens may suppress porphyrin formation. This can be confirmed by castrating male hamsters, after which there are rises both in gland enzyme activity and porphyrin content to female levels over a few months (Table 2). These increases are accompanied by changes in gland structure. Thus, Type II cells and polytubular complexes (both characteristic of male glands) disappear, while mast cell numbers rise dramatically (Payne et al. 1977). The changes in ultrastructure have been examined chronologically by Lin and Nadakavukaren (1979), who found that polytubular complexes decreased 1 week (and disappeared 2 weeks) after castration. At the same time, dark pigment began to form some 12 days after castration and membranous

Table 2. Porphyrin concentration and synthesis in intact and castrated male golden hamsters. For additional data see Payne et al. (1977) and Spike et al. (1986)

	ALA-synthase activity (nmol ALA/h/g protein)	Porphyrin (nmol/g)
Control males	392 + 98	42 + 9
Castrates 2 months	1780 + 380	1253 + 131
2 months + T[a]	376 + 80	28 + 3

[a] T = testosterone (3 × 1.5 mg/week sc)

structures (typical of cells from females, Bucana and Nadakavukaren (1972b)) appeared in the cytoplasm.

Furthermore, the administration of androgens (including testosterone, 5α-dihydrotestosterone and androstenedione) will prevent both the biochemical and morphological changes or, if these have already begun, reverse them (Table 2; Payne et al. 1977; Spike et al. 1985). The assumption that these experiments mimic natural controls is strengthened by the finding that the Harderian glands of male and female hamsters possess testosterone receptors (Vilchis and Pérez-Palacios 1989), as do those of male rats (Gustafson and Pousette 1975). Testosterone also appears to suppress porphyrin synthesis in the glands of male C3H/He mice (Shirama et al. 1981).

The results from the castration and hormone replacement studies referred to above suggest that the male hamster gland has the capacity to synthesise large quantities of porphyrin but that this capacity is normally inhibited by androgens. To that extent the presence or absence of androgens may be considered a "coarse tuning" of this particular model. The data also emphasise (1) the interdependence of synthesis and structure and (2) the extreme lability of the gland.

Changes occur in the male hamster Harderian gland during development, but it is difficult to decide whether these have a hormonal basis. Bucana and Nadakavukaren (1972c) reported that glands of both sexes showed mitotic figures from birth until 3 weeks post partum, by which time sexual dimorphism of the gland was evident; the same authors (1973) found polytubular complexes by 14 days post partum, and suggested that these arose from juxtanuclear structures and that their appearance coincided with the disappearance of porphyrin from the male gland. Our own observations are similar. At birth, Harderian gland cells possess no lipid vacuoles or polytubular complexes and few microvilli. Lipid vacuoles are in evidence by 6 days after birth, while polytubular complexes appear between 6 and 9 days. Although the polytubular complexes are androgen-dependent in the adult, the time of their appearance is clearly long before puberty (Hafez 1970). It is not clear if there are changes with post-reproductive age in the male hamster Harderian gland, or if there are seasonal changes. However, males exposed to very short photoperiods (1L:23D) show a reduction in polytubular complexes and large lipid vacuoles, presumably as a result of reduced androgen production (Nadakavukaren and Lin 1983).

2.2 The Female Hamster Harderian Gland

Androgens not only suppress porphyrin synthesis in the normal male gland, but they have the same effect in the female. Testosterone (3×1.5 mg/week) given to ovariectomised females for 5 months substantially reduced ALA-S activity and porphyrin content (Table 3). Furthermore, treatment increased Type II cells (from occurring in 0.6% of tubule profiles in normal females to 38% profiles in androgen-treated ones) and led to the formation of polytubular complexes–

Table 3. Effects of androgen administration on the female hamster Harderian gland. For additional data, see Spike et al. (1985)

	ALA-synthase activity (nmol ALA/h/g protein)	Porphyrin (nmol/g)
Control females (n = 20)	5176 ± 1492	3300 ± 302
Ovariectomised females + T[a] (n = 11)	430 ± 112[b]	706 ± 82[b]

[a] T = testosterone (3 × 1.5 mg/week).
[b] Differs from control, $P < 0.01$.

both characteristic of the glands of males. Sun and Nadakavukaren (1980) examined changes in the female gland over 7 weeks of androgen administration, and reported that polytubular complexes appeared 4 days after the start and increased with time, while membranous structures decreased in frequency. Cells with large vacuoles (Type II cells) did not appear until 35 days of treatment. When androgen treatment ceased, the ultrastructure reverted to the feminine type between 3 and 7 weeks later. Androgen administration did not alter the porphyrin content of Harderian glands in ovariectomised female rats (Ulrich et al. 1974), although it did suppress the porphyrin content of the gland in female C3H/He mice (Shirama et al. 1981).

If androgens have a "coarse tuning" effect on porphyrin synthesis and gland structure in both sexes of the hamster, does the ovary play any role in promoting synthesis in the female gland or is it simply the absence of androgens which results in both a feminine gland morphology and high levels of porphyrin biosynthesis?

In the female hamster, ovariectomy leads to decreased ALA-S activity which is progressive with time when controlled for age (Spike et al. 1986). Porphyrin content, however, remains virtually static for long periods after ovariectomy (Hoffman 1971; Spike et al. 1985, 1986). This suggests that relatively little new porphyrin is being formed, and that porphyrin stores are not being depleted. In the Sprague Dawley rat, ovariectomy leads to a reduction in porphyrin content which is reversed by combined oestrogen and progesterone treatment (Ulrich et al. 1974).

Unlike androgen administration, ovariectomy alone does not masculinise gland structure in the female hamster, but accelerates a process of tubule degeneration which also occurs in normal females with increasing age. The main changes are depicted in Fig. 1, and it is likely (Payne et al. 1985; Spike et al. 1985, 1986) that they form a sequence. Briefly:

1. Tubule walls degenerate, becoming thin with the epithelial cells reduced greatly in height. Neutrophils invade the lumen and differ from their interstitial counterparts in having many filopodial extensions and large vacuoles, suggesting active ingestion.

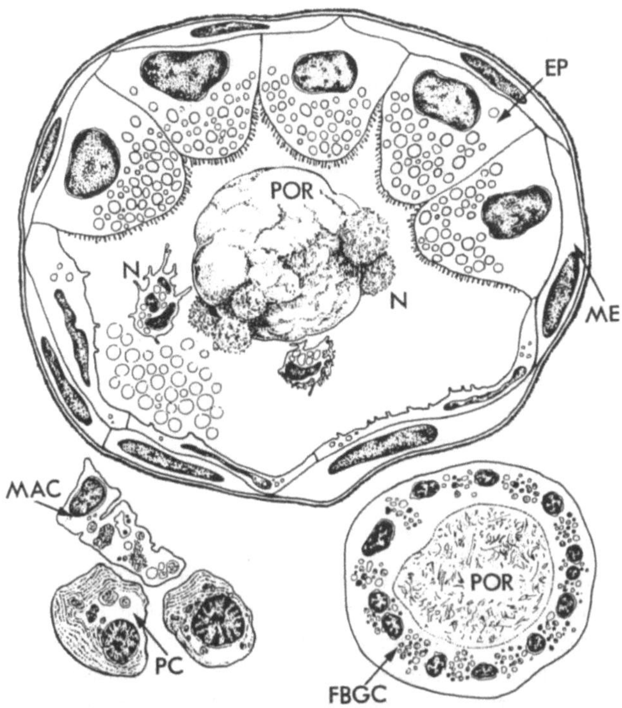

Fig. 1. This shows stages of degeneration of female hamster Harderian glands which accompany age or hormone insufficiency. At the *top*, a tubule profile consists of epithelial cells (*EP*) and myoepithelial cells (*ME*). The epithelial cells in the *lower half of the tubule* are degenerating and are greatly reduced in height. The lumen of the tubule contains a porphyrin accretion (*POR*) and lipid vacuole debris, as well as neutrophils (*N*) which become prevalent at this time. At *lower right*, a porphyrin accretion (*POR*) shown here in section to reveal its crystalline appearance, is surrounded by a foreign body giant cell (*FBGC*). At *lower left*, an individual macrophage (*MAC*) contains several crystalline deposits of porphyrin apparently in various stages of solubilisation. Such macrophages are usually surrounded by plasma cells (*PC*)

2. As tubule walls disappear, porphyrin accretions which were previously intraluminal come to lie in the interstitium of the gland. They are usually surrounded by foreign body giant cells, fused groups of macrophages which sequester solid objects.
3. Individual macrophages in the interstitium contain crystalline porphyrin in various stages of solubilisation. They are often surrounded by plasma cells; it may be that the presence of porphyrin in the interstitium is triggering an immune response. Interestingly, reduced ALA-S activity and porphyrin content within the gland following ovariectomy appear to correlate with increased blood porphyrin levels (Table 4). One possibility, although confirmatory evidence is lacking, is that porphyrin-laden cells such as macrophages are re-entering the circulation.

Table 4. Changes in Harderian gland porphyrin concentration and enzyme activity (together with blood porphyrin concentration) in intact control females and in females ovariectomised for 2 or 5 months. For additional data, see Spike et al. (1986)

	Intact females (n = 10)	Ovariectomised females 2 months (n = 10)	5 months (n = 10)
Gland ALA-synthase (nmol ALA/h/g protein)	1866 ± 231	1561 ± 219	1170 ± 109**
Gland porphyrin (nmol/g)	2634 ± 487	2341 ± 390	2244 ± 512
Blood porphyrin (nmol/l)	1585 ± 122	1670 ± 98	2293 + 219*

Differs from controls; * $P < 0.05$, ** $P < 0.01$.

While it is clear that the ovary exerts an influence on Harderian gland structure and activity in the female hamster, this is not so fundamental a control as the suppressive effect of androgens and may conveniently be thought of as a "fine tuning" of synthesis in this model. Again, the interlinking of synthesis and gland structure is noteworthy, as is the lability of the system. In the hamster, it is not clear which ovarian product is important. In female C3H/He mice, Shirama et al. (1981) found that ovariectomy produced little reduction in gland porphyrins while adrenalectomy did. However, the treatment which was most effective in raising porphyrin levels in ovariectomised, adrenalectomised females was progesterone. This may indicate something about the sources of gonadal steroids in this mouse strain.

The notion of the ovary controlling porphyrin synthesis is borne out by other findings. For example, degenerative changes in gland structure, together with reductions in ALA-S activity and porphyrin content, also occur in aged, post-reproductive female hamsters (Payne et al. 1985; Spike et al. 1988). Moreover, it is known that there are elevations in porphyrin synthesis in the gland during late pregnancy and early lactation which are particularly marked in a first (as opposed to a subsequent) pregnancy (Kelly et al. 1987). Although no changes were found during pregnancy in C3H/He mice (Shirama et al. 1981), in pregnant C57Bl mice, the Harderian gland has an enormously enhanced ability to accumulate porphyrin following injections of hematoporphyrin which was not found in the glands of non-pregnant mice (McCafferty and Mack 1964).

Changes in porphyrin content of the Harderian gland have also been reported over the oestrous cycle for the hamster (Payne et al. 1979) and mouse (Shirama et al. 1981). In both species, levels are lowest after oestrus and rise during di- and pro-oestrus.

2.3 Interrelations Between Structure and Porphyrin Synthesis

In only one species examined to date is there a robust relationship between Harderian gland structure and porphyrin biosynthetic capacity, namely in the golden hamster. Whenever experimental manipulations are performed which alter the activity of ALA-S or porphyrin content within the Harderian gland (see above), gland structure changes also. Unfortunately, it is still not clear whether the changes in synthesis and structure reveal an underlying functional relationship or if they are two quite separate consequences of hormone manipulation. If they are related, do hormones cause the change in structure which then permits altered porphyrin synthesis, or do hormones cause changes in synthesis which result in detectable changes in gland morphology? The morphological feature which has attracted most attention is the polytubular complexes which occur in all male cells and which disappear soon after castration when ALA-S activity is rising and visible porphyrin stores are begining to form. It has been suggested that these complexes either prevent the synthesis of porphyrins (Jones and Hoffman 1976) or interfere with their storage (Bucana and Nadakavukaren 1973) but confirmatory evidence is lacking.

2.4 Why Should Hormones Affect the Harderian Gland?

This is an important question to which there is no ready answer. However, a number of points may be considered.

2.4.1 Because it is a gland of ectodermal origin

A number of glands, including the Harderian gland, possess epithelia derived from embryonic ectoderm. In some cases, these epithelia show sexual dimorphisms, and in most cases they exhibit hormonal control of morphology or activity. They include the breast (Knight and Peaker 1982), sebaceous glands and modified sebaceous glands such as preputial glands, ventral glands, flank glands, sternobrachial glands, anal glands, etc. (Mykytowycz 1966; Stoddart 1972; Adachi 1974; Montagna and Parakkal 1974; Sansone-Bazzano and Reisner 1974), lacrimal glands (Paulini et al. 1972; Sullivan et al. 1984) and (probably) the salivary glands (Young and van Lennep 1978). It may be that the Harderian gland is reflecting its origin in its hormonal control. Against this, however, it must be remembered that Harderian glands occur in most terrestrial vertebrates and there are only one or two species for which hormonal control of structure or activity has been demonstrated.

2.4.2 Because it synthesises porphyrins

Since the only Harderian glands known to be controlled by hormones also produce porphyrins, this might provide a functional link. It is well established that the hepatic ("housekeeping") form of ALA-synthase (unlike erythroid ALA-synthase) is greatly influenced by steroid hormones. For example, the 5β series of the 17-oxosteroids (e.g. aetiocholanolone) are effective inducers of ALA-synthase in rat liver, while the 5α (e.g. androsterone) series are not (Moore et al. 1987). Moreover, there is good evidence to link alterations in steroid hormone synthesis with the onset of attacks of human acute porphyria (Paxton et al. 1974). Goldberg et al. (1969) noted that acute attacks were more common in women and often associated with pregnancy or particular stages of the menstrual cycle. These findings suggest either

a) that liver porphyrin synthesis is controlled by steroids,
b) that secondary control of haem synthesis is vested in the third enzyme of the pathway (porphobilinogen deaminase) which, like the controlling enzyme for steroid synthesis (adrenodoxin), is located on chromosome 11 and might thus be co-inducible, or
c) speculatively, that there is a balance between the mono-oxygenase requirements of steroid metabolism and the haem requirements for the synthesis of mono-oxygenases or cytochromes P450 such that each controls the other by feedback mechanisms based on haem availability.

2.4.3 Because it is a site of immune response

In the rat, the exorbital lacrimal gland is a rich source of secretory component (SC) present in tears. SC is believed to be an important regulatory factor in the immune system; it has a variety of functions, including control of the transfer of immunoglobulins to external secretions. Male rats have far more SC in tears than do females; moreover, levels of SC fall in males after castration and rise in both sexes after androgen administration (Sullivan et al. 1984). The Harderian gland is a site of immune response in birds and contains immunocompetent cells in rodents. While the Harderian gland appears to be a relatively poor source of SC in the rat (Sullivan et al. 1984), it is not clear if SC is hormonally controlled in this gland. Nor is it known if SC is produced by the Harderian gland of the hamster, a species where hormonal control of the gland is more marked.

2.4.4 Because it synthesises lipids

Many glands which synthesise lipids are of ectodermal origin and are usually controlled by hormones. These include many of the glands discussed above in (1), such as mammary glands, normal and modified sebaceous glands (Adachi 1974; Sansone-Bazzano and Reisner 1974). Androgens in particular are responsible for maintaining the size of sebaceous glands, the amount of sebum they produce and the activity of enzymes such as glucose-6-phosphate dehydrogenase. Since rodent Harderian glands produce lipids, hormonal control might be expected. However, it should be borne in mind (a) that there is little evidence for major changes in lipid synthesis in the Harderian gland following hormonal manipulation and (b) that most mammalian Harderian glands secrete lipids (Sakai 1981), and it is unclear at present whether hormones control the structure or activity of the gland in more than a few species of rodents.

2.4.5 Because it produces pheromones

Many mammalian glands which are known (or believed) to produce pheromones are ectodermal in origin, produce lipids and are controlled by hormones; some are also sexually dimorphic. For reviews, see Adams (1980), Albone (1984) and Drickamer (1989). These endocrinological links permit information exchange concerning sex and reproductive status and (in some cases) dominance status and age. Since the Harderian gland is considered a source of pheromones in hamsters and gerbils (Payne 1979; Thiessen 1983), it is appropriate that it should be hormonally controlled.

In considering these various factors, it is quite clear that most are inextricably interlinked. Furthermore, some relations to hormones must be secondarily acquired and are likely only to have evolved after a well-defined connection with hormones had already been established.

2.5 How Can We Use This Interrelationship?

Because we do not know the relevance of porphyrin synthesis and storage in the rodent Harderian gland, it is difficult to appreciate the functional significance of hormonal control. Nevertheless, the relationship may permit us to use the gland as a model in other areas of porphyrin synthesis which are hormonally controlled, such as human disorders of porphyrin metabolism, the porphyrias. For example, there is a relationship between gonadal hormones and acute intermittent porphyria in that the disease is rare before puberty, occurs more frequently in women than men, is precipitated by hormone changes such as the menstrual cycle, pregnancy and the contraceptive pill, and may be suppressed by LHRH antagonists such as buserelin (Brodie et al. 1977; McColl et al. 1982).

Investigators working with animals face a major problem in elevating porphyrin synthesis in a manner analogous to that seen in acute porphyria. This is usually attempted using drugs such as allyl isopropyl acetamide or succinylacetone, which interfere pharmacologically with haem synthesis (Sweeney 1985). Clearly this fails to provide a physiologically satisfactory model. It may be that the Harderian gland of species such as the golden hamster, with its high porphyrin content and enzyme activity, strong hormonal control and considerable experimental lability, will provide a useful animal model for these disorders.

References

Adachi K (1974) Receptor proteins for androgen in hamster sebaceous glands. J Invest Dermatol 62:217–233

Adams MG (1980) Odour-producing organs of mammals. Symp Zool Soc (Lond) 45:57–86

Albone ES (1984) Mammalian semiochemistry. The investigation of chemical signals between mammals. John Wiley, Chichester

Bareggi R, Crescenzi A, Narducci-Bareggi P (1979) Lipids in the Harder's gland of certain rodents. I. Neutral lipids. Basic Appl Histochem 23:155–160

Brodie MJ, Moore MR, Thompson GG, Goldberg A, Low RAL (1977) Pregnancy and the acute porphyrias. Br J Obstet Gynaecol 84:726–731

Bucana CD, Nadakavukaren MJ (1972a) Innervation of the hamster Harderian gland. Science 175:205–206

Bucana CD, Nadakavukaren MJ (1972b) Fine structure of the hamster Harderian gland. Z Zellforsch Mikrosk Anat 129:178–187

Bucana CD, Nadakavukaren MJ (1972c) Ultrastructural investigation of the postnatal development of the hamster Harderian gland. I. Cell division. Z Zellforsch u Mikrosk Anat 135:149–153.

Bucana CD, Nadakavukaren MJ (1973) Ultrastructural investigation of the postnatal development of the hamster Harderian gland. II. Male and female. Z Zellforsch Mikrosk Anat 142:1–12

Clabough JW, Norvell JE (1973) Effects of castration, blinding and the pineal gland on the Harderian glands of the male golden hamster. Neuroendocrinology 12:344–353

Davidheiser RH, Figge FHJ (1955) Enzymatic porphyrin synthesis in Harderian glands and other organs and tissues of mice. Proc Soc Exp Biol Med 90:461–462

Derrien E, Turchini J (1924) Sur l'accumulation d'une porphyrine dans la glande de Harder des Rongeurs du genre *Mus* et sur son mode d'excrétion. C R Soc Biol 91:637–639

Drickamer LC (1989) Pheromones: behavioural and biochemical aspects. In: Balthazart J (ed) Advances in comparative and environmental physiology 3. Springer, Berlin Heidelberg New York Tokyo, pp 269–348

Goldberg A, Moore MR, Beattie AD, Hall PE, McCallum J, Grant JK (1969) Excessive urinary excretion of certain porphyrinogenic steroids in human acute intermittent porphyria. Lancet 1:115–118

Gustafson J-Å, Pousette Å (1975) Demonstration and partial characterization of cytosol receptors for testosterone. Biochemistry 14:3094–3101

Hafez ESE (1970) Reproduction and breeding techniques for laboratory animals. Lea & Febiger, Philadelphia

Hoffman RA (1971) Influence of some endocrine glands, hormones and blinding on the histology and porphyrins of the Harderian glands of golden hamsters. Am J Anat 132:463–478

Hoffman RA, Jones JW (1981) Concentrations of metals in the Harderian glands of male and female hamsters. Comp Biochem Physiol 69A:153–156

Hoffman RA, Johnson LB, Reiter RJ (1985) Harderian glands of golden hamsters: Temporal and sexual differences in immunoreactive melatonin. J Pineal Res 2:161–168

Hoh JH, Lin W-L, Nadakavukaren MJ (1984) Sexual dimorphism in the Harderian gland proteins of the golden hamster. Comp Biochem Physiol 77B:729–731

Johnston HS, McGadey J, Thompson GG, Moore MR, Payne AP (1983) The Harderian gland, its secretory duct and porphyrin content in the Mongolian gerbil (*Meriones unguiculatus*). J Anat 137:615–630

Johnston HS, McGadey J, Thompson GG, Moore MR, Breed WG, Payne AP (1985) The Harderian gland, its secretory duct and porphyrin content in the plains mouse (*Pseudomys australis*). J Anat 140:337–350

Johnston HS, McGadey J, Payne AP, Thompson GG, Moore MR (1987) The Harderian gland, its secretory duct and porphyrin content in the woodmouse (*Apodemus sylvaticus*). J Anat 153:17–30

Jones CW, Hoffman RA (1976) Porphyrin concentration of the hamster (*Mesocricetus auratus*) Harderian gland: Effects of incubation with delta-aminolevulinic acid and various hormones. Int J Biochem 7:135–140

Kelly FM, Murray KJ, Maharaj JA, Stewart SL, Payne AP, Spike RC, Thompson GG, Moore MR (1987) Porphyrin synthesis during pregnancy and lactation in the golden hamster. Biochem Soc Trans 15:527–528

Knight CH, Peaker M (1982) Development of the mammary gland. J Reprod Fertil 65:521–536

Lin W-L, Nadakavukaren MJ (1979) The androgenic effect on the fine structure of the Harderian gland in the male hamster. Cell Tissue Res 198:119–127
Lin W-L, Nadakavukaren MJ (1981) Harderian gland lipids of male and female golden hamsters, (*Mesocricetus auratus*). Comp Biochem Physiol 70B:627–630
Lin W-L, Nadakavukaren MJ (1982) Delta-aminolaevulinate synthase activity in the Harderian glands of male and female golden hamsters. Comp Biochem Physiol 71B:539–540
Margolis FL (1971) Regulation of porphyrin biosynthesis in the Harderian gland of inbred mouse strains. Arch Biochem Biophys 145:373–381
McCafferty RE, Mack HP (1964) Tissue porphyrin in pregnant and non-pregnant mice injected with hematoporphyrin. Q J Exp Physiol 49:394–407
McColl KEL, Wallace AM, Moore MR, Thompson GG, Goldberg A (1982) Alterations in haem biosynthesis during the human menstrual cycle: studies in normal subjects and patients with latent and active acute intermittent porphyria. Clin Sci 62:183–191
Menendez-Pelaez A, Reiter RJ, Guerrero JM, Puig-Domingo M, Howes KA (1988) Sexual dimorphism in N-acetyltransferase activity, hydroxyindole-O-methyltransferase activity, and melatonin content in the Harderian gland of Syrian hamsters: Changes following gonadectomy. Proc Soc Exp Biol Med 187:287–291
Menendez-Pelaez A, Martinez-Telleria A, Vega JA, Coalla C, Reiter RJ (1989) Histochemical detection of monoamine oxidase and alcohol dehydrogenase activities in the Syrian hamster Harderian glands: Existence of a sexual dimorphism. Histochem J 21:125–130
Mindegaard J (1976) Studies concerning porphyrin synthesis and the origin and ultrastructure of biological porphyrin-containing pigment granules. Ph.D. Thesis, Afdelingen Biokemi Ernaering, Dept Biochem and Nutrition, Danmarks Tekniske Hojskole
Montagna W, Parakkal PF (1974) The structure and function of skin (3rd edn) Academic Press, New York, pp 281–331
Moore MR, Disler PB (1985) Chemistry and biochemistry of the porphyrins and porphyria. Clin Dermatol 3:7–23
Moore MR, McColl KEL, Rimington C, Goldberg Sir A (1987) Disorders of porphyrin metabolism. Plenum Medical Book, New York
Mykytowycz R (1966) Observations on odoriferous and other glands in the Australian wild rabbit, *Oryctolagus cuniculus* (L) and the hare, *Lepus europaeus*, P. I. The anal gland. CSIRO Wildl Res 11:11–29
Nadakavukaren MJ, Lin WL (1983) Influence of short photoperiods on the ultrastructure of hamster Harderian glands. Cell Tissue Res 233:209–213
Paulini K, Beneke G, Kulka R (1972) Age- and sex-dependent changes in glandular cells. I. Histologic and chemical investigations on the glandula lacrimalis, glandula intraorbitalis and glandula orbitalis externa of the rat. Gerontologia 18:131
Paxton JW, Moore MR, Beattie AD, Goldberg A (1974) 17-Oxosteroid conjugates in plasma and urine of patients with acute intermittent porphyria. Clin Sci Mol Med 46:207–222
Payne AP (1979) The attractiveness of Harderian gland smears to sexually naive and experienced male golden hamsters. Anim Behav 27:897–904
Payne AP (1990) The Harderian gland: its biology and role in porphyrin synthesis. Mol Aspects Med 11:145–152
Payne AP, Moore MR (1987) The hamster Harderian gland and porphyrin biosynthesis. Boll Ist Derm S Gall 13:21–28
Payne AP, McGadey J, Moore MR, Thompson GG (1977) Androgenic control of the Harderian gland in the male golden hamster. J Endocrinol 75:73–82
Payne AP, McGadey J, Moore MR, Thompson GG (1979) Changes in Harderian gland activity in the female golden hamster during the oestrus cycle, pregnancy and lactation. Biochem J 178:597–604
Payne AP, McGadey J, Johnston HS, Moore MR, Thompson GG (1982) Mast cells in the hamster Harderian gland: sex differences hormonal control and relationship to porphyrin. J Anat 135:451–461
Payne AP, McGadey J, Johnston HS (1985) Interstitial porphyrins and tubule degeneration in the hamster Harderian gland. J Anat 140:25–36
Puig-Domingo M, Guerrero JM, Reiter RJ, Peinado MA, Menendez-Pelaez A, Santana C, Webb SM (1988) Androgenic control of immunoreactive somatostatin in the Harderian gland of the Syrian hamster. J Reprod Fertil 82:753–759

Sakai T (1981) The mammalian Harderian gland: Morphology, biochemistry, function and phylogeny. Arch Histol Jpn 44:299–333
Sansone-Bazzano G, Reisner RM (1974) Steroid pathways in sebaceous glands. J Invest Dermatol 62:211–216
Shirama K, Furuya T, Takeo Y, Shimuzu K, Maekawa K (1981) Influences of some endocrine glands and of hormone replacement on the porphyrins of the Harderian glands of mice. J Endocrinol 91:305–311
Shirama K, Kohda M, Hokano M (1988) Effects of endocrine glands and hormone replacement on the mast cell count of the Harderian gland of mice. Acta Anat 131:327–331
Sinclair PR, Granick S (1976) The transport of hemin and protoporphyrin across the plasma membrane of chick embryo liver cells in culture. Ann Clin Res 8:25–258
Spike RC, Johnston HS, McGadey J, Moore MR, Thompson GG, Payne AP (1985) Quantitative studies on the effects of hormones on structure and porphyrin biosynthesis in the Harderian gland of the female golden hamster. I. The effects of ovariectomy and androgen administration. J Anat 142:59–72
Spike RC, Johnston HS, McGadey J, Moore MR, Thompson GG, Payne AP (1986) Quantitative studies on the effects of hormones on structure and porphyrin biosynthesis in the Harderian gland of the female golden hamster. II. The time course of changes after ovariectomy. J Anat 145:67–77
Spike RC, Payne AP, Moore MR (1988) The effects of age on the structure and porphyrin synthesis of the Harderian gland of the female golden hamster. J Anat 160:157–166
Spike RC, Payne AP, Thompson GG, Moore MR (1990) High-performance liquid chromatographic analyses of porphyrins in hamster Harderian glands. Biochim Biophys Acta 1034:1–3
Stoddart DM (1972) The lateral scent organs of *Arvicola terrestris* (Rodentia:Microtinae). J Zool (Lond) 166:49–54
Strong LC (1942) Sex differences in pigment content of Harderian glands of mice. Proc Soc Exp Biol Med 50:123–125
Sullivan DA, Bloch KJ, Allansmith MR (1984) Hormonal influence on the secretory immune system of the eye. Androgen control of secretory component production by the rat exorbital gland. Immunology 52:239–246
Sun CY, Nadakavukaren MJ (1980) Effect of testosterone on the female hamster Harderian gland pigmentation and ultrastructure. Cell Tissue Res 207:511–517
Sweeney GD (1985) Experimental porphyria. Clin Dermatol 3:125-143
Thiessen DD (1983) The thermoenergetics of communication and social interaction among Mongolian gerbils. In: Rosenblum L, Moltz H (eds) Symbiosis in parent–young interactions. Plenum, New York, pp 113–144
Thompson GG, Hordovatzi X, Moore MR, McGadey J, Payne AP (1984) Sex differences in haem biosynthesis and porphyrin content in the Harderian gland of the golden hamster. Int J Biochem 16:849–852
Tomio JM, Grinstein M (1968) Porphyrin biosynthesis 5. Biosynthesis of protoporphyrin IX in Harderian glands. Eur J Biochem 6:80–83
Ulrich R, Yuwiler A, Geller E, Wetterberg L (1974) Effects of sex hormones and environmental lighting on rat Harderian gland porphyrin. J Endocrinol 63:99–102
Vega-Alvarez JA, Menendez-Pelaez A, Coalla-Perez C, Martinez-Telleria A (1989) Sexual dimorphism in the cholinergic innervation of the hamster (*Mesocricetus auratus*) Harderian glands. Z Mikrosk Anat Forsch, (Leipz) 103:166–172
Vilchis F, Pērez-Palacios G (1989) Steroid hormone receptors and the sexual phenotype of the Harderian gland in hamsters. J Endocrinol 121:149–159
Young JA, van Lennep EW (1978) The morphology of salivary glands. Academic Press, London

The Harderian Gland in Birds: Histology and Immunology

R.B. Burns

ICN Biomedicals Ltd, PO Box 17, Second Avenue Industrial Estate, Irvine KA12 8NB, Scotland, UK

1 Introduction

Harderian glands occur in the orbita of all vertebrates except fishes, completely aquatic amphibians and primates. In birds, Harder's gland is the dominant orbital gland (Walls 1942) and besides a lubricating and cleaning function, it plays an important part in the local immunity of the eye and upper respiratory tract.

Since 1694, when Harder first described the gland in deer, it has been well studied in mammals, whereas it has been relatively neglected in birds. Nebel (1696) is cited by Brobby (1972) as being the first to work on the avian Harderian gland. In 1880, MacLeod described the histology and gross morphology of the gland in the duck. Ask (1913) reported on the morphology of the penguin Harderian gland and on its development in the embryo. Slonaker (1918) investigated the physiology and gross morphology of the gland in the English sparrow, *Passer domesticus*. A passing reference to the histology of Harder's gland in the western gull, *Larus argentatus*, has been made by McFarland and Sanui (1963). Some of MacLeod's (1880) work on the duck has been repeated and enlarged (Ballantyne and Fourman 1967; Fourman and Ballantyne 1967; Brobby 1972; Kühnel and Beier 1973). Harder's gland in the domestic fowl has been studied more extensively than any other avian Harderian gland. Bang and Bang's 1968 study of its histology was followed by further light microscope studies (Wight et al. 1971a; Bang et al. 1972; Burns 1974; Survashe 1976; Survashe and Aitken 1978), histochemical studies (Wight et al. 1971b; Burns and Mackenzie 1973), ultrastructural investigations (Rothwell et al. 1972) and immunological studies (Mueller et al. 1971; Bienenstock et al. 1973; Parry and Aitken 1973; Burns 1976b, 1977, 1978b; Survashe 1976).

2 The Shape of Avian Harderian Glands

Macroscopically, avian Harderian glands vary in shape although the most usual form is similar to that found in the domestic fowl (Burns 1974; Survashe 1976),

Webb et al. (Eds).
Harderian Glands

where it has been described as strap-like with hour-glass major borders (Wight et al. 1971a). In some birds, such as the duck and pelican, it is almost hemispherical with a shallow concave face ("face bulbaire" of MacLeod 1880) abutting onto the eyeball and a deeply convex surface ("face orbitaire" of MacLeod 1880) adjoining the orbit wall. It may also be roughly triangular as in the rockhopper penguin (Burns 1978a). In many of the glands, there is a groove in which an extra-ocular muscle, the pyramidalis oculi, is situated. It often divides the gland into two lobes which are not distinct in any way from each other. In contrast, bi-lobed Harderian glands are common in mammals (Sundwall 1907; Paule et al. 1955; Björkman et al. 1960; Kahán et al. 1967; Kühnel 1968) and they are often differentiated both microscopically and macroscopically into pink and white lobes.

The position of the Harderian gland in birds varies very little. Some glands might be more anterior than others, but all are ventero-medial to the eyeball. Slonaker (1918) described the gland in the sparrow and its position differs little from that of the fowl (Wight et al. 1971a) and that described in over 80 birds (Burns 1974; Survashe 1976).

The innervation and blood supply to the Harderian gland appears to be similar in all birds so far studied. The blood supply derives from the ophthalmotemporal branch of the external ophthalmic artery and innervation is by the inferior branch of the oculomotor nerve (Burns 1974).

3 Secretory Duct

There has been some dispute as to the number of secretory ducts from Harder's gland in birds. Brobby (1972) stated that the duck Harderian gland had two secretory ducts but Burns (1974), employing careful dissection and India ink injection, could find only one. Reid (1835) in his description of the embryology of the Patagonian penguin, *Aptenodytes patachonica*, said that two ducts were present, whereas Burns (1978a) could find only one in the rockhopper penguin, *Eudyptes crestatus*. Both Burns (1974) and Survashe (1976) are in agreement that the avian Harderian gland has but a single duct. It invariably leaves the anterior tip of the gland to open into the conjunctival sac at the base of the nictitating membrane.

4 Histology

Avian Harderian glands can be classified into three types based on the structure of the lobules and on the kinds of epithelial cells lining the acini and tubules (Burns 1974, 1975b; Survashe 1976; Aitken and Survashe 1977b). Type I Harderian glands are compound tubulo-acinar structures with a single lobule type composed of a single type of epithelial cell (Fig. 1). They have a large age-dependent population of plasma cells in the interstitium of the gland and are typified by the domestic fowl (cf. Bang and Bang 1968; Wight et al. 1971a; Burns 1974; Survashe 1976).

Type II Harderian glands (Fig. 2) are compound tubular structures having a single lobule type with two types of epithelial cells in the lining of the tubule. They have a much smaller population of plasma cells, which, at least in the duck, does not increase with age (Burns 1974). The duck exemplifies Type II glands and has been described by MacLeod (1880), Fourman and Ballantyne (1967), Ballantyne and Fourman (1967), Brobby (1972), Kühnel and Beier (1973), Burns (1974), and by Wight and MacKenzie (1974). The rockhopper penguin has an unusual modification to its Type II Harderian gland in the presence of PAS positive acini at the periphery of the lobules (Burns 1978a).

Type III avian Harderian glands (Fig. 3) may be regarded as "mixed" because they have two types of lobule, one of which is compound tubulo-acinar and the other is compound tubular. The rook, *Corvus frugilegus* L., typifies this class of Harderian gland (Burns 1974, 1975b). Plasma cells are found only in the compound tubulo-acinar lobules (Type I lobules). Type II lobules, which are

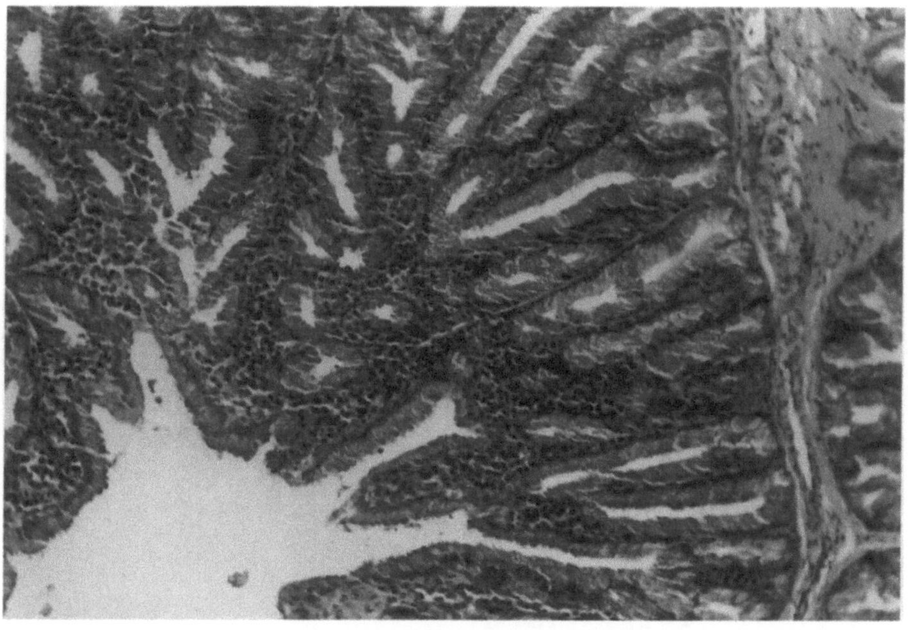

Fig. 1. Type I avian Harderian gland (Turkey, *Maleagris m. galopavo* L.). General view of a lobule. H & E × 192

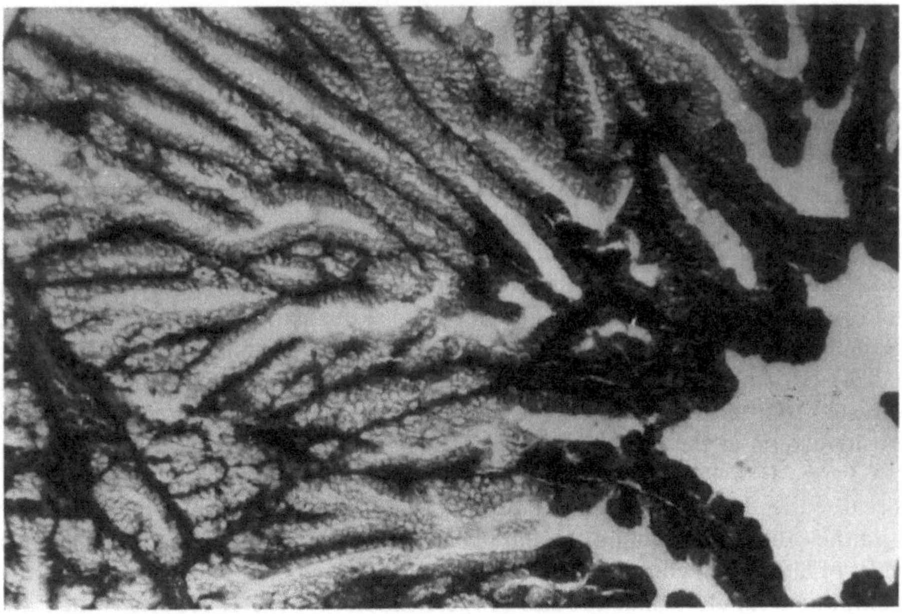

Fig. 2. Type II avian Harderian gland (Pekin duck, *Anas platyrhynchos* L.). Note two cell types in a lobule. H & E × 154

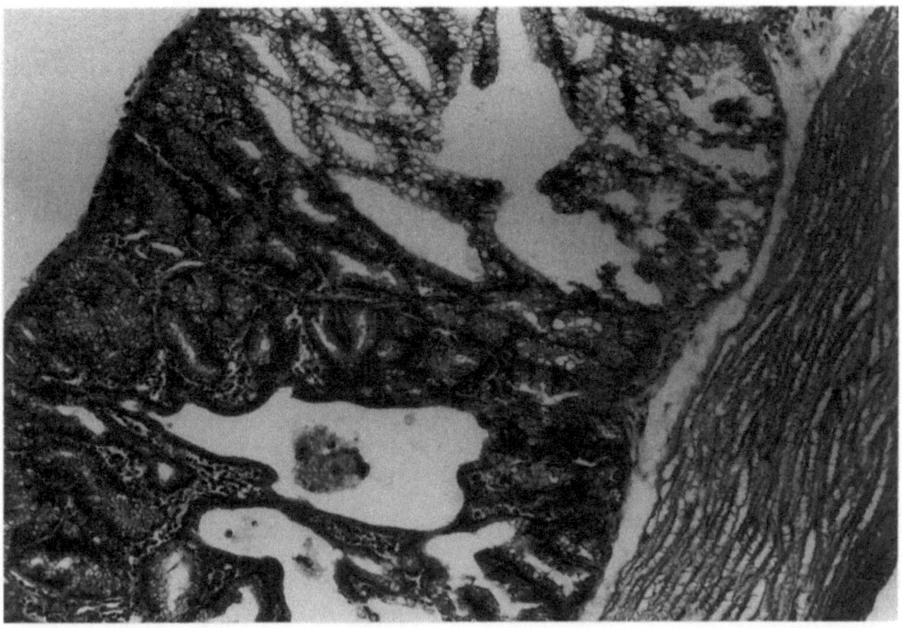

Fig. 3. Type III avian Harderian gland (blue tit, *Parus caerulus* L.). Type I lobule with plasma cells below and type II lobule above. H & E × 40

compound tubular structures, do not have plasma cells. As no age studies have been conducted on Type III avian Harderian glands, it remains to be determined whether plasma cell numbers increase with age.

It has been shown (Burns 1974; Survashe 1976; Aitken and Survashe 1977b) that the occurrence of types of avian Harderian glands falls into the following pattern:

Type I	*Type II*	*Type III*
Falconiformes	Sphenisciformes	Gruiformes
Galliformes	Pelicaniformes	Charadriiformes
Columbiformes	Ciconiformes	Piciformes
Psittaciformes	Anseriformes	Passeriformes

Burns (1974, 1975b) has claimed that the type of Harderian gland that a bird possesses does not seem to be related to its mode of life but rather to its evolutionary status. However, Survashe (1976) and Aitken and Survashe (1977b) are of the opinion that a segregation of gland type is evident when birds are assorted according to habitat. Certainly, no terrestrial species has a Type II gland whereas aquatic birds can have either a Type II or a Type III gland, suggesting that Type II lobules might be related to an aquatic environment.

5 Secretory Mechanism

The principal method of secretion found in avian Harderian glands is merocrine (Burns 1974). In none of the glands studied by either Burns (1974) or by Survashe (1976) was there any evidence of mitotic activity in the secretory cells so holocrine secretion seems unlikely. Wight et al. (1971a) ascribed merocrine secretion to the fowl Harderian gland. Ballantyne and Fourman (1967), Brobby (1972) and Kühnel and Beier (1973) classified secretion in the anatid Harderian gland as merocrine. Type III glands also appear to have a merocrine mode of secretion (Burns 1974, 1975b).

6 Secretory Products

Brobby (1972) described secretory granules within the epithelial cells of the duck Harderian gland and showed that the gland secreted a mixture of acid and neutral mucosubstances. He found no glycogen in the gland. Burns (1974, 1978b) found a similar sort of secretory product in other Type II glands. Ballantyne and Fourman (1967) concluded that the principal secretion from the duck Harderian gland was mucoid with a high mucopolysaccharide and lipid content.

Type I glands, as typified by the domestic fowl, have been shown to secrete weakly acidic sulphated and neutral mucosubstances together with a small proportion of sialic acid and hyaluronic acid (Wight et al. 1971b). Neutral lipids have been described in the fowl Harderian gland and in other Type I glands (Burns 1974).

It can be concluded that avian Harderian glands secrete mucus and fatty substances to lubricate the movement of the nictitating membrane and to nourish the avascular cornea.

7 Plasma Cells and Lymphocytes

One of the more striking features of the Harderian gland in the domestic fowl is the constant presence of a large population of plasma cells in the interstitium of the gland (Bang and Bang 1968; Wight et al. 1971a; Bang et al. 1972). Wight et al. (1971a) have shown that the numbers of plasma cells increase with age. In its normal state, no other avian organ appears to have such an extensive population of plasma cells. Although present in Harder's glands of all species of birds examined (Burns 1974, 1975b; Survashe 1976), plasma cells were least numerous in the duck and this probably accounted for their being overlooked by MacLeod (1880), Brobby (1972) and by Kühnel and Beier (1973). There are few reports of plasma cells in mammalian Harderian glands, though Burns (1975b) claimed to have found them in small numbers in the secretion from rabbit, guinea pig and rat.

8 Immune Status of the Fowl Harderian Gland

The presence of a large, age-dependent population of plasma cells probably means that the Harderian gland in the domestic fowl plays a part in local immunity of the eye and upper respiratory tract. This has been emphasised by the work of Albini and Wick (1973) who showed that surface immunoglobulin-bearing cells are present in the gland. Using a plaque-forming technique and assay, Mueller et al. (1971) found antibody-producing cells in specimens of Harderian gland from birds stimulated by topical application of sheep red blood cells. Systemic antigenic stimulus did not affect the gland.

Evidence has been presented that the lymphoid cells in the fowl Harderian gland are mainly bursa-dependent (Sundick et al. 1973), suggesting that the gland is active in humoral immune responses.

9 Immunoglobulin Production

The course of immunoglobulin production in Harderian glands of non-infected fowls has been studied by Albini et al. (1974). They found that up to the 4th week after hatching, most of the plasma cells produced IgM, from the 4th–9th week, both IgG and IgA were present, and after this, IgA was the main immunoglobulin encountered. These results differ from those of Bienenstock et al. (1973), who found mainly IgG and IgM with a trace amount of IgA in the glands of 5-month-old birds. Rothwell et al. (1973) and Burns (1974) reported a predominance of IgG in the glands of 9-week-old chickens. Koch and Jongenelen (1988) have shown that 60% of the lymphoid cells in the fowl Harderian gland are immunoglobulin-secreting. They claim a predominance of IgM production at 8 and 16 weeks of age.

Parry and Aitken (1973, 1977) have shown that the virus-neutralising activity of respiratory tract secretions is principally associated with the lymphoid and plasma cells of the upper respiratory tract and of the paraocular and paranasal tissues of the fowl. They have shown, using class-specific antisera, that the Harderian gland plasma cells synthesise mainly IgA during the early post-infection stages of Newcastle disease. Davelaar and Kouwenhoven (1976) have shown that infectious bronchitis virus has a marked effect on the plasma cells of the gland in the fowl. They found an increase in plasma cell numbers after infection and also the formation of germinal centres in the interstitium of the gland.

10 Specific Antibody Production

Exogenous antigens that enter the eye of the domestic fowl can reach the Harderian gland via the secretory duct (Burns 1977, 1978b) and stimulate a local and specific antibody response. Mueller et al. (1971) found antibody-producing cells in the glands of birds immunised by topical applications of sheep red blood cells but not in those immunised by systemic application. Burns (1975a, 1976b) immunised domestic fowls with a soluble antigen, bovine serum albumin, and induced an immune response in the Harderian gland. He (Burns 1976b, 1978b) also induced anti-BSA antibodies in the Harderian glands of topically stimulated turkeys and duck.

References

Aitken ID, Survashe BD (1977a) Plasma cells in vertebrate paraocular glands. Int Arch Allergy Appl Immunol 53:62–67

Aitken ID, Survashe BD (1977b) Lymphoid cells in avian paraocular glands and paranasal tissues. Comp Biochem Physiol 58:235–244

Albini B, Wick G (1973) Immunoglobulin surface determinants on chicken lymphoid cells. Int Arch Allergy Appl Immunol 44:804–822

Albini B, Wick G, Rose E, Orlans E (1974) Immunoglobulin production in chicken Harderian glands. Int Arch Allergy Appl Immunol 47:23–34

Ask F (1913) Über die Entwicklung der orbitalen Drüsen bei *Pygoscellis papua*. Lunds Univers Arsskt N. F. 9:1–12

Ballantyne B, Fourman J (1967) The histology and histochemistry of the Harderian gland of the domestic duck. J Anat 101:194–199

Bang BG, Bang FB (1968) Localized lymphoid tissues and plasma cells in paraocular and paranasal organ systems in chickens. Am J Pathol 53:735–751

Bang BG, Bang FB, Foard MA (1972) Lymphocyte depression induced in chickens on diets deficient in vitamin A and other components. Am J Pathol 68:147–162

Bienenstock J, Gauldie J, Perey DYE (1973) Synthesis of IgG, IgA, IgM by chicken tissues: immunofluorescent and ^{14}C amino acid incorporation studies. J Immunol 111:1112–1118

Björkman N, Nicander L, Schantz B (1960) On the histology and ultrastructure of the Harderian gland in rabbits. Z Zellforsch Mikrosk Anat 52:93–104

Brobby GW (1972) On the Harderian gland of the domestic duck (*Anas. platyrhinchos*). Z Zellforsch Mikrosk Anat 133:223–230

Burns RB (1974) The Harderian gland in birds. M Phil Thesis, University of London

Burns RB (1975a) Specific antibody production to BSA in the domestic fowl Harderian gland. Folia Biol Praha 21:360

Burns RB (1975b) Plasma cells in avian Harderian glands and the morphology of the gland in the rock. Can J Zool 53:1258–1269

Burns RB (1976a) The structure of the lachrymal glands of the domestic fowl and of the duck. Res Vet Sci 21:292–299

Burns RB (1976b) Specific antibody production against a soluble antigen in the Harderian gland of the domestic fowl. Clin Exp Immunol 26:371–374

Burns RB (1977) Possible route of antigen uptake by the Harderian gland of the domestic fowl. Br Poult Sci 18:407–409

Burns RB (1978a) Histology of Harder's gland of the rockhopper penguin, *Eudyptes crestatus*. Zool J Linn Soc 64:129–133

Burns RB (1978b) Immunological studies on the avian Harderian gland with particular reference to the domestic fowl. PhD Thesis, University of London

Burns RB, MacKenzie GM (1973) Cholinesterase activity in the Harderian gland of the domestic fowl (*Gallus domesticus*). Res Vet Sci 14:261–263

Davelaar FG, Kouwenhoven B (1976) Changes in the Harderian gland of the chicken following conjunctival and intranasal infection with infectious bronchitis virus in 1- and 20-day-old chickens. Avian Pathol 5:39–50

Fourman J, Ballantyne B (1967) Cholinesterase activity in the Harderian gland of *Anas domesticus*. Anat Rec 159:17–27

Kahán IL, Jugasz K, Mindszenti S (1967) Lipids of the Harderian gland of rabbits. Acta Biol Hung 18:295–301

Koch G, Jongenelen IMCA (1988) Quantification and class distribution of immunoglobulin-secreting cells in mucosal tissues of the chicken. In: Fossum S, Relstad B (eds) Histophysiology of the immune system. Plenum, New York pp 633–639

Kühnel W (1968) Comparative histological, histochemical and electron microscopical studies on the lacrimal glands. I. Rabbit and cat. Z Zellforsch Mikrosk Anat 119:384–404

Kühnel W, Beier HM (1973) Morphologie und Cytochemie der Harderschen Drüse von Anatiden. Z Zellforsch Mikrosk Anat 141:255–283

MacLeod J (1880) Sur la structure de la glande de Harder du canard domestique. Arch Biol Paris 1:45–57

McFarland LZ, Sanui H (1963) Sodium and potassium binding by microsomes from the nasal salt gland, Harder's gland, kidney and liver of the sea gull. Biol Med 113:105–107

Mueller AP, Sato K, Glick B (1971) The chicken lacrimal gland, gland of Harder, caecal tonsil and accessory spleen as sources of antibody-producing cells. Cell Immunol 2:140–152

Nebel (1696) De glandula lachrymali Harderiana non tantum in arvis sed etiam in aliis diversi generis animalibus reperta in miscellanus curiosa decem. III Lipsiae. Cited by Brobby (1972)

Parry SH, Aitken ID (1973) Immunoglobin A in the respiratory tract of the chicken following exposure to Newcastle disease virus. Vet Rec 93:258–260
Parry SH, Aitken ID (1977) Local immunity in the respiratory tract of the chicken. II The secretory immune response to Newcastle disease virus and the role of IgA. Vet Microbiol 2:143–165
Paule WJ, Hayes ER, Marks BH (1955) The Harderian gland of the Syrian hamster. Anat Rec 121:349–350
Reid J (1835) Anatomical description of the patagonian penguin (*Aptenodytes patachonica*). Proc Zool Soc Lond 3:132–148
Rothwell B, Wight PAL, Burns RB, MacKenzie GM (1972) The Harderian glands of the domestic fowl. III Ultrastructure. J Anat 112:233–250
Rothwell B, Wight PAL, Burns RB (1973) The Harderian gland of domestic avian species. Proc III Eur Anat Congr, Manchester
Slonaker R (1918) A physiological study of the anatomy of the eye and its accessory parts of the English sparrow (*Passer domesticus*). J Morphol 31:351–456
Sundick RS, Albini B, Wick G (1973) Chicken Harder's gland: evidence for a relatively pure bursa-dependent lymphoid cell population. Cell Immunol 7:332–335
Sundwall J (1907) The structure of the Harderian gland of the ox. Anat Rec 1:72–73
Survashe BD (1976) Immunological aspects of avian paraocular glands with particular reference to the Harderian gland of the domestic fowl. PhD Thesis, University of Liverpool
Survashe BD, Aitken ID (1978) Immunocompetent cells in avian paraocular glands. Res Vet Sci 24:182–190
Walls GL (1942) The Vertebrate Eye and its Adaptive Radiation. Cranbrook Inst Sci (Bull 11)
Wight PAL, MacKenzie GM (1974) Mucosubstances in the Harderian gland of the domestic duck. Res Vet Sci 17:114–121
Wight PAL, Burns RB, Rothwell B, MacKenzie GM (1971a) The Harderian gland of the domestic fowl. I Histology with reference to the genesis of plasma cells and Russell bodies. J Anat 110:307–315
Wight PAL, MacKenzie GM, Rothwell B, Burns RB (1971b) The Harderian glands of the domestic fowl. II Histochemistry. J Anat 110:323–333

Porphyrins and Their Possible Significance in Harderian Glands

ROSEMARY C. SPIKE[1], ANTHONY P. PAYNE[1] and MICHAEL R. MOORE[2]

[1]Department of Anatomy, University of Glasgow, Glasgow G12 8QQ, UK
[2]Department of Medicine and Therapeutics, Western Infirmary, Glasgow G11 6NT, Scotland UK

1 Introduction

Porphyrins are highly coloured compounds which fluoresce red in light of wavelength around 400 nm. They are widely distributed throughout the animal and plant kingdom. Protoporphyrin IX occurs in the pigmented eggshells of many species of birds while uroporphyrin is found in the shells of molluscs. However, their most important role is when complexed to metals to form metalloporphyrins, which are essential for all biological oxidation reactions. Haem is the iron complex of protoporphyrin. The rodent Harderian gland is remarkable in its ability to synthesise porphyrins, which in some species exceeds that of the liver.

The arrangement of the four pyrrole nitrogen atoms within the ring structure of the porphyrin molecule allows the combination with metal atoms such as iron, copper, zinc and magnesium. Protohaem, the iron complex of protoporphyrin IX, is the most important biologically (Fig. 1). It is the prosthetic group of both haemoglobin and myoglobin and combines with various proteins to form the oxidative enzymes catalase, peroxidase, the cytochromes and tryptophan pyrrolase. Protohaem is therefore central to biological oxygen use. The chlorophylls, magnesium porphyrins, are essential for the biological utilisation of solar energy and carbohydrate synthesis in most higher plants, algae and photosynthetic bacteria.

2 Haem Biosynthesis

Although haem biosynthesis can occur in all body tissues, it takes place predominantly in immature red blood cells, for the synthesis of haemoglobin, and in the liver, to produce haem for a number of haemoproteins. Eight enzymes are involved in the haem biosynthesis (Fig. 2).

Webb et al. (Eds).
Harderian Glands

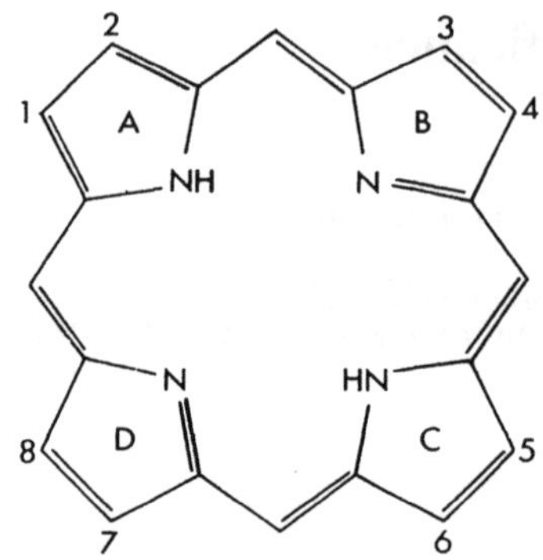

Fig. 1. The structure of the porphyrin molecule showing the Fischer nomenclature. The types of porphyrin are defined by the substitutents of the side chains of positions 1 to 8 in Fischer nomenclature.

	1	2	3	4	5	6	7	8
Uroporphyrin III	A^H	P^R	A^H	P^R	A^H	P^R	P^R	A^H
Coproporphyrin III	Me	P^R	Me	P^R	Me	P^R	P^R	Me
Harderoporphyrin	Me	V	Me	P^R	Me	P^R	P^R	Me
Iso-Harderoporphyrin	Me	P^R	Me	V	Me	P^R	P^R	Me
Protoporphyrin	Me	V	Me	V	Me	P^R	P^R	Me

Me = $-CH_3$ (methyl); A^H = $-CH_2CO_2H$ (acetyl); P^R = $-CH_2CH_2CO_2H$ (propionyl)
V = $-CH=CH_2$ (vinyl)

2.1 5 Aminolaevulinate (ALA) Synthase (EC 2.3.1.37)

The first of these enzymes, 5-aminolaevulinate synthase (ALA-S), is mainly located in the mitochondrion. ALA-S is the rate-limiting enzyme in porphyrin biosynthesis (Moore et al. 1987). Pyridoxal phosphate is required as a cofactor. In this reaction, succinate, activated as succinyl CoA, reacts with glycine bound to pyridoxal phosphate. This produces alpha-amino beta-keto adipic acid, which is then decarboxylated to form ALA. ALA-S is a relatively unstable enzyme with a short half-life in vivo of approximately 1 h in mammalian species and low basal activity (Moore and Disler 1985), which is greatly elevated in porphyria or by inducers of porphyrinogenesis (Granick 1966). The hepatic or "housekeeping" form of this enzyme is strongly inhibited by protohaem and protohaemin, the end product of the pathway, and this feedback may allow an organism to control porphyrin synthesis and avoid generating unnecessary biosynthetic intermediates (Sinclair and Granick 1976).

2.2 ALA Dehydratase (EC 4.2.1.24)

The next four enzymatic steps take place in the cytoplasm. The first of these is the combination of 2 mol of ALA to form the monopyrrole porphobilinogen

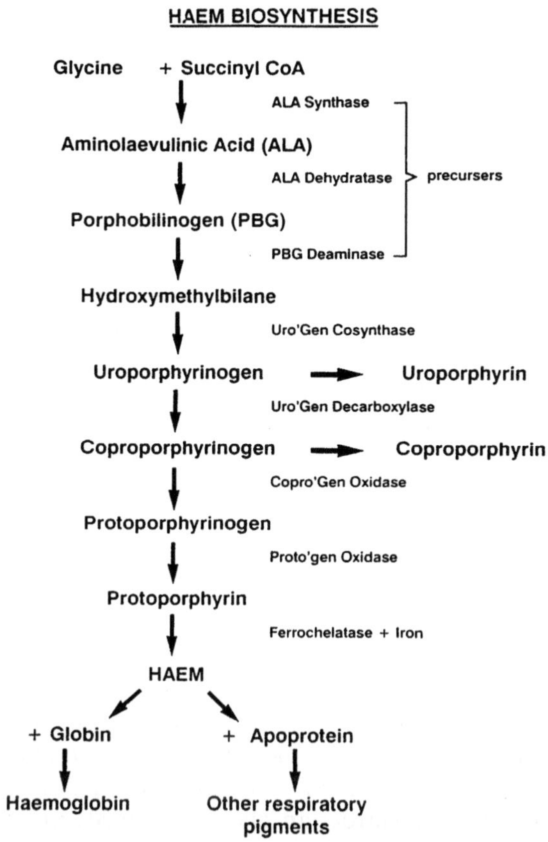

Fig. 2. A schematic diagram of the pathway of haem biosynthesis

(PBG) by the enzyme 5-aminolaevulinic acid dehydratase (Shemin 1976). PBG is the unique precursor of porphyrins. The activity of ALA dehydratase is much higher than that of ALA-S and it is thought to play little part in regulating the overall rate of haem synthesis.

2.3 PBG Deaminase (EC 4.3.1.8) and Uroporphyrinogen 3-Cosynthase (EC 4.2.1.75)

These two enzymes condense four molecules of PBG to form the linear tetrapyrrole hydroxymethylbilane and cyclize them to form the first of the porphyrins, uroporphyrinogen III. In theory, four isomers of uroporphyrinogen can be formed, but the important porphyrin derivatives such as protohaem, chlorophylls and cytochromes are all based on the asymmetrical type III isomer. If the cosynthase activity is reduced, as in congenital porphyria, large amounts of the superfluous type I isomers are formed which are excreted in the urine (Moore et al. 1979; Rimington 1975). Porphobilinogen deaminase has the second lowest endogenous activity in the haem pathway and is thus defined as a secondary control point in the pathway (Brodie et al. 1977).

2.4 Uroporphyrinogen Decarboxylase (EC 4.1.1.37)

The basic ring structure of the porphyrins has now been formed and there follows a series of decarboxylations and oxidations to form protoporphyrin. Uroporphyrinogen III, an octacarboxylic molecule, is sequentially decarboxylated by uroporphyrinogen decarboxylase to give heptacarboxylic, hexacarboxylic and pentacarboxylic intermediates prior to the formation of coproporphyrinogen III, a tetracarboxylic structure. This sequence is followed by both the type III and type I isomers, but because of substrate requirements the series 3 isomer is decarboxylated more rapidly than the series 1 isomer (Smith and Francis 1981).

2.5 Coproporphyrinogen Oxidase (EC 1.3.3.3)

The pathway then re-enters the mitochondrion, where coproporphyrinogen oxidase catalyses the formation of protoporphyrinogen. The propionyl groups in positions 2 and 4, respectively, of coproporphyrinogen III are oxidised and decarboxylated to yield the two vinyl groups of protoporphyrinogen IX (Smith 1975). The intermediate of this reaction is the monovinyl Harderoporphyrinogen, which was first isolated from rodent Harderian glands (Kennedy et al. 1970).

2.6 Protoporphyrinogen Oxidase (EC 1.3.3.4)

Protoporphyrinogen IX is then oxidised by the removal of six hydrogen atoms to form protoporphyrin IX, by protoporphyrinogen oxidase. This enzyme is embedded in the inner mitochondrial membrane and has a high substrate specificity (Dailey 1990).

2.7 Ferrochelatase (EC 4.99.1.1)

The final step in the pathway involves the insertion of ferrous iron into the protoporphyrin molecule by mitochondrial ferrochelatase to form haem or protohaem (Goldberg et al. 1956). Ferrochelatase is not substrate specific and will catalyse the chelation of a number of different bivalent metals into various dicarboxylic porphyrins, such as mesoporphyrins (Dailey et al. 1986).

3 Control of Haem Biosynthesis (Fig. 3)

3.1 The Role of ALA Synthase

The initial enzyme, ALA synthase, is the primary controller of the haem biosynthetic pathway. There is negative feedback regulation of this enzyme by intracellular haem, at both transcriptional and translational levels. The short half-life and the low endogenous activity of this enzyme enhance its ability to control the pathway (Bishop and Desnick 1982). Haem concentrations as low as 10^{-7} m will suppress the formation of ALA synthase; inhibition of the enzyme itself occurs about 10^{-5} m. Figure 3 shows the role that haem plays by assuming the existence of a "free haem pool", which is haem either newly synthesised or from exogenous sources which is unbound to any form of protein. This pool is small and has a rapid turnover, necessary prerequisites for a controlling substance.

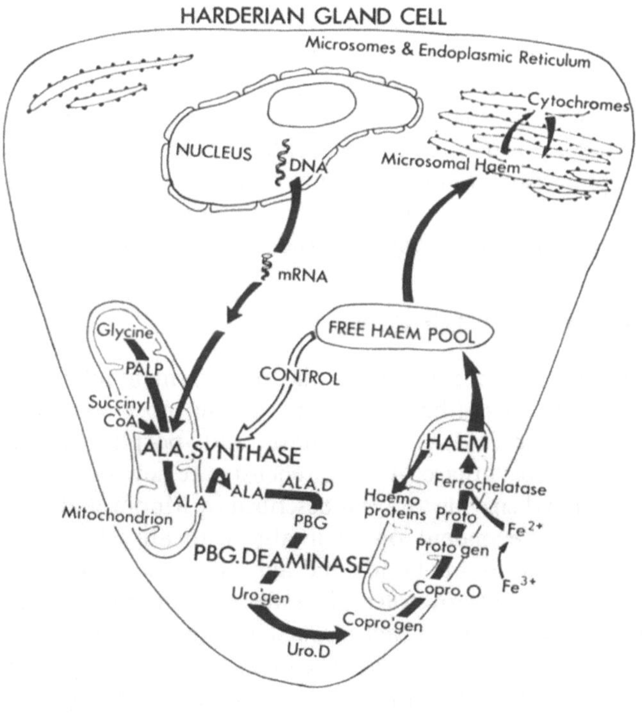

Fig. 3. A schematic diagram showing the control of haem biosynthesis. The enzyme ALA synthase and the free haem pool are the central components. Biosynthesis may be affected by substrate availability and by the presence or absence of the end product, haem. Haem acts on ALA synthase not only by limiting its formation, but also by inhibiting its activity and transmembrane transfer. Haem concentrations in the cell are maintained by the balance between synthesis and degradation

3.2 Glycine

Substrate availability may also control ALA-S activity. This enzyme has a very low affinity for one of its substrates, glycine, with K_m values of between 5 and 20 nM. The concentration of glycine in the liver, however, is less than these values. When compounds known to complex glycine, such as para-aminobenzoic acid or sodium benzoate, are given to animals, experimental porphyrias induced by chemicals can be reversed. These results suggest the possibility of control of the pathway through availability of glycine substrate for ALA-S (Piper et al. 1973; Neuberger 1980).

3.3 Control in Normal Subjects and in Porphyrias

In normal subjects, where the biosynthetic pathway is functioning efficiently, porphyrin production greater than that required for haem synthesis results in the excretion of minute quantities of porphyrins and their precursors. However, in metabolic disorders of haem biosynthesis (the porphyrias), there is an increase in the activity of the initial and rate-limiting enzyme, ALA-S, which results in the over-production of porphyrins and their precursors. In acute porphyrias there is excessive production of ALA and PBG unlike the non-acute porphyrias, in which there is no excess of these precursors. In acute porphyrias PBG-D activity is either normal or depressed so that ALA and PBG accumulate and are subsequently excreted, whereas in the non-acute porphyrias the activity of this enzyme is raised so these precursors do not accumulate (Brodie et al. 1977). The difference in the activity of PBG deaminase in the acute and non-acute porphyrias therefore defines the role of this enzyme as the secondary control point.

3.4 Measurement of Porphyrins and their Enzymes

Standardised techniques have been devised for the measurement of all the various enzymes in the haem biosynthetic pathway. Those most commonly measured in human tissue are ALA dehydratase and porphobilinogen deaminase/uroporphyrinogen cosynthase. A comprehensive review of quantitative porphyrin measurement and enzyme assays is described by Moore et al. 1987.

Porphyrin quantitation is dependent on their ability to absorb light at around 400 nm. The conjugated double bonds and alternating series of single and double bonds in the cyclic tetrapyrroles gives these compounds a high degree of resonance resulting in a series of well-marked absorption bands in the visible spectrum and a characteristic red colour. The intense absorption band at approximately 400 nm, known as the Soret band, is used in spectrophotometric determinations, and porphyrins can be measured at concentrations as low as 10^{-9} M. In alkaline and acid solutions other peaks appear between 480 and 650 nm. The nature of the substituents on the porphyrin ring alters the intensity of these peaks.

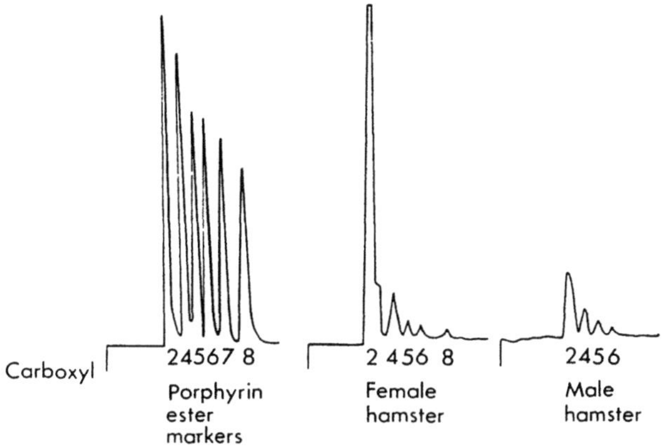

Fig. 4. High pressure liquid chromatography (HPLC) chromatograms of the carboxyl groups in porphyrin ester markers, female hamster Harderian glands and male hamster Harderian glands. Protoporphyrin (2 carboxyl groups) occurs predominantly in both female and male Harderian glands. Harderoporphyrin (3 carboxyl groups), when present, appears as a shoulder on the protoporphyrin peak. Coproporphyrin (4 carboxyl groups) is the other main constitutent porphyrin of the gland

The most characteristic feature of porphyrins is their bright red fluorescence. Light is absorbed within the Soret absorption band of 400 nm and is re-emitted at longer wavelengths, around 600 nm, as fluorescence. Fluorimetric determination of porphyrin concentration can be measured with a high degree of specificity and with greater sensitivity than is possible with spectrophotometry.

Porphyrins may be measured in biological material such as liver, kidney, Harderian glands, urine, faeces, plasma and blood. The classical method of separation is by extraction. Alternatively, chromatography can be used to separate porphyrins; this includes paper chromatography, thin layer chromatography, and high pressure liquid chromatography (Fig. 4). Subsequently, detection is normally performed by spectrophotometry or spectrofluorimetry.

4 Porphyria

Although type I and type III isomers of coproporphyrin and uroporphyrin, and protoporphyrin IX are often found in relatively small amounts in tissue fluids and urine, large amounts of porphyrins in body tissues usually only occur as a result of porphyria or lead poisoning (Moore et al. 1987). Porphyria is usually hereditary in origin. Sporadic and toxic porphyrias also occur. The porphyrias are each characterised by derangement and partial blockade of haem biosynthesis, leading to a build-up of the precursor preceding the block. All have raised

5-ALA-S activities. A number of common drugs may induce a porphyric attack in patients. Steroids have been demonstrated to be porphyrinogenic; oral contraceptives or hormonal changes in the menstrual cycle or pregnancy may trigger an attack in porphyrics. Attacks are more common in women and rarely occur before puberty. Each porphyria has a characteristic pattern of porphyrin and porphyrin precursor production and excretion.

5 Porphyrin Synthesis in the Rodent Harderian Gland

Rodents are the only mammalian group in which porphyrins have been located within the Harderian gland. Porphyrin pigment was first described in the gland of the white mouse (Loewenthal 1892), although no attempt was made to identify it until the work of Derrien and Turchini in 1924. The gland was originally thought to be a passive storage organ of porphyrin formed elsewhere in the body (Derrien and Turchini 1924; Grafflin 1942; Towbin et al. 1945; Orten and Keller 1946) or to act as an excretory organ for porphyrins (Figge 1944). However, later experiments comparing the capacity of various homogenised mouse tissues to produce porphyrin from ALA showed that the Harderian gland has the greatest porphyrin-producing enzyme activity (Davidheiser and Figge 1955). Kennedy (1954) showed that rat Harderian gland tissue could synthesise uroporphyrin III from porphobilinogen, and Formgren and Wetterberg (1973) demonstrated the presence of the enzyme PBGD in the hamster Harderian gland. Tomio and Grinstein (1968) and Tomio et al. (1968) provided experimental evidence for the de novo biosynthesis of protoporphyrin IX from glycine and succinyl-Co-A by the rat Harderian gland. The presence of 5-ALA-S in the mitochondrial fraction was demonstrated by the formation of 14C-ALA. The presence of ALA-D, PBG deaminase, uroporphyrinogen decarboxylase, and coproporphyrinogen oxidase systems, and the similarity of the biosynthetic pathway with other living cells was also demonstrated by these authors. The entire enzymatic complement required to synthesise porphyrins from the precursors was also demonstrated in the Harderian gland of the mouse (Margolis 1971) and the golden hamster (Thompson et al. 1984).

In the hamster gland the levels of ferrochelatase are very low (Thompson et al. 1984) and consequently there are no metal complexes of porphyrins in the gland. In the gland of the female hamster, high activity of protoporphyrinogen oxidase coupled with low ferrochelatase activity is the probable cause of the deposition of solid intraluminal porphyrin accretions.

It has been generally assumed that the control of haem synthesis in the gland is similar to that of the liver. To test this, a mixing experiment was carried out. The activities of ALA synthase and PBG deaminase were measured in male and female hamster gland homogenates and in a 50:50 mixture of these two. The results show that although the mean PBG deaminase activity of the mixture

Table 1. The mixing experiment[a]

	ALA-S (nmol/g/h)	PBG-D (nmol/g/h)
Female	1767	1224
Male	542	201
50:50 mix	525	655

[a]In this experiment, male and female Harderian glands were homogenised to equal protein concentrations. The activity of 5-aminolaevulinate synthase (ALA-S) and porphobilinogen deaminase (PBG-D) was measured. In the last part of the experiment, equal volumes of male and female homogenates were mixed (50:50) and the enzyme activities measured once again.

was, as expected, mid-way between the male and female activities, the activity of ALA synthase in the mixture was depressed to a figure similar to that of the male (Table 1). This suggests firstly that there are controlling substance(s) in the male gland which inhibit ALA synthase and secondly that there is primary control of the biosynthetic pathway at this enzyme as there is in the liver.

6 Histological Appearance of Porphyrins in the Harderian Gland

Porphyrins are synthesised by the epithelial cells of the Harderian gland. The red fluorescence of the gland, showing the presence of porphyrins, was first observed in rats by Derrien and Turchini (1924). There is still some dispute as to which type of epithelial cells synthesise porphyrin. Porphyrin release into the lumen is thought to occur by incorporation into intracellular lipid droplets (Brownscheidle and Niewenhuis 1978; Wooding 1980). However, holocrine secretion may also be involved in some cases (Carriere 1985).

Once released into the lumen, the porphyrins are oxidised (Cohn 1955) and combine with other secretory products such as lipids. These secretions are transformed into dense red-brown pigment that is no longer fluorescent (Strum and Shear 1982a). Porphyrin is usually deposited in the form of large intraluminal accretions in the hamster (Fig. 5a); Christensen and Dam 1953; Woolley and Worley 1954), mouse (Woodhouse and Rhodin 1963), rat (Kanwar 1960; Brownscheidle and Niewenhuis 1978), and gerbil (Johnston et al. 1983). Where tubule walls have degenerated, porphyrins may also occur as large interstitial accretions surrounded by foreign body giant cells, and subsequently as small accretions within individual macrophages (Fig. 5c, d; Payne et al. 1982a, 1985; Spike et al. 1985, 1986a, 1988).

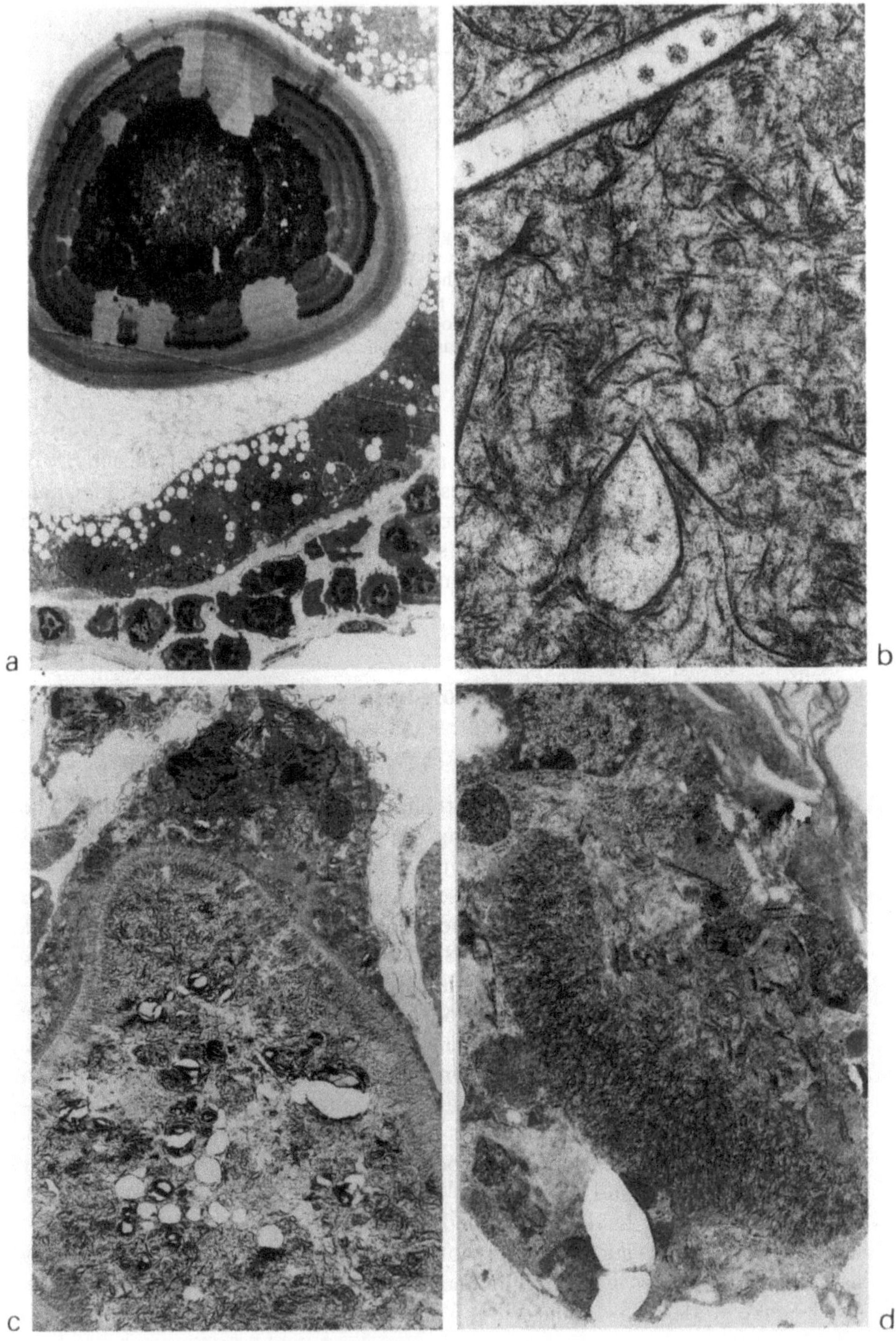
a
b
c
d

Cohn (1955) suggested that some lipid is attached to porphyrin as a lipoprotein complex. Fibrogranular components are also seen in the lipid vacuoles of secretory A cells and in the lumen of the rat (Brownscheidle and Niewenhuis 1978) and are similar in appearance to aggregates of protoporphyrin crystals within the cytoplasm and lipid droplets of hepatocytes in some cases of porphyria (Fig 5b; Matilla and Molland 1974; James et al. 1980).

6.1 Histological Preparation

Porphyrins are soluble in a wide range of organic solvents, so histological routines must minimise porphyrin loss during processing as well as giving good tissue preservation. Porphyrin loss during routine histological processing may therefore be determined by the behaviour of fixative towards lipids. A study of four commonly used fixatives showed that there was least loss of porphyrin when using dichromate, which fixes both lipids and phospholipids, and glutaraldehyde, which fixes most lipids but has a slower penetration time than dichromate. Both alcohol-formaldehyde-acetic acid and Bouin's routines produced very high losses of tissue porphyrin. Neither of these routines fixes lipids or phospho-lipids (Spike et al. 1986b). These findings substantiate earlier observations that porphyrin is not fixed by formaldehyde alone (Woolley and Worley 1954; Kanwar 1960).

6.2 Effects of drugs

Porphyrin synthesis in the Harderian and in other tissues differ: porphyrinogenic compounds such as 3,5-diethoxycarbonyl-1,4-dihydro-2,4,6-trimethylpyridine (DDC) and allylisopropylacetamide (AIA), which induce hepatic porphyria, had no effect on the porphyrin content of the mouse Harderian gland. This was thought to reflect a different form of regulation of the porphyrin pathway in the mouse Harderian gland, and genetic studies showed that the locus regulating porphyrin content in the gland of inbred mice was distinct from the liver locus (Margolis 1971). More recently, Janousek et al. (1990) found that griseofulvin elevated porphyrin content in the liver and kidneys by a hundred times and ten times respectively, but had no effect on the Harderian gland porphyrin content in young female nu/nu mice. Further differences in the effect of porphyrinogenic drugs on the rat liver and Harderian gland were illustrated when liver ALA-S activity increased sevenfold compared to a twofold increase in

Fig. 5a–d. Transmission electron micrographs of female hamster Harderian glands; these show: **a** Intraluminal porphyrin accretion showing laminar nature surrounded by type 1 secretory cells (× 750). **b** Crystalline nature of porphyrin accretion (× 10000). **c** Large interstitial porphyrin accretion surrounded by macrophages forming a giant cell (× 1400). **d** Small interstitial porphyrin accretion in a macrophage. The nucleus lies at the *top left* (× 2800) the gland activity following a single injection of AIA. Difference in the time taken to reach peak activity also occurred. The injection resulted in an increase in the total porphyrin content of the gland which returned to normal values in 48 h: excess porphyrin, which fluoresced red under UV illumination, was excreted onto the nose and whiskers (Wetterberg et al. 1971). Distinct mechanisms may also exist in the Harderian gland due to differences in co-factor availability between it and the liver (Lin and Nadakavukaren 1982; Gross and Hutton 1971).

7 Factors Influencing Porphyrin Synthesis in the Harderian Gland

7.1. Species Differences

It has long been observed that the Harderian glands of rats, hamsters and most mice fluoresced red while those of rabbits and guinea pigs did not (even though they do synthesise porphyrins) (Derrien and Turchini 1924; Thomas 1938; Figge et al. 1942). A quantitative study by Davidheiser and Figge (1958), measuring the amount of porphyrin synthesised by gland homogenates, showed that there was a great variation in enzyme activity in these species, which parallelled the degree of red fluorescence observed. More porphyrin was present in the mouse, rat and hamster glands than in the rabbit and guinea pig glands; the white fluorescent lobe of the rabbit contained considerably greater quantities of porphyrin than the grey-red lobe. This study also showed that synthesis in the Harderian gland was considerably greater than that of the liver in rats and hamsters: the reverse was true of rabbits and guinea pigs.

Porphyrins are present in the Harderian glands of all rodent species assayed so far, although the amount varies considerably (Table 2). In the woodmouse, *Apodemus sylvaticus*, there are no intraluminal or interstitial stores and levels of porphyrin in both sexes are very low (Johnston et al. 1987). Solid intraluminal porphyrin accretions and interstitial deposits are observed in the plains mouse, *Pseudomys australis* (Johnston et al. 1985), the rat (Grafflin 1942), and the female hamster (Payne et al. 1982a, 1985; Spike et al. 1985, 1986a). Both sexes of the Mongolian gerbil, *Meriones unguiculatus*, have low levels of porphyrin and intraluminal deposits of porphyrin are sparse (Johnston et al. 1983). Species differences also occur in the amount of different forms of porphyrins produced, although protoporphyrin usually predominates (Table 3).

7.2 Age and Strain

Studies of the porphyrin content and porphyrin-synthesising enzyme activity of the rodent Harderian gland show that these parameters vary with age and

Table 2. Porphyrin concentration (nmol/g tissue) in the Harderian glands of various rodent species (adult)

	Male	Female[a]
Woodmouse (*Apodemus sylvaticus*)	61 ± 22	125 ± 64(1)
Mongolian gerbil (*Meriones unguiculatus*)	89 ± 35	213 ± 69(2)
Plains mouse (*Pseudomys australis*)	237 ± 83	1932 ± 772(3)
Golden hamster (*Mesocricetus auratus*)	42 ± 9	3300 ± 302(4)
Rat (albino)	350 ± 112	288 ± 81(5)
Mice C57L/J	1.6 ± 0.4	1.4 ± 0.8(6)
A/HRJ	5.9 ± 3.6	10.1 ± 6.4
C3H/HeJ	34 ± 7	44 ± 32
AKR/J	83 ± 13	130 ± 46
(nmol/pair of glands)		
Mice C3H/He	99 ± 53	800 ± 89(7)

[a](1) Johnston et al. (1985); (2) Johnston et al. (1983); (3) Johnston et al. (1987); (4) Payne (1990); (5) Rohony and Kélenyi (1962); (6) Margolis (1971); (7) Shirama et al. (1981a).

Table 3. Porphyrin profiles of Harderian glands of various rodent species, detected by high performance liquid chromatography

	Proto-porphyrin	Hardero-porphyrin	Copro-porphyrin	Penta-carboxylic porphyrin	Hexa-carboxylic porphyrin	Hepta-carboxylic porphyrin	Uro-porphyrin
(1)[a]Woodmouse	93	nd	7	nd	nd	nd	nd
(2) Plains mouse	91.1–96.7	nd	0.3–5.5	0.6–2.6	1.5–5	nd	nd
(3) Rat	64	29	9	nd	nd	nd	nd
(4) Golden hamster							
Female	92.9 ± 2.3	1.9 ± 0.6	2.8 ± 0.8	0.8 ± 0.3	1.2 ± 0.6	0.5 ± 0.4	< 0.1
♀ 5 months	78.9 ± 4.4	10.1 ± 3.6	3.8 ± 1.6	1.0 ± 0.4	1.5 ± 0.5	2.9 ± 0.9	1.7 ± 1.1
Male	71.9 ± 13.1	nd	10.7 ± 5.7	8.8 ± 9.5	8.5 ± 9.2	nd	nd
♂ 6 weeks	84.3 ± 4.1	nd	15.3 ± 4.1	0.4 ± 0.3	nd	nd	nd

[a](1) Johnston et al. (1985); (2) Johnston et al. (1987); (3) Kennedy et al. (1970); (4) Spike et al. (1990c)
♀ = ovariectomised female
♂ = castrated male
nd = not detected

strain. The porphyrin content of glands of rats, mice and hamsters remains low until about 2 weeks old, at which time the porphyrin content increases rapidly (Wetterberg et al. 1970a; Shirama et al. 1981a; Bucana and Nadakavukaren 1973). Enzyme activities are almost absent in the glands of very young mice but as the animals become sexually mature (2–5 weeks) the enzymes appear. There is then a rapid increase in activity which reaches a maximum at 8 weeks and remains constant until 52 weeks (Figge and Davidheiser 1957). In rats, visible porphyrin appears on day 13 after birth (Rapin 1952) coinciding with the eyes opening (Wetterberg et al. 1972a), and there is a rapid alteration in cellular morphology and increased secretory activity (Müller 1969). In the albino rat, pink fluorescence was observed at day 8, and pigment granules seen in the lumen at day 9 (Müller 1969); porphyrin content increases as weight increases (Ono et al. 1956; Rohonyi and Kelényi 1962). In gerbils, porphyrin is rarely seen at birth but is abundant when the eyes open and increases up to 3 weeks of age (Arvy and de Lerma 1961). In other desert rodents porphyrin content also increases from birth (Abou-Harb and Abou-Harb 1964).

The number of intraluminal porphyrin deposits has increased in female hamsters by 4 weeks, but has almost disappeared in males. The disappearance of porphyrin in males coincides with the appearance of polytubular complexes, which are thought to contain enzymes which break down porphyrin (Bucana and Nadakavukaren 1973). In an age series of male hamsters ranging from 2 to 30 months, there was no significant difference in the very low levels (25–70 nmol/g) of porphyrin present (Spike 1986). In an age series of female hamsters (2–24 months), the porphyrin content was greatest in the 2-month-old animals which are sexually mature, and thereafter decreased. The ALA-S activity showed a similar pattern, peaking at 6 months and then decreasing. These parameters appear to be dependent on the levels of circulating ovarian hormones since they are significantly reduced in the post-reproductive senescent females. Although not reflecting porphyrin formation, the pattern of porphyrin stores within the gland also alters as the animals age; intraluminal deposits are greatest in the 2-month females, while there is a progressive increase in the number of interstitial deposits as the animals age (Table 4) (Spike et al. 1988).

Table 4. Changes occurring in total porphyrin content, ALA-S activity, and some morphological parameters in the Harderian glands of female hamsters from post-puberty until senescence. (Spike et al. 1988)

	Age in months 2 (n = 10)	6 (n = 10)	12 (n = 10)	18 (n = 10)	24 (n = 10)	Analyses of variance (F)
Total porphyrin content (nmol/g tissue)	6012 ± 542	5383 ± 743	3649 ± 507	1860 ± 313	1677 ± 334	16.07 $P < 0.001$
ALA-S activity (nmol ALA formed/g/h)	937 ± 173	5545 ± 1856	2873 ± 247	1457 ± 435	2010 ± 464	4.6 $P < 0.01$
Intraluminal porphyrin accretions/mm^2	28.58 ± 1.70	22.70 ± 5.70	16.94 ± 1.65	8.54 ± 1.43	8.37 ± 1.02	10.43 $P > 0.001$
Large interstitial porphyrin deposits/mm^2	0.03 ± 0.02	0.04 ± 0.02	0.12 ± 0.05	0.10 ± 0.04	0.54 ± 0.16	7.84 $P < 0.001$
Small interstitial porphyrin deposits/mm^2	0.01 ± 0.01	0.13 ± 0.09	1.04 ± 0.32	0.88 ± 0.18	4.30 ± 1.29	8.96 $P < 0.001$

All values are means ± SEM

Porphyrin synthesis in the Harderian glands of mice (Davidheiser and Figge 1958) and hamsters (Bucana and Nadakavukaren 1973) appears to be dependent on developing gonadal hormones, but it has not yet been demonstrated if the enzymes in the haem pathway are sequentially induced as hormone levels rise, as occurs in perinatal rat liver (Woods and Dixon 1970; Moore et al. 1972).

The mouse Harderian gland shows strain variations in porphyrin content. Early studies by Figge (1944) demonstrated that the red fluorescence of the glands of mice of different strains vary in intensity. Quantitative experiments confirmed these findings (Strong 1942; Figge and Davidheiser 1957; Margolis 1971; Shirama et al. 1981a, b).

7.3 Sex Differences

In some rodent species, such as C57, A and LCSa strains of mice (Figge and Davidheiser 1957; Shirama et al. 1981a), the white rat (Müller 1969) and the woodmouse (Johnston et al. 1987) the amount of porphyrin synthesised by the two sexes is comparable, but in others the female produces more than the male, as in the gerbil (Johnston et al. 1983), the Plains mouse (Johnston et al. 1985) and C3H strains of mice (Strong 1942; Shirama et al. 1981a). The most extreme sex differences occurs in the golden hamster (Christensen and Dam 1953; Bucana and Nadakavukaren 1972; Payne et al. 1978). Gonadal hormones appear to be responsible for these sex differences as castration of males results in increased porphyrin synthesis in both mice (Shirama et al. 1981b) and hamsters (Woolley and Worley 1954; Hoffman 1971; Clabough and Norvell 1973; Payne et al. 1977a). Increased activity of porphyrin-synthesising enzymes accounts for this (Spike et al. 1986c). In the hamster, there is also loss of type II cells and polytubular complexes (Lin and Nadakavukaren 1979), and an increase in mast cell numbers (Payne et al. 1977a, 1982b). Testosterone administration prevented these changes (Payne et al. 1977a; Shirama et al. 1981a). Conversely, ovariectomised female hamsters given androgen treatment show greatly decreased porphyrin concentrations and ALA-S activity drops to male levels; morphological changes include the appearance of type II cells and polytubular complexes, and a decrease in the number of intraluminal porphyrin accretions accompanied by an increase in the number of large interstitial porphyrin deposits (Sun and Nadakavukaren 1980; Spike et al. 1985). Ovariectomised female hamsters do not develop these male morphological features: instead there is progressive degeneration of the tubules in the gland which becomes more pronounced with increasing time post-operatively. Porphyrin concentrations remain similar to those of intact females for many weeks after ovariectomy, although the activity of ALA-S shows a significant progressive decrease. This suggests that porphyrin is neither being formed nor lost (Payne et al. 1985; Spike et al. 1985, 1986a; Table 5). Similarly, ovariectomy in the rat does not result in lowered Harderian gland porphyrin levels (Ulrich et al. 1974).

Table 5. Changes occuring in porphyrin content, ALA-S activity and some morphological parameters in intact control females and female golden hamster ovariectomised for 10, 20, 30 or 40 weeks (Spike et al. 1986a)

	Control (n = 20)	10 week (n = 20)	20 week (n = 21)	30 week (n = 23)	40 week (n = 17)	Analyses of variance (F)
Total porphyrin content (nmol/g tissue)	3649 ± 507	3340 ± 406	3236 ± 521	3130 ± 246	3995 ± 584	0.62 n.s.
ALA-S activity (nmol ALA formed/g protein/h)	1795 ± 227	1413 ± 246	** 922 ± 89	1309 ± 222	** 896 ± 105	3.59 $P < 0.01$
Intraluminal porphyrin accretions/(mm^2)	16.94 ± 1.64	12.67 ± 3.74	15.62 ± 1.31	14.61 ± 0.87	13.00 ± 1.52	2.33 n.s
Large interstitial porphyrin deposits (mm^2)	0.12 ± 0.05	0.34 ± 0.33	0.35 ± 0.11	0.51 ± 0.16	** 1.66 ± 0.44	9.88 $P < 0.001$
Small interstitial porphyrin deposits (mm^2)	1.04 ± 0.32	2.57 ± 0.55	2.89 ± 0.49	* 4.28 ± 0.85	**10.89 ± 1.57	22.44 $P < 0.001$

Differs significantly from control females: *$P < 0.05$, **$P < 0.01$.
All values are means ±SEM.

However, although porphyrin concentrations remained fairly constant after ovariectomy in the hamster, there is a clear redistribution of porphyrin stores with a decrease in the number of intraluminal accretions being accompanied by an increase in both the number of large and small interstitial deposits. The increase in blood porphyrin levels at 20–40 weeks post-operatively may be due to the blood-borne dispersal of these small deposits. Porphyrin synthesis also alters over the oestrous cycle and during pregnancy, though there may be species differences. The porphyrin content and ALA-S activity of the female hamster gland shows changes over the 4 day oestrous cycle with porphyrin content greatest at oestrus (day 1), decreasing on day 2 and increasing again on days 3 and 4. ALA-S activity is lowest on day 1 and then gradually increases to reach maximum activity on day 4. There is also a significant variation in the number of intraluminal deposits, the most being found on day 1 (Moore et al. 1977; Payne et al. 1977b, 1979). This differs from the cyclic variation described in mice where porphyrin levels are lowest during metoestrous and highest at dioestrous (Shirama et al. 1981a). In pregnant mice, the porphyrin concentration of the gland was nearly the same as in intact oestrous mice (Shirama et al. 1981b), whereas in hamsters the porphyrin content and ALA activity of the gland rise during a first pregnancy, reaching a peak just after birth and then decrease during the lactation period. These changes do not occur during subsequent pregnancies. During a first pregnancy there is an increase in abnormally located porphyrin stores, as both large and small interstitial deposits increase, as do mast cell numbers. These histological changes are maintained in non-pregnant multiparous females. ALA-S activity was significantly greater in primsi than in multiparous females during late pregnancy in liver and kidney: faecal porphyrin content increased during lactation in primagravidae (Spike et al. 1990a).

Jones and Hoffman (1976) found that testosterone and oestrogen had no effect on porphyrin synthesis in the glands of male hamsters in vitro. These authors also suggested that increased porphyrin synthesis in the male could be produced by the addition of ALA to the incubation medium and that, normally, production of ALA in the male gland might be limited by substrate availability.

Sex differences have also been determined in the activities of the haem biosynthetic enzymes in the hamster Harderian gland, firstly in uro-cosynthase (Formgren and Wetterberg 1973) and then in ALA-S (Mindegaard 1976; Lin and Nadakavukaren 1982). These results were later confirmed and the research extended by Thompson et al. (1984), who demonstrated that there were sex differences in five of seven enzymes in the hamster Harderian gland (ALA-S, PBG-D, URO-D, COPRO-O, and PROTO-O). In each case, female tissue exhibited higher activity than the male.

7.4 Effects of Temperature and Lighting

In the mouse, exposure to heat leads to an increase in the accumulation of porphyrin in the Harderian gland. In hot, humid conditions albino strains show visible signs of fluorescence on the face. Rats exposed to the same high temperature and humidity as the mice also exhibited red fluorescence around the eyes and nose which was not observed in animals at lower temperatures, although the porphyrin content was the same in the two groups. The rat gland normally contains more porphyrin than that of the mouse. The author (Collins 1957) suggested that, in the mouse, heat exposure resulted in porphyrin accumulation, but pigment was discharged only when the porphyrin level was similar to that of the rat gland.

The golden hamster is a seasonal breeder and variations in photoperiod regulate the annual growth and atrophy of the sex organs. Studies have shown that hamsters normally require long photoperiods (i.e. more than 12.5 h of light/24 h) for the maintenance of testicular size and function (Hoffman et al. 1965; Elliot 1981). In the female hamster, the porphyrin content of the Harderian gland varies throughout the year, being significantly greater in summer than in winter. ALA-S activity parallelled these seasonal fluctuations (Payne et al. 1977b; Moore et al. 1980). Diurnal variations in porphyrin content have been observed in rat Harderian glands, being minimal at noon and increasing to a maximum at midnight (Joó and Kahán 1975).

Lighting conditions also alter the pattern of porphyrin synthesis. In the female golden hamster, 72 days of continuous light produced a fourfold increase in gland porphyrin content compared to animals kept in total darkness. Histologically, the light-exposed animals had a higher amount of luminal pigment than those kept in darkness, while the latter exhibited male Type II cells (Wetterberg et al. 1972b). Conversely, Harderian glands of rats and mice exposed to continuous lighting showed degenerative changes. Histological degeneration of the gland occurred during exposure to light for periods of 1 week in female Albino Swiss rats, but some proliferation was also noted among epithelial cells (O'Steen et al. 1978). In pigmented and albino rats exposed to continuous lighting for 65 days, the Harderian glands regressed. Histologically, the tubular lumina were dilated and many tubules contained cellular debris (Reiter 1973; Reiter and Klein 1971). Exposure of female rats to continuous light for 35 days caused a significant decrease in porphyrins (Ulrich et al. 1974), while Shirama et al. (1977) found that either constant light or constant darkness had an inhibitory effect on the porphyrin content of the Harderian gland of female rats. A more recent study (Shirama et al. 1987) confirmed these findings: constant exposure to light for 60 days produced a 60% decrease in gland porphyrin content of intact female rats compared to animals kept under diurnal lighting conditions. Exposure to constant darkness produced a smaller decrease in porphyrin content. Ovariectomised females had gland porphyrin content similar to that of intact control females when kept under constant darkness or diurnal lighting conditions, but those kept in constant light had decreased porphyrin levels. Male rats exposed to constant light also had lower gland porphyrin levels, but constant darkness induced an increase in these levels. Castration resulted in increased porphyrin concentrations under each lighting regime, although there was no difference in gland weight or morphological appearance. Constant exposure to light in albino mice caused damage to secretory cells within the Harderian gland and induced some of them to undergo squamous metaplasia. Porphyrin content of the gland, assayed after 1, 3 and 7 days of exposure, showed a progressive and significant decrease compared to the control animals kept in 12hL:12hD (Strum and Shear 1982b).

Environmental lighting conditions affect the diurnal rhythm in secretion pattern of the rat Harderian gland in neonatal males and females. Under diurnal lighting conditions, there was a statistically significant difference in gland secretion patterns between males and females at both 1300 and 2300 h. In continuous light or continuous darkness, this diurnal rhythm in secretion pattern was lost and there was no significant difference between data from male and female neonates (Feria-Velasco et al. 1983).

In male hamsters, the porphyrin content of the Harderian gland is less than 10% of the female, but castration of males results in the disappearance of Type II cells and polytubular complexes (Lin and Nadakavukaren 1979) and an increase in ALA synthase activity and porphyrin content (Payne et al. 1977a; Spike et al. 1986c). However, the length of light exposure and visual function also influenced the porphyrin-synthesising ability of the gland in castrates. This increase in gland porphyrin occurred in hamsters kept under diurnal lighting conditions (12h L: 12h D) or in constant light for 72 days, but not in animals kept in constant darkness for this period. The gland porphyrin content of blinded castrated male hamsters kept in constant light was as low as in castrated males kept in total darkness (Wetterberg 1972). Nadakavukaren and Lin (1983) showed that short photoperiods (1h L: 23h D) influenced the ultrastructure of the hamster Harderian gland after 4 to 10 weeks of exposure. Changes in the male gland included reduction in the number of large vacuoles and the number of polytubular complexes and an increase in the number of membrane formations, as occurs following castration. The opposite occurs in female hamster Harderian glands. These changes are most likely due to changes in testosterone levels induced by gonadal regression. Further experiments showed that pinealectomy prevented these ultrastructural changes induced by short photoperiod. Atrophy of the testes and uteri also occurred during exposure to short photoperiods but not in pinealectomised animals (DiIorio and Nadakavukaren 1984).

Finally, the wavelength of light also affects the porphyrin concentration in the Harderian gland of the golden hamster. Castrated males kept in constant blue light for 72 days had higher Harderian gland concentrations than those kept in green, white, UV or red light. Harderian porphyrins in castrates exposed to red light were very low, and the testes of normal hamsters kept in red light had lower Harderian gland porphyrin concentration compared to the four other groups kept in white, UV, blue and green light. Females kept in red or blue light conditions had small uteri than those kept in white light (Formgren and Wetterberg 1976).

Table 6. The porphyrin content and ALA-S activity of liver, kidney and Harderian gland in the female golden hamster. (Payne 1990)

	Porphyrin content (nmol/g)	ALA-S activity (nmol ALA/h/g protein)
Liver	25 ± 2	180 ± 29
Kidney	27 ± 5	380 ± 75
Harderian gland	3400 ± 539	1800 ± 511

8 Harderianectomy and Porphyrin Synthesis

The anatomical route of porphyrin secretion from the Harderian gland can be followed by observing the red fluorescence. The secretion collected from the lacrimal sac passes into the nasal cavity via the nasolacrimal duct and then passes either to the alimentary tract via the naso-pharynx or to the nose and whiskers through the external nares. In normal rats the secretion is removed and spread onto the fur during grooming, and red pigment is occasionally seen on the fur. Harderian gland ablation prevented the accumulation and incrustation of a red fluorescent porphyrin-containing exudate in rats fed a panthothenic acid deficient diet. The Harderian gland is, therefore, the source of porphyrin in the encrusted material observed on the nose, whiskers and fur of rats on such a diet (Figge and Salomon 1942).

In the Harderian gland of the female golden hamster the porphyrin content is several hundred times greater than that of the liver or kidney, while in the male hamster gland the porphyrin content and enzyme activity are low (Table 6). The effects of the gland on porphyrin synthesis elsewhere in the body has received little attention. However, Spike et al. (1990b) showed that liver porphyrin content decreased in males and females which had been Harderianectomised for 5 months. ALA-S activity of the liver also decreased in both sexes and was accompanied by elevated blood porphyrin levels. Kidney porphyrin content and ALA-S activity were not affected by Harderianectomy. The data suggested some link between porphyrin biosynthesis in the liver and the Harderian gland.

9 Porphyrins and Cancer Susceptibility

Early findings on mice indicated a match between the amount of protoporphyrin in the Harderian glands and intrinsic susceptibility to cancer of the mammary gland. Inbred strains of mice with a high incidence of spontaneous mammary tumours had a very high concentration of protopophyrin, which emitted deep red fluorescence under ultraviolet light; mice of strains with intermediate degrees of tumour susceptibility showed an intermediate pink fluorescence, while mice of cancer-resistant strains exhibited little or no fluorescence (Strong and Figge 1941; Figge et al. 1942). Female mice of a cancer-susceptible strain had higher protoporphyrin concentrations than age-matched males of the same strain and these females developed breast cancer spontaneously (Strong 1942). Consistent with these observations is the association between hepatocellular carcinoma and both Porphyria Cutanea Tarda and Acute Intermittent Porphyria (Moore et al. 1987; Lithner and Wetterberg 1984). A further parallel was demonstrated between cancer susceptibility and maternal inheritance of Harderian gland protoporphyrin. In F1 crosses between C3H strain mice (high protoporphyrin and high cancer susceptibility) and JK strain mice (low protoporphyrin and low cancer susceptibility), the protoporphyrin content lay between the two ancestral strains, but was closer to that of the ancestral female than the male (Strong 1944). An examination of numerous birds, reptiles and mammals showed that red-fluorescent Harderian glands were present only in mice, rats and hamsters. These three species are very prone to the induction of tumours by carcinogenic agents. A relationship between excess porphyrins and susceptibility to carcinogens was postulated (Figge 1944), but at that time it was believed that the Harderian gland excreted porphyrins synthesised in other organs and tissues.

Porphyrins injected into the peritoneal cavity of rats became concentrated in the skin, subcutaneous tissues and Harderian glands, as might occur during excess porphyrin synthesis. Porphyrin excreted by the Harderian gland is mixed with saliva and smeared onto the forepaws, ears, nose, and around the eyes during grooming; these areas developed tumours when the animals were exposed to UV light (Figge 1944). Rats fed a diet deficient in pantothenic acid have decreased saliva production caused by changes in water metabolism. In these animals, excreted porphyrin accumulated and became more abundant around the nose: in normal rats this would have been smeared onto the head and ears (Figge and Atkinson 1941; Figge and Salomon 1942). Porphyrins appear to sensitise cells to the action of carcinogenic or growth stimuli, but would not induce atypical growth in the absence of porphyrin sensitisation (Figge 1944). Albino rats exposed to UV light developed multiple skin carcinomas, while hairless rats reacted with an accentuation of their normal hyperkeratotic condition. Porphyrin content of the haired rats was four times greater than that of hairless rats, again suggesting that the excretion of larger amounts of porphyrin by the Harderian gland is a factor influencing the susceptibility to light-induced cancer in rats (Hueper and Figge 1945).

10 The Possible Significance of Harderian Gland Porphyrins

As yet we do not know the role that porphyrins play in the Harderian gland. Their function can only be speculated upon in light of the potential functions of the gland. These include the following.

10.1 Lubrication

In the lower vertebrates, the Harderian gland secretion is important in lubricating the eye and may also be involved in the maintenance of moisture in the nasal apertures (Davis 1929; Sakai 1981). In mammals, however, the Harderian gland may no longer have such an important role in lubrication since:

a) the Harderian gland is always accompanied by a well developed lacrimal gland,
b) there is little correlation between the sizes of the Harderian gland, eye and the nictitating membrane (Walls 1942; Balemans et al. 1980; Sakai 1981),
c) in rodents, the excretory duct may open on the nasal side of the nictitating membrane rather than the bulbar surface (Sakai 1981).

Furthermore, in some mammals the gland produces a lipid secretion as in rodents (Hais et al. 1968; Bucana and Nadakavukaren 1972; Yamazaki et al. 1981) and lagomorphs (Wooding 1980). Lipids may prevent the evaporation of tears or prevent them falling over the lower lid (Cohn 1955; Getty 1975). Although many porphyrins are lipid-soluble, it is unlikely that they play any significant role in lubrication.

10.2 Pheromones and Thermoregulation

Although lipids are the major secretory product of the mammalian Harderian gland, the large amount of porphyrins synthesised in addition to lipids in the rodent gland suggest alternative functions to lubrication alone. During normal grooming in rodents, the forepaws pick up both saliva and secretions from the Harderian gland which exit onto the face at the external nares. The latter can be confirmed because the secretion fluoresces red under UV light and can be seen to be spread onto the pelage (Thiessen et al. 1976). It has been suggested that the spread of lipid-borne Harderian gland secretion onto the pelage during normal grooming may have a thermoregulatory functions (Thiessen and Kittrell 1980; Harlow 1984) while spread during "social grooming" may have a pheromonal role (Payne 1977, 1979; Thiessen et al. 1976). Porphyrins have not been found to have behavioural effects on conspecifics and this is also an unlikely role for them.

10.3 Phototransduction

Since porphyrins are photoactive molecules, Harderian gland porphyrins may provide a photoprotective role by acting as a light filter in rodents adapted to living in dark environments, which are suddenly exposed to light. Sudden lighting resulted in an increase in porphyrin secretion into the conjunctival sac in the rat, while gland porphyrin content decreased significantly after 15 or 30 min of exposure to intense light compared to animals in a dark room (Hugo et al. 1987). Porphyrin secretion in the rat Harderian gland also increased in response to light. Porphyrins have also been detected in high concentrations in both the exorbital and infraorbital lacrimal glands in young and old rats (Hugo et al. 1990), protoporphyrin being the major component (Hugo et al. 1989). However, porphyrins have not been detected in hamster exorbital glands (Mhâtre et al. 1988).

Indeed, since porphyrins are phototransducing molecules, they may affect retinal photoreception in some way, and play a role in modifying light energy in nocturnal animals or animals adapted to living in dark environments. Wetterberg et al. (1970b) have suggested that the Harderian gland may act as a reflector and transducer of light energy as the visual elements of the mammalian retina face towards the gland. However, an alternative concept is that of a monolayer of porphyrin or porphyrinogen dissolved in lipid solution being spread over the surface of the cornea. This thin film would be capable of absorbing light and causing some wavelength shift of shorter wavelength light. This could increase the amount of usable light reaching the retina by converting ultra violet to visible red. This would be of greatest importance in low light conditions such as in early morning and late evening.

10.4 Reproductive Function

The rodent Harderian gland has also been postulated as a link in the retinal-pineal-gonadal chain. This link could be achieved through effects of the gland on photic stimulation, as described above. However, it has been demonstrated that indoles may be involved in porphyrin biosynthesis in the rodent Harderian gland. Melatonin, which has antigonadal effects, doubled the amount of porphyrin in rat Harderian glands only 20 min after intra-peritoneal administration (Joó and Kahán 1975). In the rat Harderian gland, the administration of the serotonin precursor, 5-hydroxytryptophan caused a temporary increase in gland porphyrin content. Serotonin may also be involved in the secretion and/or biosynthesis of Harderian gland porphyrins (Shirama et al. 1979). Alternatively, porphyrin synthesis may control the availability of precursors for melatonin formation. Tryptophan pyrrolase is a haemoprotein which catalyses the rate-limiting step of tryptophan degradation and thus controls the availability of tryptophan for serotonin and melatonin synthesis. This enzyme is uniquely sensitive to haem availability and rapidly loses activity in conditions of haem depletion (Badawy and Morgan 1980). This is more speculative than phototransduction as a functional role for the gland porphyrins but has merit because of the effects on haem biosynthesis.

These remain tentative suggestions as to the function(s) of porphyrins in the Harderian gland. To date there are no data on turnover or rate of accumulation of porphyrin stores within the gland. The possible contribution of Harderian gland porphyrins to the economy of porphyrin biosynthesis in the rest of the body and the relationship to the pineal-retinal gonadal axis requires further analysis.

References

Abou-Harb M, Abou-Harb N (1964) La glande de Harder chez certains rongeurs sauvages désertiques et tropicaux et l'élaboration des porphyrines. C R Soc Biol 158:2268–2270

Arvy L, de Lerma B (1961) Sur la présence d'un pigment fluorescent de type porphyrinique dans la glande de Harder de *Meriones crassus*, Sundevall. C R. Acad Sci 253:1012–1014

Badawy AAB, Morgan CJ (1980) Tryptophan pyrrolase in haem regulation Biochem J 186:763–772

Balemans MGM, Pévet P, Legerstee WC, Nevo E (1980) Preliminary investigations on melatonin and 5-methoxytryptophol synthesis in the pineal, retina, and Harderian gland of the mole rat and in the pineal of the mouse "Eyeless" J Neural Transm 49:249–255

Bishop DF, Desnick RJ (1982) Assays of the haem biosynthetic enzymes. Enzyme 28:91–231

Brodie MJ, Moore MR, Thompson GG, Campbell BC, Goldberg A (1977) Is porphobilinogen deaminase activity a secondary control mechanism in haem biosynthesis in humans? Biochem Soc Trans 5:1466–1468

Brownscheidle CM, Niewenhuis RJ (1978) Ultrastructure of the Harderian gland in male albino rats. Anat Rec 190:735–754

Bucana CD, Nadakavukaren MJ (1972) Fine structure of the hamster Harderian gland. Z Zellforsch Mikrosk Anat 129:178–187

Bucana CD, Nadakavukaren MJ (1973) Ultrastructural investigation of the postnatal development of the hamster Harderian gland. II. Male and female. Z Zellforsch Mikrosk Anat 142:1–12

Carrière R (1985) Ultrastructural visualization of intracellular porphyrin in the rat Harderian gland. Anat Rec 213:496–504

Christensen F, Dam H (1953) A sexual dimorphism of the Harderian glands in hamsters. Acta Physiol Scand 27:333–336

Clabough JW, Norvell JE (1973) Effects of castration, blinding, and the pineal gland on the Harderian glands of the male golden hamster. Neuroendocrinology 12:344–353

Cohn SA (1955) Histochemical observations on the Harderian gland of the albino mouse. J Histochem Cytochem 3:342–353

Collins KJ (1957) Pigment in the Harderian glands of mice exposed to a hot environment. B J Exp Physiol 42:24–32

Dailey HA (1990) Biosynthesis of heme and chlorophylls. McGraw-Hill, New York

Dailey HA, Fleming JE, Harbin BM (1986) Ferrochelatase from *R. spheroides*; substrate-specificity and role of sulfhydryl and arginyl residues. J Bacteriol 165:1–5

Davidheiser RH, Figge FHJ (1955) Enzymatic porphyrin synthesis in Harderian glands and other organs and tissues of mice. Proc Soc Exp Biol Med 90:461–462

Davidheiser RH, Figge FHJ (1958) Comparison of porphyrin producing enzyme activities in Harderian glands of mice and other rodents. Proc Soc Exp Biol Med 97:775–777

Davis FA (1929) The anatomy and histology of the eye and orbit of the rabbit. Trans Am Opthalmol Soc 27:401–441

Derrien E, Turchini J (1924) Sur l'accumulation d'une porphyrine dans la glande de Harder des rongeurs du genre *Mus* et sur son mode d'excrétion. C R Soc Biol 91:637–639

DiIorio DP, Nadakavukaren MJ (1984) Prevention by pinealectomy of short photoperiod-induced ultrastructural changes in the hamster Harderian gland. Anat Rec 210:449–452

Elliot JA (1981) Circadian rhythms, entrainment and photoperiodism in the Syrian hamster. In: Follet BK, Follet DE (eds) Biological clocks in seasonal reproductive cycles. Scientechnica, Bristol, pp 203–218

Feria-Velasco A, Meléndez C, Jaramillo J (1983) Sex differences in secretion patterns of neonatal rat Harderian gland under various environmental lighting conditions. Comp Biochem Physiol 74B:709–714

Figge FHJ (1944) Fluorescence studies on cancer. I. Porphyrin metabolism, Harderian gland fluorescence, and susceptibility to carcinogenic agents. Cancer Res 4:465–471

Figge FHJ, Atkinson WB (1941) Relation of water metabolism to porphyrin incrustations in pantothenic acid-deficient rats. Proc Soc Exp Biol Med 48:112–114

Figge FHJ, Davidheiser RH (1957) Porphyrin synthesis by mouse Harderian gland extracts: sex, age, and strain variation. Proc Soc Exp Biol Med 96:437–439

Figge FHJ, Salomon K (1942) Prevention of porphyrin incrustations on pantothenic acid-deficient rats by Harderian gland ablation. J Lab Clin Med 27:1496–1501

Figge FHJ, Strong LC, Strong, Jr LC, Shanbrom A (1942) Fluorescent porphyrins in Harderian glands and susceptibility to spontaneous mammary carcinoma in mice. Cancer Res 2:335–342

Formgren B, Wetterberg L (1973) Uroporphyrinogen-III-cosynthetase activity in the Harderian gland of the golden hamster. Enzyme 16:146–150

Formgren B, Wetterberg L (1976) Effects of light of different wavelengths on the porphyrin concentration in the Harderian gland of the golden hamster. In: Doss M (ed) Porphyrins in Human diseases, 1st Int Porphyrin Meet Freiburg/Br 1975. S. Karger, Basel, pp 380–384

Getty R (ed) (1975) Sisson and Grossman's: The anatomy of domestic animals, vols 1, 2, 5th edn. WB Saunders, Philadelphia, pp 229–230, 1410–1411

Goldberg A, Ashenbrucker H, Cartwright GE, Wintrobe MM (1956) Studies on the biosynthesis of heme in vitro by avian erythrocytes. Blood 11:821–833

Grafflin AL (1942) Histological observations upon the porphyrin-excreting Harderian gland of the albino rat. Am J Anat 71:43–64

Granick S (1966) The induction in vitro of the synthesis of 5-aminolevulinic acid synthetase in chemical porphyria: a response to certain drugs, sex hormones and foreign chemicals. J Biol Chem 241:1359–1375

Gross SR, Hutton JJ (1971) Induction of hepatic d-amino-levulinic acid synthetase activity in strains of inbred mice. J Biol Chem 246:606–614

Hais IM, Strych A, Chmelar V (1968) Preliminary thin layer chromatographic characterization of the rat Harderian gland lipid. J Chromatogr 35:179–191

Harlow HJ (1984) The influence of Harderian gland removal and fur lipid removal on heat loss and water flux to and from the skin of muskrats, *Ondatra zibethicus*. Physiol Zool 57:349–356

Hoffman RA (1971) Influence of some endocrine glands, hormones and blinding on the histology of the Harderian glands of golden hamster. Am J Anat 132:463–478

Hoffman RA, Hester RJ, Towns C (1965) Effect of light and temperature on the endocrine system of the golden hamster (*Mesocricetus auratus*, Waterhouse). Comp Biochem Physiol 15:525–533

Hueper WC, Figge FHJ (1945) Porphyrin excretion of Harderian glands in its relation to actinic carcinogenesis in hairless rats. Cancer Res 5:328–330

Hugo J, Krijt J, Vokurka M, Janousek V (1987) Secretory response to light in rat Harderian gland: possible photoprotective role of Harderian porphyrin. Gen Physiol Biophys 6:401–404

Hugo J, Vokurka M, Sanitrák J, Melková Z, Janousek V (1989) first demonstration of porphyrins in the lactrimal glands: similarity to the Harderian gland. Biomed Biochem Acta 48:K1–K4

Hugo J, Vokurka M, Krijt J, Janousek V (1990) Photoprotection as a possible function of the rodent Harderian gland porphyrins: first description of porphyrins in the lacrimal gland. Mol Aspects Med 11:150

James KR, Cortes JM, Paradinas FJ (1980) Demonstration of intracytoplasmic needle-like inclusions in hepatocytes of patients with porphyria cutanea tarda. J Clin Pathol 33:899–900

Janousek V, Sanitrak J, Krijt J, Holub M (1990) Tissue porphyrins in athymic nude (nu/nu) mice. Mol Aspects Med 11:150–151

Johnston HS, McGadey J, Thompson GG, Moore MR, Payne AP (1983) The Harderian gland, its secretory duct and porphyrin content in the Mongolian gerbil (*Meriones unguiculatus*). J Anat 137:615–630

Johnston HS, McGadey J, Thompson GG, Moore MR, Breed WG, Payne AP (1985) The Harderian gland, its secretory duct and porphyrin content in the plains mouse (*Pseudomys australis*). J Anat 140:337–350

Johnston HS, McGadey J, Payne AP, Thompson GG, Moore MR (1987) The Harderian gland, its secretory duct and porphyrin content in the woodmouse (*Apodemus sylvaticus*). J Anat 153:17–30

Jones CW, Hoffman RA (1976) Porphyrin concentration of the hamster (*Mesocricetus auratus*) Harderian gland: effects of incubation with delta-aminolevulinic acid and various hormones. Int J Biochem 7:135–139

Joó I, Kahán A (1975) The porphyrin content of Harderian glands in rats and the melatonin-melanocyte stimulating hormone system. Endokrinologie 65:308–312

Kanwar KC (1960) Morphological and cytochemical studies on the Harderian glands of rats. Cellule 61:129–143

Kennedy GY (1954) The chemistry and metabolism of some porphyrins and porphyrin derivatives. PhD Thesis, University of Sheffield, England

Kennedy GY, Jackson AH, Kenner GW, Suckling CJ (1970) Isolation, structure and synthesis of a tricarboxylic porphyrin from the Harderian glands of the rat. FEBS Lett 6:9–12

Lin WL, Nadakavukaren MJ (1979) The androgenic effect on the fine structure of the Harderian gland in the male hamster. Cell Tissue Res 198:119–127

Lin WL, Nadakavukaren MJ (1982) Delta-aminolevulinate synthase activity in the Harderian glands of male and female golden hamsters. Comp Biochem Physiol 71B:539–540

Lithner F, Wetterberg L (1984) Hepatocellular carcinoma in patients with acute intermittent porphyria. Acta Med Scand 215:271–274

Loewenthal N (1892) Notiz über die Harder'sche Drüse des Igels. Anat Ang 7:48–54

Margolis FL (1971) Regulation of porphyrin biosynthesis in the Harderian gland of inbred mouse strains. Arch Biochem Biophys 145:373–381

Matilla A, Molland EA (1974) A light and electron microscopic study of the liver in case of erythrohepatic protoporphyria and in griseofulvin-induced porphyria in mice. J Clin Pathol 27:698–709

Mhâtre MC, van Jaarsevela AS, Reiter RJ (1988) Melatonin in the lacrimal gland, first demonstration and experimental manipulation. Biochem Biophys Res Commun 153:1186–1192

Mindegaard J (1976) Studies concerning porphyrin synthesis and the origin and ultrastructure of biological porphyrin-containing pigment granules (in Danish). PhD Thesis, Afdelingen Biokemi Ernaering, Dept Biochem Nutr, Danmarks Tekniske Hojskole

Moore MR, Thompson GG, Low RAL, Cumming RLC (1972) The perinatal development of haem biosynthesis in rats, guinea pigs and humans. Proc Biochem Soc 129:43

Moore MR, Thompson GG, Payne AP, McGadey J (1977) Cyclical oscillations in the activity of δ-aminolaevulinate synthase and porphyrin synthesis in the Harderian gland during the oestrous cycle of the golden hamster (*Mesocricetus auratus*). Biochem Soc Trans 5:1475–1478

Moore MR, McColl KEL, Goldberg A (1979) The porphyrias. Diabete Metab (Paris) 5:323–336

Moore MR, Thompson GG, Payne AP, McGadey J (1980) Seasonal variation in δ-aminolaevulinate synthase and porphyrin content in the Harderian gland of the female golden hamster (*Mesocricetus auratus*). Int J Biochem 12:501–504

Moore MR, Disler PB (1985) Chemistry and biochemistry of the porphyrins and porphyria. Clin Dermatol 3:7–23

Moore MR, McColl KEL, Rimington C, Goldberg A (1987) Disorders of porphyrin metabolism. Plenum, New York

Müller HB (1969) Die postnatale Entwicklung der Harderschen Drüse der weissen Ratte. I. Lichtmikroskopische Befunde. Z Zellforsch Mikrosk Anat 100:421–438

Nadakavukaren MJ, Lin WL (1983) Influence of short photoperiods on the ultrastructure of hamster Harderian glands. Cell Tissue Res 233:209–213

Neuberger A (1980) The regulation of chlorophyll and porphyrin biosynthesis. Int J Biochem 12:787–789

Ono T, Umeda M, Sugimura T (1956) Porphyrin metabolism in tumor bearing animals. Free porphyrin in liver, Harderian gland and urine and the effect thereon of toxohormone. Gann 47:171–180

Orten JM, Keller JM (1946) Dietary protein and protoporphyrin formation in the rat. J Biol Chem 165:163–167

O'Steen WK, Kraeer SL, Shear CR (1978) Extraocular muscle and Harderian gland degeneration and regeneration after exposure of rats to continuous fluorescent illumination. Invest Opthalmol Visual Sci 17:847–856

Payne AP (1977) Pheromonal effects of Harderian gland homogenates on aggressive behaviour in the hamster. J Endocrinol 73:191–192

Payne AP (1979) Attractant properties of the Harderian gland and its products on male golden hamsters of differing sexual experience. J Endocrinol 80:69–70

Payne AP (1990) The Harderian gland: its biology and role in porphyrin synthesis. Mol Aspects Med 11:145–152

Payne AP, McGadey J, Moore MR, Thompson G (1977a) Androgenic control of the Harderian gland in the male golden hamster. J Endocrinol 75:73–82
Payne AP, McGadey J, Moore MR, Thompson GG (1977b) Cyclic and seasonal changes in Harderian gland activity in the female golden hamster. J Endocrinol 72:41
Payne AP, McGadey J, Moore MR, Thompson GG (1978) The Harderian gland in the golden hamster: its histology, physiological control and behavioural role. J Anat 127(3):662
Payne AP, McGadey J, Moore MR, Thompson GG (1979) Changes in Harderian gland activity in the female golden hamster during the oestrous cycle, pregnancy and lactation. Biochem J 178:597–604
Payne AP, McGadey J, Johnston HS (1982a) The observation of interstitial porphyrin in the Harderian gland of normal and ovariectomised female hamsters. J Anat 135:831–832
Payne AP, McGadey J, Johnston HS, Moore MR, Thompson GG (1982b) Mast cells in the hamster Harderian gland: sex differences, hormonal control and relationship to porphyrin. J Anat 135:451–461
Payne AP, McGadey J, Johnston HS (1985) Interstitial porphyrins and tubule degeneration in the hamster Harderian gland. J Anat 140:25–36
Piper WN, Condie IW, Tephly TR (1973) The role of substrates for glycine acyltransferase in the reversal of chemically induced porphyria in the rat. Arch Biochem Biophys 159:671–677
Rapin M (1952) Apercus sur la glande de Harder du rat. Son dévelopment, sa secretion et son histologie. Bull Soc Zool Fr 77:239
Reiter RJ (1973) Comparative effects of continual lighting and pinealectomy on the eyes, the Harderian glands and reproduction in pigmented and albino rats. Comp Biochem Physiol 44:503–509
Reiter RJ, Klein DC (1971) Observations on the pineal gland, the Harderian glands, the retina and the reproductive organs of adult female rats exposed to continuous light. J Endocrinol 51:117–125
Rimington C (1965) Porphyrin metabolism In: The comparative physiology and pathology of the skin. Blackwell Scientific, Oxford, pp 309–316
Rimington C (1985) A review of the enzymic errors in the various porphyrias. Scand J Clin Lab Invest 45:291–301
Rohonyi B, Kelényi G (1962) Porphyrin content of the Harderian gland of the rat. Acta Biol Acad Sci Hung 13:241–245
Sakai T (1981) The mammalian Harderian gland: Morphology, biochemistry, function and phylogeny. Arch Histol Jpn 44:299–333
Shemin D (1976) Delta-aminolaevulinic dehydratase: structure, function and mechanism. Philos Trans R Soc Lond B273:109–115
Shirama K, Takeo Y, Shimizu K, Maekawa K (1977) Influence of environmental lighting and sex hormones on the development of Harderian gland in the rat. Annot Zool Jpn 50:70–80
Shirama K, Furuya T, Takeo Y, Shimizu K, Maekawa K (1979) Neural control of the Harderian gland in the rat (in Japanese). Zool Mag (Tokyo) 88:149–154
Shirama K, Furuya T, Takeo Y, Shimizu K, Maekawa K (1981a) Sex and strain differences in porphyrin content of the Harderian gland of mice. Zool Mag (Tokyo) 90:247–250
Shirama K, Furuya T, Takeo Y, Shimizu K, Maekawa K (1981b) Influences of some endocrine glands and of hormone replacement on porphyrins of the Harderian glands of mice. J Endocrinol 91:305–312
Shirama K, Kohda M, Kohda M, Hokano M (1987) Effects of lighting conditions and of hormone replacement on the level of porphyrins in the rat Harderian gland. J Endocrinol Invest 10:79–82
Sinclair PR, Granick S (1976) The transport of hemin and protoporphyrin across the plasma membrane of chick embryo liver cells in culture. Ann Clin Res 8:250–258
Smith AG, Francis JE (1981) Investigations of rat liver uroporphyrinogen decarboxylase – comparisons of porphyrinogens I and III as substrates and the inhibition by porphyrins. Biochem J 195:241–250
Smith KM (1975) Porphyrins and metalloporphyrins. Elsevier Scientific, Amsterdam
Spike RC (1986) The hamster Harderian gland: regulation of morphology and porphyrin biosynthesis by gonadal steroids. PhD Thesis Glasgow Univ, Glasgow, Scotland
Spike RC, Johnston HS, McGadey J, Moore MR, Thompson GG, Payne AP (1985) Quantitative studies on the effects of hormones on structure and porphyrin biosynthesis in the Harderian gland of the female golden hamster. I. The effects of ovariectomy and androgen administration. J Anat 142:59–72

Spike RC, Johnston HS, McGadey J, Moore MR, Thompson GG, Payne AP (1986a) Quantitative studies on the effects of hormones on structure and porphyrin biosynthesis in the Harderian gland of the female golden hamster. II. The time course of changes after ovariectomy. J Anat 145:67–77

Spike RC, McGadey J, Johnston HS, Payne AP, Thompson GG, Moore MR (1986b) The prevention of porphyrin loss from tissues during routine histological processing: Quantitative studies on the Harderian gland. Histochem J 18:245–250

Spike RC, Payne AP, Moore MR (1986c) The effects of castration and steroid administration on porphyrin biosynthesis in the Harderian gland of the male golden hamster. Biochem Soc Trans 14:140–141

Spike RC, Payne AP, Moore MR (1988) The effects of age on the structure and porphyrin synthesis of the Harderian gland of the female golden hamster. J Anat 160:157–166

Spike RC, Stewart SL, Murray K, Kelly FM, Maharaj JA, Payne AP, Moore MR (1990a) Porphyrin synthesis in the Harderian gland and other tissues of the golden hamster during pregnancy and lactation. Mol Aspects Med 11:151–152

Spike RC, Payne AP, Moore MR (1990b) Harderianectomy and porphyrin synthesis. Mol Aspects Med 11:151

Spike RC, Payne AP, Thompson GG, Moore MR (1990c) High-performance liquid chromatographic analysis of porphyrins in hamster Harderian glands. Biochem Biophys Acta 1034:1–3

Strong LC (1942) Sex differences in pigment content of Harderian glands of mice. Proc Soc Exp Biol Med 50:123–125

Strong LC (1944) Maternal inheritance of protoporphyrins of the Harderian glands in mice. Proc Soc Exp Biol Med 57:78–79

Strong LC, Figge FHJ (1941) Fluorescence of Harderian glands in mice of cancer-susceptible and cancer-resistant strains. Science 94:331

Strum JM, Shear CR (1982a) Harderian glands in mice: fluorescence, peroxidase activity and fine structure. Tissue Cell 14:135–148

Strum JM, Shear CR (1982b) Constant light exposure induces damage and squamous metaplasia in Harderian glands of albino mice. Tissue Cell 14:149–161

Sun CY, Nadakavukaren MJ (1980) Effect of testosterone on the female hamster Harderian gland pigmentation and ultrastructure. Cell Tissue Res 207:511–517

Thiessen DD, Kittrell EMW (1980) The Harderian gland and thermoregulation in the gerbil (*Meriones unguiculatus*). Physiol Behav 24:417–424

Thiessen DD, Clancy A, Goodwin M (1976) Harderian gland pheromone in the Mongolian gerbil (*Meriones unguiculatus*). J Chem Ecol 2:231–238

Thomas MJ (1938) Contributions à l'étude des porphyrines en biologie et en pathologie. I. Genèse du noyau tetrapyrrolique. Br Soc Chim Biol 20:471

Thompson GG, Hordovatzi X, Moore MR, McGadey J, Payne AP (1984) Sex differences in haem biosynthesis and porphyrin content in the Harderian gland of the golden hamster. Int J Biochem 16:849–852

Tomio JM, Grinstein M (1968) Porphyrin biosynthesis 5. Biosynthesis of protoporphyrin IX in Harderian glands. Eur J Biochem 6:80–83

Tomio JM, Tuzman V, Grinstein M (1968) 5-Aminolevulinate dehydratase from rat Harderian gland. Purification and properties. Euro J Biochem 6:84–87

Towbin EJ, Fanta PE, Hodge HC (1945) The porphyrin of Harder's gland. Proc Soc Exp Biol Med 60:228–231

Ulrich R, Yuwiler A, Geller E, Wetterberg L (1974) Effects of sex hormones and environmental lighting on rat Harderian gland porphyrin. J Endocrinol 63:99–102

Walls GL (1942) The vertebrate eye and its adaptive radiation. Cranbrook Inst Sci Bull, Bloomfield Hills, MI, 19

Wetterberg L (1972) Increase of Harderian gland porphyrin content in castrated male hamsters dependent on light and visual function. Life Sci 11:541–546

Wetterberg L, Yuwiler A, Geller E, Shapiro S (1970a) Harderian gland: development and influence of early hormonal treatment on porphyrin content. Science 168:996–998

Wetterberg L, Geller E, Yuwiler A (1970b) Harderian gland: an extraretinal photoreceptor influencing the pineal gland in neonatal rats? Science 167:884–885

Wetterberg L, Marver HS, Swanson AL (1971) Delta-aminolaevulinic acid synthetase in the Harderian gland. S Afr J Lab Clin Med (Spec Issue) 45:189–190

Wetterberg L, Ulrich R, Yuwiler A (1972a) Light, the Harderian gland and the rodent pineal. In: Proc 4th Int Cong Endocrinol, Washington DC. Am Elsevier, New York, pp 268–272

Wetterberg L, Nyström B, Nilsson B (1972b) Effect of lighting conditions on the porphyrin content and morphology of the Harderian gland and on uterine morphology in the golden hamster. Experientia 28:1084–1085

Woodhouse MA, Rhodin JAG (1963) The ultrastructure of the Harderian gland of the mouse with particular reference to the formation of its secretory product. J Ultrastr Res 9:76–98

Wooding FBP (1980) Lipid droplet secretion by the rabbit Harderian gland. J Ultrastruct Res 71:68–78

Woods JS, Dixon RL (1970) Perinatal differences in delta-aminolevulinic acid synthetase activity. Life Sci 9:711–719

Woolley GW, Worley J (1954) Sexual dimorphism in the Harderian gland of the hamster (*Cricetus auratus*). Anat Rec 118:416–417

Yamazaki T, Seyama Y, Otsuka H, Ogawa H, Yamakawa T (1981) Identification of alkyldiacylglycerols containing saturated methyl branched chains in the Harderian gland of guinea pig. J Biochem 89:683–691

Lipids in Harderian Glands and Their Significance

YOUSUKE SEYAMA[1], TAKESHI KASAMA[2], ETSUKO YASUGI[3], SUN-HEE PARK[1] and KAZUTAKA KANO[1]

[1] Department of Physiological Chemistry and Nutrition, Faculty of Medicine, The University of Tokyo, Bunkyo-ku, Tokyo, Japan
[2] Laboratory for Biomedical Analysis, Tokyo Medical and Dental University, Bunkyo-ku, Tokyo, Japan
[3] Division of Clinical Biochemistry, Clinical Research Center, National Medical Center, Shinjuku, Tokyo, Japan

1 Introduction

The Harderian gland, discovered by Harder (1694), is well developed in rodents (rat, mouse, golden hamster, Mongolian gerbil, guinea pig, etc.), lagomorphs (rabbit, pika) and cetaceans, and may have been lost in primates, terrestrial carnivores, and some other species. The Harderian gland occupies a considerable part of the orbit and is located around the posterior half of the eyeball. Histologically, a single layer of columnar epithelial cells surrounds the lumen, with the single excretory duct directly opening onto the eyelid. The surface of the gland is smooth and is covered with endothelium of the orbital venous sinus (Sakai and Yohro 1981).

One of the functions ascribed to this gland is the secretion of an oily substance which facilitates the movement of the eyelid (Cohn 1955; Kennedy 1970). The secretory material is a simple lipid which differs from species to species. The major lipid component of the rabbit Harderian gland is 1-alkyl-2, 3-diacylglycerols (Fig. 1, lane 1, major spot) containing esterified, hydroxy-substituted alkyl chains instead of saturated or unsaturated chains (Kasama et al. 1973; Rock and Snyder 1975a). The upper spot of this lane 1 in Fig. 1 is a mixture of 2-*O*-acylhydroxy fatty acid alkyl esters (Rock et al. 1976). The upper spot of the lipids in rat Harderian gland (Fig. 1, lane 3) is wax ester (Murawski and Jost 1974). 1-Alkyl-2, 3-diacylglycerol is also the main lipid component (Fig. 1, lane 5) of the mouse (Watanabe 1980) and golden hamster (Lin and Nadakavukaren 1981). The lipid from the Harderian gland of guinea pig is also 1-alkyl-2, 3-diacylglycerol (Fig. 1, lane 6), but containing a large amount of branched alkyl and acyl chains (Yamazaki et al. 1981). The excretory lipid of the Mongolian gerbil (*Meriones unguiculatus*) is 2,3-alkanediol diacyl esters (2,3-diol waxes) (Otsuru et al. 1983). These secretory lipids are classified as wax and glycerolipid with ether bonds, and account for approximately 20% of fresh tissue weight.

As the Harderian gland excretes a large amount of characteristic lipids, this gland contains the active components of enzymes related to lipid metabolism.

Webb et al. (Eds).
Harderian Glands

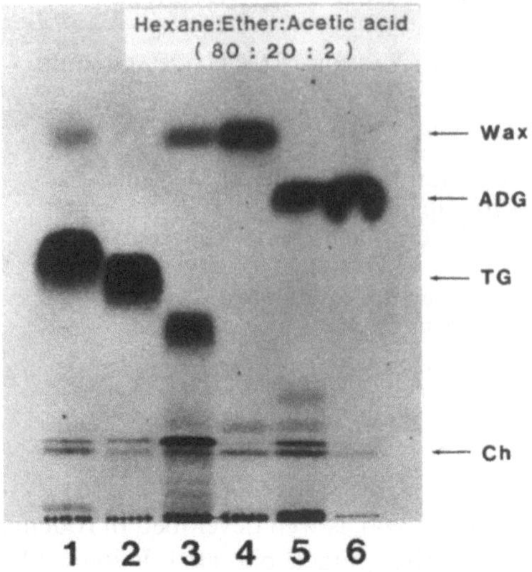

Fig. 1. Thin layer chromatogram of lipids in Harderian glands. Samples: lipids in the Harderian glands of rabbit (*lane 1*), rat (*lane 3*), squirrel (*lane 4*), mouse (*lane 5*), and guinea pig (*lane 6*), and human orbital fat (*lane 2*). Standards: *Wax*; *ADG* alkyldiacylglycerol; *TG* triacylglycerol; *Ch* cholesterol. (After Yamazaki et al. 1981)

Hence, this organ seems to be an appropriate model for investigating the mechanism of lipid metabolism. Fatty acid synthase, acyl-CoA ligase, and acyltransferase are examples of these extensively studied enzymes obtained from the Harderian gland.

2 Chemical Diversity of Lipids

2.1 Secretory Lipids

2.1.1 Rabbit

The Harderian gland of rabbit consists of two distinct lobes, an upper (white) and a lower (red, pink) portion (Björkman et al. 1960). Analyses of the lipids of each portion showed that the pink and white portions had the same components on thin layer chromatography (TLC) (Kahán et al. 1967); these authors did not refer to the secretory nature of the lipids. Precise histological studies revealed that the secretory cells in the tubular endpieces of both portions contained lipids (Kühnel 1971). The lipid droplets were stained with Sudan IV and Sudan Black B. Before release into the glandular lumen, the limiting membranes of the adjacent lipid droplets fused, forming large vacuoles. Secretory lipids of the pink portion

Table 1. Chemical structure of secretory lipids in Harderian glands of various animals

Animal	Secretory lipid	Structure
Rabbit (white portion)	2-*0*-Acylhydroxy fatty acid alkyl ester	(a)
(pink portion)	2-Acyl-1-(*0*-acylhydroxyalkyl)-3-(3-methylbutyryl)-*sn*-glycerol	(b)
Guinea pig	1-Alkyl-2,3-diacylglycerol	(c)
Mouse	1-Alkyl-2,3-diacylglycerol	(c)
Mouse tumor	1-Alkyl-2,3-diacylglycerol	(c)
	2-Acyl-1-alkyl-3-(2′,3′-diacylglycerol)glycerol	(d)
Rat	Fatty acid alkyl ester (wax)	(e)
Mongolian gerbil	(2*S*,3*R*)-2,3-Diacylalkanediol	(f)
Golden hamster	1-Alkyl-2,3-diacylglycerol	(c)

(a)

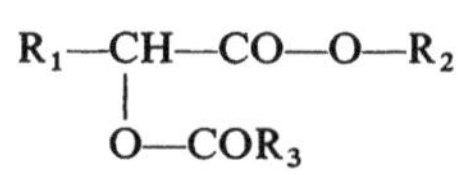

(b)

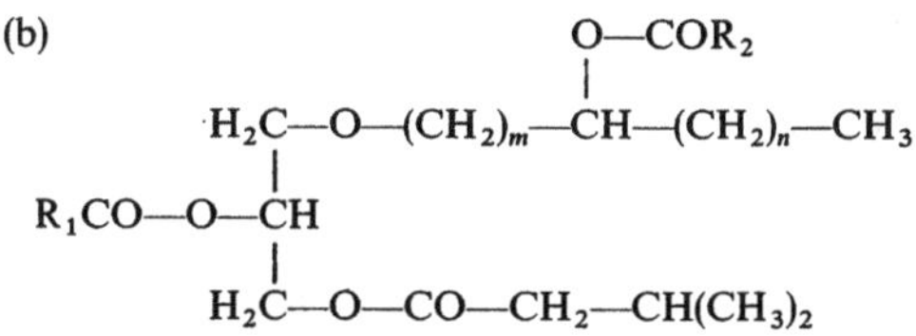

(c)

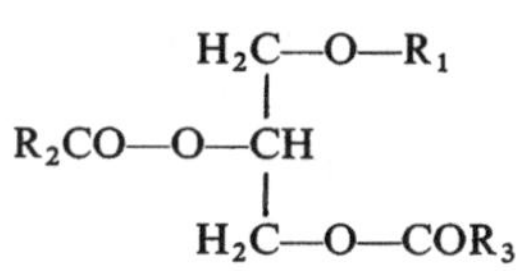

(d)

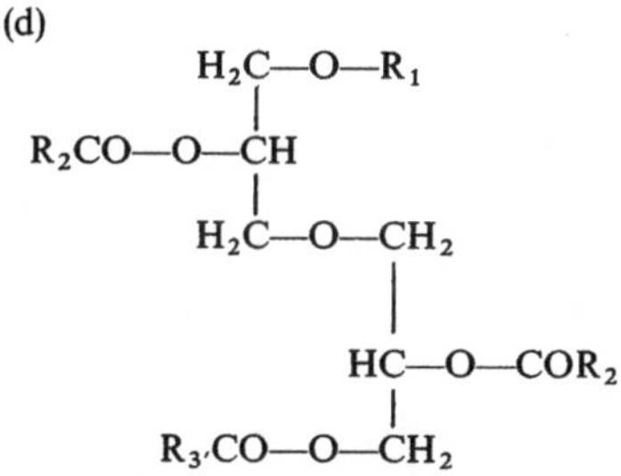

(e)

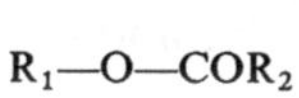

(f)

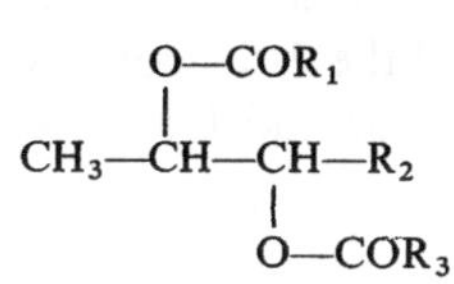

were analyzed and one of the major lipids proved to be an 1-alkyl-2, 3-diacylglycerol containing 1 mol of isovaleric acid (Blank et al. 1972, Table 1b). More than 50% of the *O*-alkyl moieties consisted of 16:0 and 18:0 carbon chains, whereas the other major *O*-alkyl moieties were 15:0 and 17:0 branched chain (~ 30%). The long chain acyl groups of the 1-alkyl-2, 3-diacylglycerol consisted primarily of saturated fatty acids (60% 16:0 and 30% 18:0) and a small amount of branched chain fatty acids (~ 5%), whereas the 3-position was occupied by isovaleric acid. The major secretory lipid of the pink portion was later purified and reduced by $LiAlH_4$. The esterified lipid yielded a hydroxyalkylglycerol (Kasama et al. 1973). Identification was based on chromatography, infrared (IR) and gas chromatography-mass spectrometry (GC-MS). The isomers of the hydroxyalkylglycerols consisted of *O*-hexadecyl moieties with hydroxy-groups at C10 and C11 and of *O*-octadecyl moieties with hydroxy-groups at C11 and C12.

Structure of the major lipid in the pink portion of the rabbit Harderian gland was incorrectly reported to be 1-alkyl-2, 3-diacylglycerol containing unsaturated alkyl chains (Jost 1974). The fatty acid composition of the lipid from the pink portion was mainly composed of C18:0 and C16:0. Rock and Snyder (1975a) demonstrated that the pink portion of the rabbit Harderian gland contained primarily of 1-(*O*-acyl) hydroxyalkyl-2, 3-diacyl-*sn*-glycerols and smaller amounts of 1-alkyl-2, 3-diacyl-*sn*-glycerols. They also indicated that the 1-(*O*-acyl) hydroxyalkyl-2, 3-diacyl-*sn*-glycerols and 1-hydroxyalkyl-2, 3-diacyl-*sn*-glycerols contain a short chain acyl moiety identified as 3-methylbutanoic acid (isovaleric acid). Two additional minor lipid components in this gland were identified to be 1-hydroxyalkyl-2-acyl-*sn*-glycerols and 1-hydroxyalkyl-2, 3-diacyl-*sn*-glycerols. The long chain acyl groups occurring in these two components were almost exclusively hexadecanoic acid, whereas the 1-(*O*-acyl)hydroxyalkyl-2, 3-diacyl-*sn*-glycerol has a ratio of hexadecanoic acid to octadecanoic acid of 2:1. They also suggested that the isovaleric acid at the 3-position may be incorporated from short chain acyl-CoA by microsomal 1-alkyl-2-acyl-*sn*-glycerol acyltransferase (Rock and Snyder 1975b).

Detailed studies on the lipid components in the white portion of the gland were done by Rock et al. (1976). The lipid compositions of the white portion of the rabbit Harderian gland differed from those of the pink portion. The major lipid of the white portion was 2-*O*-acylhydroxy fatty acid alkyl ester (Table 1a). The chain lengths of the aliphatic component of this lipid were determined as follows; fatty acids from C12:0 to to C22:0 including 65% of C16:0, 13.5% of C18:0, 8.9% of C22:0, 4.7% of C20:0, and 4.3% of C17:0. Fatty alcohols accounted for 68.8% of C22:0, 18.6% of C20:0, and 12.5% of C21:0. The chain lengths of the 2-hydroxy fatty acid were comprised of C14:0 to C16:0 in which C16:0, C14:0 and C15:0 were about 60%, 25%, and 14%, respectively.

2.1.2 Guinea Pig

The lipid components of the Harderian gland of guinea pig were first analyzed by Valeri et al. (1974). They reported the presence of triacylglycerol (3.43 mg/ 100 mg wet weight) among other neutral lipids, determined by TLC. They explained that the high content of triacylglycerol gave to this gland a characteristic common to lipid-storage organs. However, Yamazaki et al. (1981) rectified the incorrect results. They extracted lipids with chloroform-methanol and purified the major lipid by Iatrobeads column chromatography. The major lipid was identified to be 1-alkyl-2,3-diacylglycerol (Table 1c). More detailed analysis using TLC, gas liquid chromatography (GLC), GC-MS, IR and nuclear magnetic resonance (^{1}H-NMR) revealed that the alkyl and acyl moieties, 70 mol% of 1-alkyl moieties, 80 mol% of 2-acyl moieties and 40 mol% of 3-acyl moieties, were methyl-branched. Most of the branched chains had mono- and dimethyl branches. GC-MS study further revealed that the methyl branches occurred at 4, 8, 10, 14 and terminal anteiso positions. 8-Methyl-C16:0, 14-methyl-C16:0 (anteiso), and n-C17:0 aliphatic chains were the major alkyl chains. The 2-acyl moieties contained larger amounts of short and dimethyl-branched chain fatty acids than did those of the 3-acyl moieties. The physiological roles of the branched chain fatty acids in these lipids have not been clarified. However, it is conceivable that the branched fatty acids with a low melting temperature can replace the unsaturated fatty acids which are commonly observed in glycerolipids.

2.1.3 Rat

Hais et al. (1968) reported that a considerable proportion of ^{14}C-labeled acetate was incorporated into the major lipid fraction of the rat Harderian gland. About 83% of the radioactivity in the lipid extract migrated with the sterol ester band on TLC. The unsaponifiable material gave a single radioactive band which they considered to be a sterol ester. However, it was unfortunate that the wax and sterol esters cochromatographed in their solvent system (petroleum ether:ethyl ether:acetic acid, 90:10:1, v/v). Murawski and Jost (1974) identified the major lipid in the rat Harderian gland to be mixture of wax esters (Table 1e). They further reported the presence of (n-7) unsaturated alcohols and (n-7) monoenoic fatty acids in the wax ester by GC-MS. The major components are the C24:1(n-7) alcohol and C20:1(n-7) fatty acid. Zackor et al. (1976) reported that the major lipid in infant rats was 1-alkyl-2,3-diacylglycerol and wax ester was not detected. Bareggi et al. (1979a) independently suggested that the major lipid class of the rat Harderian gland was esterified cholesterol, which accounted for more than 40% of neutral lipids. Tvrzicka et al. (1988) separated the lipid of the rat Harderian gland using TLC. Wax ester as well as polar lipids were detected as the major components. GLC and GC-MS analysis showed the fatty acid composition in the wax. The acyl moieties of the wax predominantly consisted of C20:1(n-7) (86 mol%), C18:1(n-9) (12 mol%) C18:1(n-7) (1 mol%). From these results, they suggested that the position of a double bond was not only n-7 but also n-9 and n-5.

2.1.4 Mouse

The lipids in the mouse Harderian gland have been examined by biochemical, autoradiographic, and electron microscope techniques. Watanabe (1980) observed in the DD mouse that the radioactive carbon from [^{14}C] glucose injected into the abdominal cavity was incorporated into the acid insoluble fraction within 30 min. More than 90% of the incorporated radioactivity was detected in the lipid fraction 1h after injection. The total lipids of the Harderian gland were separated into five fractions on TLC. IR Spectroscopy of the samples from the TLC indicated that one of the five bands was glycerol ether ester (Table 1c). More than 60% of the incorporated radioactivity was detected in the glycerol ether ester.

Fatty acid compositions of the total lipids and the major lipid fraction in the Harderian gland of the ICR mouse were determined by Tvrzicka et al. (1988). The Harderian gland contained more than 20% (wet weight) of lipids. Neutral lipids occupied 55–57% of the total lipids. No difference in the lipid contents was observed between the sexes except that the male gland was slightly larger. The major nonpolar lipid in mice was 1-alkyl-2,3-diacylglycerol. Fatty acids of C20:1 (60 mol%), C20:0 (20 mol%), C22:0 (3.7 mol%) and C18:1 (2.9 mol%) were predominant. In contrast to those of the guinea pig Harderian gland, the content of methyl-branched fatty acids was much lower. They observed iso and anteiso acids, but fatty acids with branching in the center part of aliphatic chain, which were common in guinea pig lipids, were not evident.

2.1.5 Mouse Tumor

Spontaneous neoplasms of the Harderian gland have been detected incidentally in mice (Deringer 1959, 1962; Madison et al. 1968; Peters et al. 1972; Lindsay 1976; Krieg et al. 1978; Ward et al. 1979), and in Syrian hamsters (Krieg et al. 1978). The Harderian gland tumor can be readily induced by a single intraperitoneal injection of urethane (Deringer 1962). When the lipid components of the mouse Harderian gland tumor were analyzed (Itoh and Kasama 1970; Kasama et al. 1970; Kasama et al. 1989), a tumor in a female mouse of 129/Jax strain was detected and was maintained by transplantation into subcutaneous or intramuscular organs. Numerous large lipid vacuoles inside of the epithelial cells were well stained with Sudan III and Sudan Black B. The tumor contained a large amount of lipid (25.5% on wet weight tissue). The 1-alkyl-2,3-diacylglycerols were identified in the Harderian gland tumors of mice (59% of total lipid and 66% of neutral lipid) (Table 1c). The ether-bound side chain of the 1-alkyl-2,3-diacylglycerol was determined after formation of the isopropylidene derivative by GLC (C16:0 12.9%, C18:0 21.8%, C20:0 22.4%, C20:1 8.7%, and C22:1 22.9%). The ratio of saturated to unsaturated alcohol was 63:39. A chain length exceeding C20 was predominant and the content of selachyl alcohol was remarkably low. The fatty acid of 1-alkyl-2,3-diacylglycerol was also found to be predominant in the case of a chain length over C20 (C16:0 12.2%, C18:1 18%, C20:1 36.6%). The content of unsaturated fatty acids was 69%. The structure of the lipid in the mouse Harderian gland tumor was analyzed by Kasama et al. (1989). The transplanted tumor from a 129/Jax mouse contained 19.2% lipids (fresh tissue) and 88.9% of the total lipid was neutral. Nearly 40% of the neutral lipid was 1-alkyl-2,3-diacylglycerol (Table 1c) and 20.4% was 2-acyl-1-alkyl-3-(2′,3′-diacylglycerol) glycerol (Table 1d). The latter was scanty in the normal Harderian gland of mice. This was the first report suggesting the presence of bis-glyceryl ether lipid in mammals. Structural analysis of the tumor-specific lipid was performed with GC-MS, ^{1}H-NMR, IR, and chromatography, using synthesizsed 3-(1′-glycerol)-1-octadecylglycerol and 3-(2′,3′-dioctadecanoyl-glycerol)-1-octadecylglycerol as standards. The alkyl moieties of this unique lipid consisted of saturated aliphatic chains with a chain length of 14 to 20 carbon atoms. The alkyl and acyl moieties of 2-acyl-1-alkyl-3-(2′,3′-diacylglycerol) glycerol were similar to those of 1-alkyl-2,3-diacylglycerols present in the same tissue, except for the presence of monounsaturated alkyl-moieties in the latter. They also compared lipids from the normal Harderian gland and those from Harderian gland tumor in DBA mice; the results are summarized in Table 2. It is noteworthy that the chain length and ratio of saturated and unsaturated alkyl or acyl moieties in 1-alkyl-2,3-diacylglycerol and 2-acyl-1-alkyl-3-(2′,3′-diacylglycerol) glycerol of Harderian gland tumors showed similarities.

2.1.6 Golden Hamster

Histochemical observations revealed a distinct sexual dimorphism of the Harderian gland of golden hamsters (Bucana and Nadakavukaren 1972). When the lipid contents of the hamster Harderian gland were studied (Lin and Nadakavukaren 1981), TLC findings suggested the presence of 1-alkyl-2,3-diacylglycerol (Table 1c). Although the chloroform-methanol extractable lipids

Table. 2. Alkyl and acyl moieties in ether lipids of the normal Harderian gland and its tumor from DBA mice. (After Kasama et al. 1989)

	Normal gland		Tumor			
	Alkyldiacyl glycerols		Alkyldiacyl glycerols		1-Alkyl-2-acyl-3-(2′,3′ diacylglycerol)glycerols	
	Alkyl	Acyl	Alkyl	Acyl	Alkyl	Acyl
Major chain length	18:0	20:1	16:0	20:1	16:0	20:1
Weight % of each alkyl and acyl moieties	52	73	58	43	60	43
Saturated chains	94	15	87	22	100	22
Unsaturated chains	6	85	13	78	0	78

of male and female hamster were similar, the fatty acid compositions of these lipids showed significant sexual differences. The major fatty acids in glands of females were C16:0 and C18:1, which accounted for more than 70%, whereas the lipids in glands of males had a greater proportion of fatty acids with a chain length ranging from C10 to C15. Whether or not the fatty acid compositions of the 1-alkyl-2, 3-diacylglycerol differ between female and male organs is unknown. They also reported the absence of triacylglycerol and suggested that the major lipids were secretory rather than storage products.

2.1.7 Mongolian Gerbil

The histological structure of the Harderian gland of Mongolian gerbil was investigated by means of light and electron microscopy (Sakai and Yohro 1981). The glandular cells had numerous clear secretory vacuoles containing lipids.

The major lipid class in the Harderian gland of Mongolian gerbils was identified to be a wax-like compound, 2, 3-diacylalkenediol (Otsuru et al. 1983, Table 1f). Methanolysis of these lipids gave fatty acid methyl esters having a chain length ranging from C10 to C22. These were palmitic (29.5%), myristic (16.5%), behenic (9.8%), stearic (5.1%), arachidic (16.4%), capric (3.0%), and lauric acid (16.3%). The structure of the alcohol moiety was determined to be 2, 3-alkanediol with a carbon chain length from C12 to C22 by GC-MS of the TMS and isopropylidene derivatives. The predominant diol was 2, 3-hexadecanediol, which accounted for 52.3% of the total diols. Stereochemical analyses demonstrated that the five major 2, 3-alkanediol, C14-C22 (even carbon numbers), had the *erythro* configuration as determined by GC-MS and ^{13}C-NMR in comparison with synthetic materials. The absolute configurations of the C-2 and C-3 asymmetric centers were assigned as 2*S* and 3*R* (Otsuka et al. 1986). Differences in the 2, 3-alkanediol or fatty acid constituents between the sexes are unknown.

2.2 Nonsecretory Neutral Lipids

2.2.1 Rabbit

Mono-, di- and triacylglycerols and cholesterol were detected by TLC (Kahán et al. 1967). The major portion of cholesterol was nonesterified and a minor portion was esterified. They also reported that the white and pink portions had the same lipid composition on TLC. Later, Jost (1974) and Jost et al. (1974) analyzed nonesterified cholesterol (2% of total lipid in the white portion, 7.9% in the pink portion) and esterified cholesterol (2% in the white portion, 1.6% in the pink portion) by TLC and densitometry.

2.2.2 *Guinea Pig*

Valeri et al. (1974) characterized the lipids of the Harderian gland of the guinea pig using histochemical and biochemical techniques. They reported the presence of cholesterol (0.126% of tissue), triacylglycerol (3.43%), free fatty acids (0.30% as palmitic acid) and an unidentified material which was later identified as 1-alkyl-2,3-diacylglycerol by Yamazaki et al. (1981). Trace amounts of nonesterified cholesterol were also detected.

2.2.3 *Rat*

Small amounts of triacylglycerol, cholesterol, and free fatty acids were detected by TLC (Hais et al. 1968; Murawski and Jost 1974; Bareggi et al. 1979a; Tvrzicka et al. 1988). Hais et al. (1968) studied the incorporation of radioactivity from [^{14}C] labeled acetate into the rat Harderian gland lipid. Nonsecretory neutral lipids were detected as the nonesterified cholesterol, diacylglycerol and triacylglycerol on TLC. But only a small amount of nonesterified cholesterol was labeled in the autoradiogram. The fatty acid compositions of triacylglycerol were analyzed by Tvrzicka et al. (1988); 74.3% of C18:1, 12.9% of C16:0, 4.6% of C20:1, 4.3% of C16:1 and others (4.0%).

2.2.4 *Mouse*

The lipid compositions of mouse Harderian gland were analyzed by TLC (Bareggi et al. 1979a; Tvrzicka et al. 1988). Free fatty acids, nonesterified cholesterol, triacylglycerol, and esterified cholesterol were detected. Tvrzicka et al. (1988) analyzed the fatty acid composition of triacylglycerol; 37.1% of C18:1, 23.9% of C16:0, 14.9% of C20:1, 12.3% of C18:2, 6.8% of C16:1 and others (4.9%).

2.2.5 *Mouse Tumor*

The lipid compositions of mouse Harderian gland tumor were reported by Itoh and Kasama (1970), Kasama et al. (1970) and Kasama et al. (1989). Small amounts of triacylglycerol were detected as a minor neutral lipid on TLC. The fatty acid composition of triacylglycerol was 32.8% of C18:1, 21.3% of C16:1, 13.3% of C20:1, 12.5% of C18:2 and others (20.1%), which differed considerably from that of 1-alkyl-2, 3-diacylglycerols (secretory lipid). The C18:2 content was relatively higher in triacylglycerol than in 1-alkyl-2, 3-diacylglycerols.

2.2.6 *Golden Hamster*

Lin and Nadakavukaren (1981) detected small amounts of cholesterol and fatty acids as nonsecretory neutral lipids in the Harderian gland of golden hamster. But triacylglycerol and esterified cholesterol were not detected.

2.2.7 *Mongolian Gerbil*

The lipid compositions of the Harderian gland of Mongolian gerbil were analyzed by Otsuru et al. (1983). They detected cholesterol, triacylglycerol and other simple lipids on TLC.

2.3 Phospholipids

2.3.1 *Rabbit*

The compositions of phospholipid in the rabbit Harderian gland (white and pink portions) were examined using TLC (Kahán et al. 1967) or autoradiography (Jost et al. 1974). It was suggested that the major components of phospholipid were phosphatidylcholine and phosphatidylethanolamine. In addition to these diacylglycerophospholipids, Radominska-Pyrek et al. (1979) demonstrated the existence of glycerophospholipids with ether bonds. Less than 20% of the choline glycerophospholipids and 37% of ethanolamine glycerophospho lipids had ether bonds in the white and pink portions. In the ether-linked glycerophospholipids, both alkenylacyl and alkylacylglycerophospholipids were observed. Although the rabbit Harderian gland had a large amount of neutral lipids with ether bonds, ether types did not constitute a large part of the choline and ethanolamine glycerophospholipids.

2.3.2 *Guinea Pig*

Valeri et al. (1974) found a relatively high quantity of phospholipids in the guinea pig Harderian gland. As the secretory lipid of guinea pig Harderian gland contained a large amount of methyl branches (Yamazaki et al. 1981), Seyama et al. (1983, 1984) examined the fatty acid compositions of phosphatidylethanolamine, phosphatidylcholine, and cardiolipin. Oleic acid proved to be the major component of these three phospholipids, but a large amount of saturated branched fatty acids was also found: 34.5% for phosphatidylethanolamine, 42.9% for phosphatidylcholine, and 72% for cardiolipin. On the other hand, polyunsaturated fatty acids such as linoleic and arachidonic acids were not detected. These characteristics were found only in the Harderian gland, but not in the liver of the same animal. Sphingomyelin from the guinea pig Harderian gland was isolated and characterized (Yasugi et al. 1990). The purified sphingomyelin gave a broad spot on TLC. The fatty acid composition was 71% nonhydroxy acids and 29% 2-hydroxy acids. Methyl-branched fatty acids were only 2% of the total acids. The long chain bases were composed of straight-chain sphingenines (50%) and sphiganines (6%). Methyl-branched long chain bases were 44% of the bases. The methyl-branched fatty acids of sphingomyelin were only a minor component in contrast to those of glycerophospholipids.

2.3.3 *Rat*

Phospholipids in the rat Harderian gland were investigated by autoradiography (Hais et al. 1968) and TLC (Bareggi et al. 1979b). The phospholipids in male rats were higher than in those of females (Bareggi et al. 1979b; Tvrzicka et al. 1988). Tvrzicka et al. (1988) carefully examined the compositions of phospholipid: in phosphatidylcholine and phosphatidylethanolamine, the C16:0, C16:1, C18:1 predominated and at least two or three positional isomers of monoenoic fatty acids were always present.

2.3.4 *Mouse*

Phospholipids in the mouse Harderian gland were analyzed using TLC (Bareggi et al. 1979b) or autoradiography (Watanabe 1980). These workers considered the phosphatidylcholine and phosphatidylethanolamine to be the major components. Contrary to findings in the rat, the phospholipid contents in males were less than those of females (Bareggi et al. 1979b; Tvrzicka et al. 1988). Oleic acid predominated in both phosphatidylethanolamine and phosphatidylcholine, and a large amount of palmitic acid was also present in the phosphatidylcholine (Tvrzicka et al. 1988). Kasama et al. (1989) compared the fatty acid compositions of the phospholipids in the normal Harderian gland and in tumors, and found considerable differences. The total amounts of saturated acyl chain of phospholipids from the tumor were much higher than those of the normal Harderian gland.

2.3.5 *Golden Hamster*

Phospholipids were detected in the golden hamster Harderian gland by TLC (Lin and Nadakavukaren 1981). Most of these phospholipids migrated as phosphatidylcholine and a minor fraction migrated as phosphatidylethanolamine. Lipid extracts from Harderian glands of both sexes showed identical compositions.

2.4 Glycolipids

2.4.1 *Guinea Pig*

Yasugi et al. (1987, 1988) identified cerebrosides in the guinea pig Harderian gland and found that they contained a large amount of methyl-branched sphinganines. The cerebrosides consisted of 75% of glucosylceramide and 25% of galactosylceramide. The fatty acid composition of these cerebrosides was 0.7% of nonhydroxy fatty acids and 99.3% of 2-hydroxy fatty acids. Among these 2-hydroxy fatty acids, a small amount of methyl-branched acids was detected. The long chain bases were composed of sphinganines (78%) and sphingenines (22%). The most remarkable feature of the long chain bases of cerebrosides in the Harderian gland was the presence of a large amount of methyl-branched sphinganine. The structures of methyl-branched bases were analyzed by GLC and GC-MS after conversion to aldehydes, alcohols, and fatty acids. The predominant branched long chain bases were found to be sphinganines with methyl substitutions at carbon numbers 8, 9, or 10, of which the last is dominant. These branched long chain bases, in which the substituted positions exist in the middle part of aliphatic chain, are novel long chain bases in mammals.

2.4.2 Rat and Mouse

Bareggi et al. (1979c), using TLC, proposed the existence of glycolipids, such as ceramide monohexoside, ceramide dihexoside, ceramide trihexoside and sulphatide, in Harderian glands of rats and mice.

3 Enzymes Related to Lipid Synthesis

The Harderian gland is not an adipose tissue like the orbital fat of human, but is a sort of sebaceous gland which secretes large amounts of lipids, the chemical structures of which are species-specific. In most cases, the lipids belong to a category of 1-alkyl-2,3-diacylglycerol as described in the previous sections. Consequently, most investigations of lipid metabolism in Harderian glands have been concentrated on the enzymes related to the ether lipid synthesis. As shown in Fig. 2, the pathway to produce alkyldiacylglycerol is quite different from that to produce triacylglycerol. In both pathways, the fatty acid synthase is a key enzyme to supply acyl-CoA as an acyl donor. As the Harderian gland contains a relatively high amount of fatty acid synthase, it is thought to be a good model organ to study the characters of this enzyme.

3.1 Fatty Acid Synthase

Fatty acids, the common components of lipids, are synthesized in the cytosol fraction by the two enzymes, acetyl-CoA carboxylase (Fig.2, 1) and fatty acid synthase (FAS; Fig. 2, 2). Acetyl-CoA carboxylase catalyzes the carboxylation of acetyl-CoA to yield malonyl-CoA, the activated donor of two-carbon units for the elongation of fatty acids catalyzed by fatty acid synthase. The rate of fatty acid production is regulated by the acetyl-CoA carboxylase step, according to nutritional, hormonal, developmental, and genetic conditions (Numa and Yamashita 1974). There seems to be no documentation on this enzyme from the Harderian gland, while several investigations of fatty acid synthase have been made. Usually the major product is palmitic acid, but several other fatty acids,

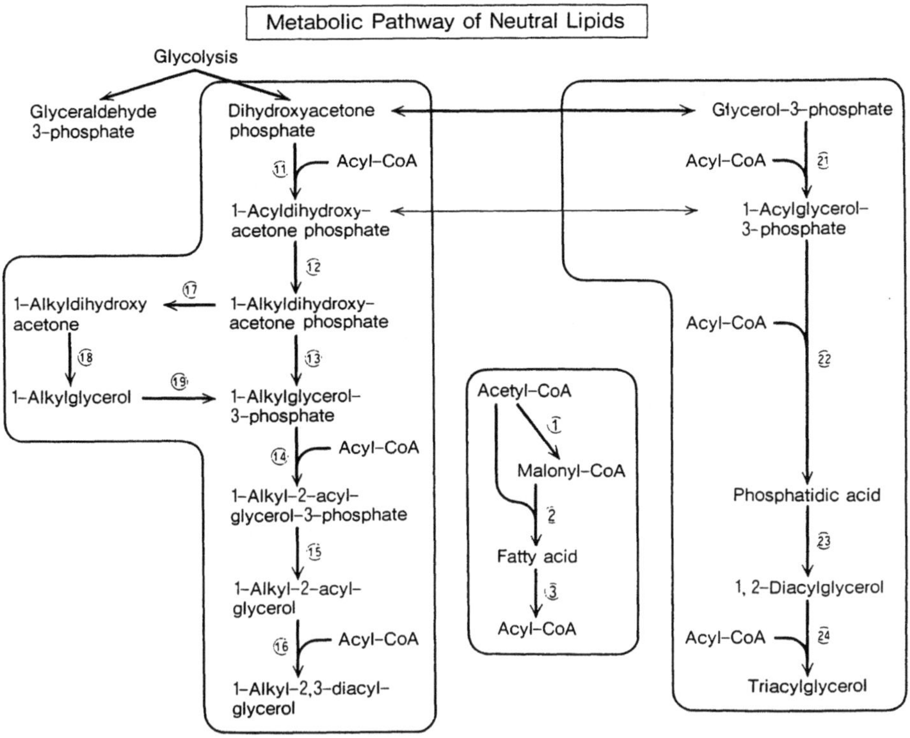

Fig. 2. Metabolic pathway of neutral lipids. *1* Acetyl-CoA carboxylase (EC 6.4.1.2). *2* Fatty acid synthase (FAS) (enzyme complex). *3* Acyl-CoA synthase (EC 6.2.1.3). *11* Dihydroxyacetone phosphate acyltransferase (DHAP AT) (EC 2.3.1.42). *12* 1-Alkyldihydroxyacetone phosphate synthase (EC 2.5.1.26). *13* 1-Alkyldihydroxyacetone phosphate reductase. *14* 1-Alkylglycerol-3-phosphate acyltransferase (Alkyl-G3P AT). *15* 1-Alkyl-2-acylglycerol-3-phosphate phosphatase. *16* 1-Alkyl-2-acylglycerol acyltransferase. *17* 1-Alkyldihydroxyacetone phosphate phosphatase. *18* 1-Alkyldihydroxyacetone oxidoreductase. *19* 1-Alkylglycerol phosphotransferase (EC 2.7.1.93). *21* Glycerol-3-phosphate acyltransferase (G3P AT) (EC 2.3.1.15). *22* 1-Acylglycerol-3-phosphate acyltransferase (acyl-G3P AT) (EC 2.3.1.51). *23* Phosphatidic acid phosphatase (EC 3.1.3.4). *24* Diacylglycerol acyltransferase (DG AT) (EC 2.3.1.20)

including methyl-branched fatty acids are also produced. The composition of the molecular species of fatty acids is determined at the step of fatty acid synthase. The regulatory mechanism of the determination of the products is a main interest of lipid biochemists. The lipids of the guinea pig Harderian gland contain a large amount of branched chain fatty acids (Yamazaki et al. 1981); thus, the fatty acid synthase from this organ functions under regulations which produce straight chain fatty acids together with branched ones. We purified a fatty acid synthase from this gland (Seyama et al. 1981).

The purified fatty acid synthase of the guinea pig Harderian gland is a homodimer consisting of two subunits of $Mr = 2.5 \times 10^5$ (Kitamoto et al. 1985).

Electron micrograph examinations revealed an ellipsoidal contour with a continuous middle cleft along the major axis. The subunit has a rod-like structure of about 220 Å long and 50 Å wide, and the dimension of the dimer is 220 × 150 × 7 Å (Kitamoto et al. 1988). Two monomers lie together in an antiparallel fashion (Fig. 3). There is a groove between the two subunits, giving the appearance of the Greek letter theta. The two subunits are in strong contact at their centers. The outer two attachment sites are not often in contact, thereby indicating a certain flexibility of subunits at their ends. Limited proteolysis by *Staphylococcus aureus* V8 protease produced four fragments (120, 70, 35, 30 *k*Da), aligned in this order from the NH_2 terminus. Using fragment-specific antibodies, three subregions of the subunit were determined. Treatment of this enzyme with phenylmethanesulfonyl fluoride indicated a negative functional interference between two active centers of the native enzyme (Kyushiki and Ikai 1989).

The fatty acids synthesized by the crude or purified fatty acid synthase of the guinea pig Harderian gland were analyzed using mass fragmentography (Seyama et al. 1981; Kitamoto et al. 1985). This enzyme had an inherent capacity to utilize methylmalonyl-CoA and to synthesize methyl-branched fatty acids. These fatty acids are characteristic components of the lipid secreted from this gland (Yamazaki et al. 1981). In the presence of an appropriate concentration

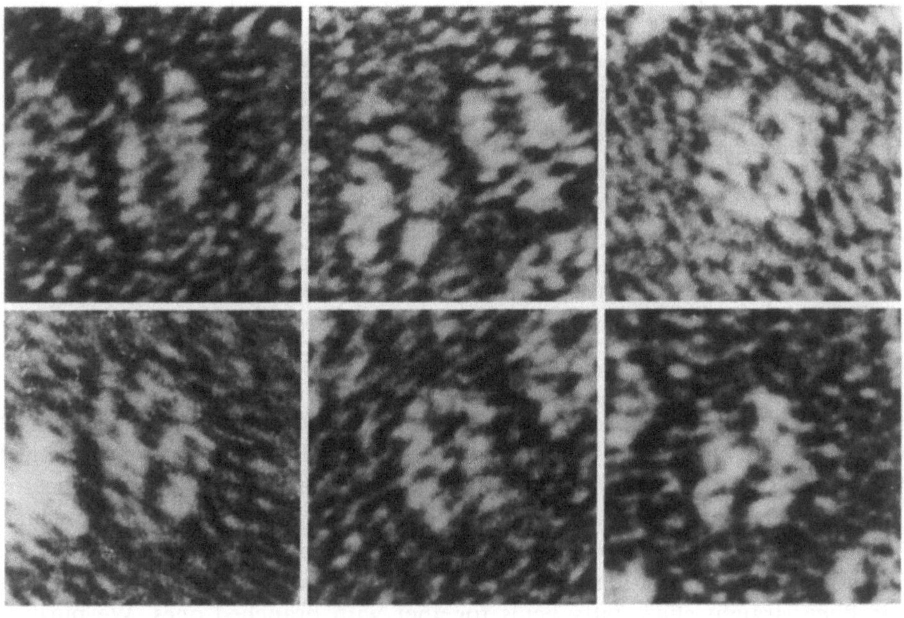

Fig. 3. Electron micrographs of Harderian gland fatty acid synthase. The native enzyme negatively stained by methylamine tungstate. *Scale bars* 500 Å. (Data from Kitamoto et al. 1985)

of acetyl-CoA, propionyl-CoA, malonyl-CoA, and methylmalonyl-CoA, the enzyme can produce fatty acids needed for the synthesis of the 1-*O*-alkyl-2,3-diacylglycerol, especially for the acylation at the C-2 position. Under standard incubation conditions with acetyl-CoA and malonyl-CoA as substrates, a large amount of palmitic acid together with small amounts of myristic and stearic acids were synthesized (Fig. 4a). When methylmalonyl-CoA was added together with malonyl-CoA, more than 19 different fatty acids were produced (Fig. 4b). The effect of the methylmalonyl-CoA concentration on this phenomenon was examined by changing the amount of methylmalonyl-CoA relative to malonyl-CoA (Fig. 5). Production of even-numbered fatty acids was strongly inhibited by methylmalonyl-CoA, but formation of odd-numbered and methyl-branched fatty acids increased to some extent with increasing amounts of methylmalonyl-CoA. Thus, the presence of methylmalonyl-CoA regulates the composition of fatty acids produced by the fatty acid synthase from the Harderian gland, as was seen in the case of the uropygial gland of birds (Buckner et al. 1978). This characteristic was first thought to be specific to the enzyme from the Harderian gland, based on the finding that the fatty acid synthase from the liver of the same animal could not utilize methylmalonyl-CoA (Seyama et al. 1981). But later, Kolattukudy et al. (1987) found that the partially purified liver enzyme also utilized methylmalonyl-CoA to produce branched fatty acids and concluded that the production of methyl-branched acids only by the Harderian gland is not due to any unique specificity of the fatty acid synthase of this gland.

Two hydrogen atoms bound to odd-numbered carbons were incorporated from the reduced pyridine nucleotide (NADPH), in two steps. Gas chromatography-mass spectrometric analysis revealed that the β-ketoacyl reductase in the fatty acid synthase from the guinea pig Harderian gland utilizes pro-4*S* hydrogen of NADPH and the enoyl reducatase uses the pro-4*R* hydrogen (Seyama et al. 1981). These results are similar to those observed with the fatty acid synthase from the rat liver (Seyama et al. 1978).

3.2 Enzymes in the Ether-Lipid Synthesis

Harderian glands of several animals secrete lipids, which are classified as ether-lipids, 1-alkyl-2,3-diacylglycerols. These lipids are synthesized from dihydroxyacetone phosphate (DHAP) by sequential actions of several enzymes such as dihydroxyacetone phosphate acyltransferase (DHAP AT), alkyl-DHAP synthase, alkyl-DHAP reductase, 1-alkyl-G3P acyltransferase, 1-alkyl-2-acyl-G3P phosphatase, and 1-alkyl-2-acyl-*sn*-glycerol acyltransferase (Fig. 2). Rock and Snyder (1974a) described an enzyme, 1-alkyl-*sn*-glycerol phosphotransferase, which enabled 1-alkyl-*sn*-glycerol to enter the pathway that forms 1-alkyl-2,3-diacylglycerol. Another key enzyme producing phospholipids is glycerol-3-phosphate acyltransferase (G3P AT), which is also present in Harderian glands.

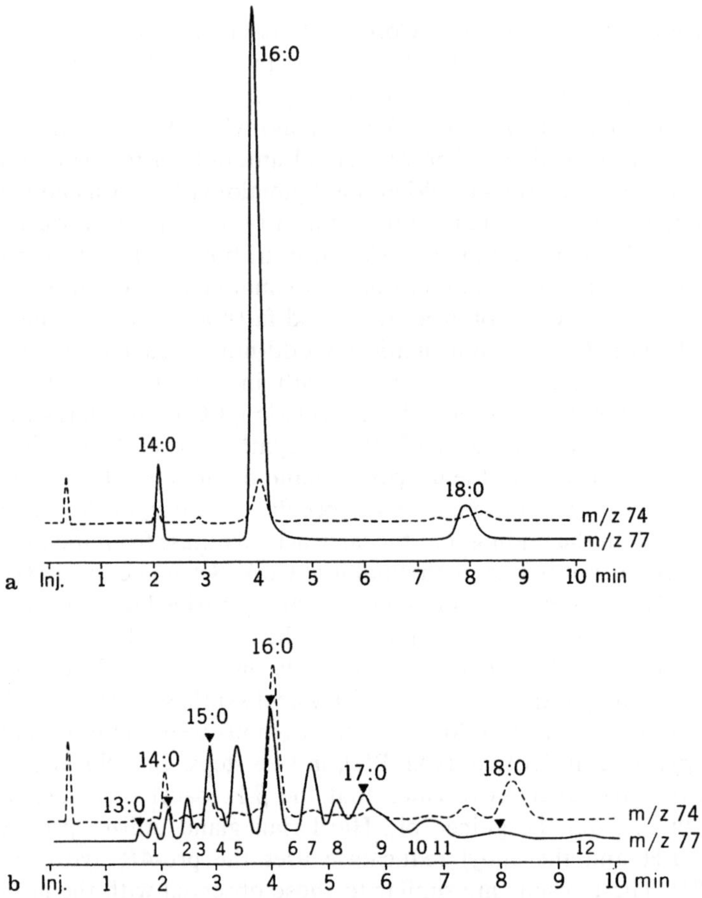

Fig. 4a,b. Fatty acids produced by fatty acid synthase from guinea pig Harderian gland. The substrates used in experiments **a** and **b** were as follows: **a** acetyl-CoA and malonyl-CoA, **b** acetyl-CoA, malonyl-CoA, and methylmalonyl-CoA. Monitoring at m/z 77 ion (——) indicates newly synthesized fatty acids, and the m/z 74 tracing (– – –) shows the endogenous fatty acids in the enzyme preparations. Identification of each peak is shown in both **a** and **b**. In **b**, together with six kinds of straight chain fatty acids (C13:0 to C18:0), at least 12 methyl-branched fatty acids were detected as follows: *1* 4-methyl-C13:0; *2* 4-methyl-C14:0; *3* 12-methyl-C14:0; *4* 4,13-dimethyl-C15:0; *5* 8-methyl-C15:0; *6* 4,14-dimethyl-C16:0; *7* 4-methyl-C16:0; *8* 14-methyl-C16:0; *9* 6,15-dimethyl-C17:0 or 8,15-dimethyl-C17:0; *10* 8-methyl-C17:0; *11* 4-methyl-C17:0; *12* 4-methyl-C18:0. (Data from Seyama et al. 1981)

3.2.1 Dihydroxyacetone Phosphate Acyltransferase

The first extensive study on dihydroxyacetone phosphate acyltransferase (DHAP AT; Fig. 2, 11) was performed by Rock et al. (1977a), who used microsomal preparations from the pink portion of the rabbit Harderian gland. The addition of the detergents, deoxycholate and cholate, stimulated the enzymatic activity,

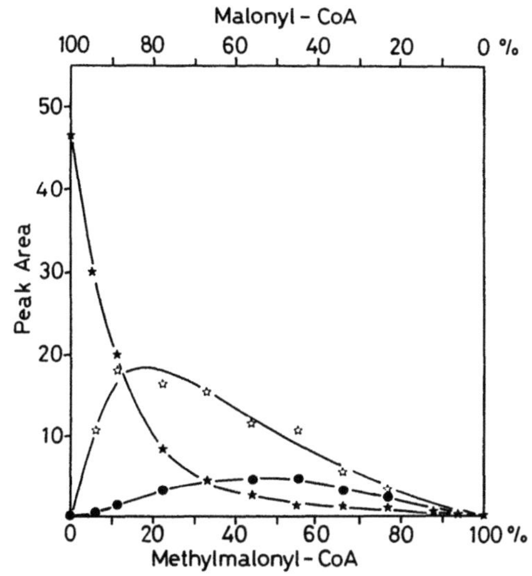

Fig. 5. Effect of methylmalonyl-CoA concentration on fatty acid synthesis. Peak areas (arbitrary unit) of even-numbered fatty acids (★), odd-numbered fatty acids (☆), and methyl-branched fatty acids (●) were plotted against the concentration ratio of malonyl-CoA and methylmalonyl-CoA. (Data from Seyama et al. 1981)

though the enzyme was not solubilized. They suggested that the active site of this enzyme is exposed to the luminal surface of microsomal vesicles. The results of protease digestion, treatment by a sulfhydryl reagent and *sn*-glycerol-3-P, suggested that this enzyme is distinct from the glycerol-3-phosphate acyltransferase (G3P AT; Fig. 2, 15). As the glycerol-phosphate did not inhibit DHAP AT activity, Kume et al. (1987) thought that the G3P AT is an enzyme different from DHAP AT. Schlossman and Bell (1976) reported that the acylation of glycerol-3-phosphate and dihydroxyacetone-phosphate were dual catalytic functions of the same enzyme in case of adipose cells. The cellular localization of this enzyme in the guinea pig Harderian gland was studied by Horie and Suga (1989); they noted the existence of high activities of DHAP AT in peroxisomes. They also suggested that DHAP AT and G3P AT are different enzymes in the Harderian gland.

3.2.2 Alkyl-DHAP Synthase

The acyl residue of acyldihydroxyacetone phosphate (acyl-DHAP) is replaced by the alkyl portion of the long-chain alcohol to produce alkyldihydroxyacetone phosphate (alkyl-DHAP). This reaction catalyzed by alkyl-DHAP synthase (Fig. 2, 12) was investigated by Rock et al. (1977b). As an enzyme source, they used microsomes of the pink portion of the rabbit Harderian gland. Treatment of the microsomes with detergents under critical micellar concentrations, or trypsin digestion, revealed that the catalytic portion of this enzyme is exposed to the luminal surface of the microsomal vesicles. They proposed a localization of the ether-lipid biosynthetic pathway in the lumen of the microsomes. This enzyme was also found in the Harderian gland of guinea pig (Horie and Suga 1989).

3.2.3 Alkylacylglycerol Acyltransferase

1-Alkyl-2,3-diacylglycerol is synthesized by the acylation of 1-alkyl-2-acylglycerol at the 3-position. 1-Alkyl-2-acylglyerol acyltransferase (Fig. 2, 16) in the microsome catalyzes this reaction. Substrate specificities of the enzyme from rabbit Harderian gland were examined by Rock and Snyder (1975b). The enzyme is most active in the presence of CoA esters with chain lengths similar to isovaleric acid and inactive in the presence of acetyl-CoA and long-chain acyl-CoA's. This would account for the evidence that the 3-position of alkyldiacylglycerol produced in the pink portion of this gland is occupied exclusively by isovaleric acid. As the 3-position of alkyldiacylglyerol of other sources contains long chain fatty acids, this substrate specificity is characteristic to the enzyme only from the pink portion of rabbit Harderian gland.

3.2.4 Alkylglycerol Phosphotransferase

The presence of 1-alkyl-*sn*-glycerol phosphotransferase (Fig. 2 ,19) enables the 1-alkyl-*sn*-glycerol to enter pathways leading to alkyldiacylglycerol, via 1-alkyl-glycerol 3-phosphate. The pink portion of the rabbit Harderian gland is a rich source of this enzyme (Rock and Snyder 1974), which is microsomal and requires ATP and Mg^{2+} for functions.

3.2.5 Alkylglycerol-3-Phosphate Acyltransferase

When microsomes of the pink portion of rabbit Harderian gland were incubated with alkylglycerol, ATP and CoA, 1-alkyl-2-acylglycerol-3-phosphate was produced (Rock and Snyder 1974). This finding suggested the presence in microsomes of 1-alkylglycerol-3-phosphate acyltransferase (Alkyl-G3P AT; Fig. 2, 14) together with acyl-CoA ligase.

4 Significance of Lipids in the Harderian Gland

Lipids, excreted from the Harderian glands, act as lubricants (Cohn 1955; Lin and Nadakavukaren 1981) and probably as the solvent for pheromones or other biologically active substances (Sakai 1981; Sakai and Yohro 1981) (Fig. 6). From the Harderian gland of guinea pig, Yokoyama et al. (1989) purified a novel

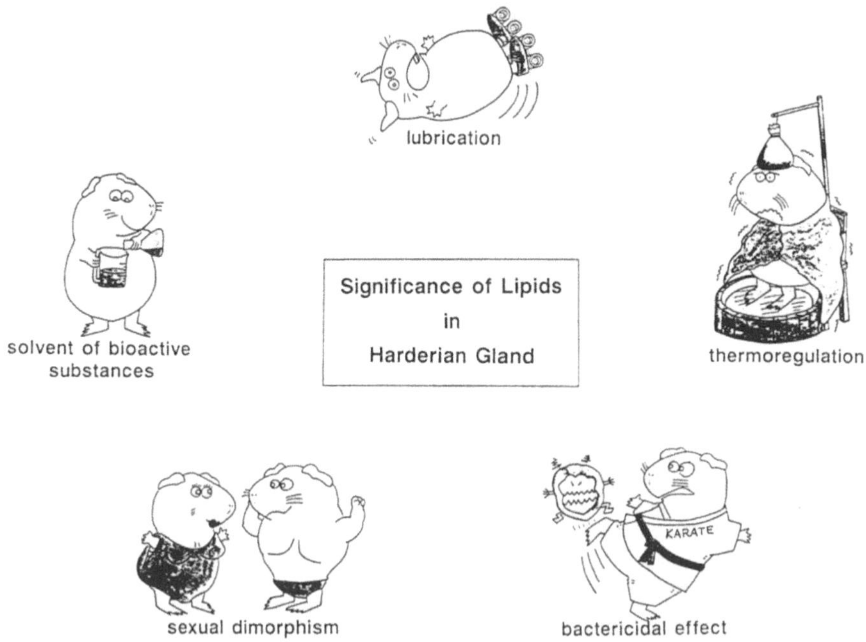

Fig. 6. Significance of lipids in Harderian gland

growth factor, which also dissolves in the excretory lipids. Excretory lipids also serve a thermoregulation function in gerbils (Thiessen et al. 1977; Thiessen 1988). The lipids spread on the pelage act to insulate the animal against cold and wetness, and the lipids and pigments (porphyrins) darken the pelage and increase radiant absorption. Sandbathing removes excessive amounts of pelage lipids, and lowers the body temperature. Also, removal of Harderian gland or fur lipids can result in an increased heat loss as well as water flux to and from the skin (Harlow 1984). Removal of Harderian materials from the pelage by sandbathing aids osmoregulation through lowering rate of insensible perspiration (Harriman and Thiessen 1983). Water and fat content of Harderian glands was determined in the exophthalmic animals (Smelser 1943). An increase in water content was observed but no change occurred in the lipid content. In a transient exophthalmos induced by exposure to high-intensity illumination, the lipid in the Harderian gland decreased significantly while the water content increased in accordance with the weight increase of this gland (Johnson et al. 1979). In toads, Harderian gland of only females contains numerous lipid droplets (Minucci et al. 1989). This finding suggests a pheromonal function, which may influence the sexual behavior of the male. Sexual dimorphism of fatty acids in the golden hamster Harderian gland was reported by Lin and Nadakavukaren (1981). In this respect, this gland could offer a model system to study the role of sex hormones involved in lipid metabolilsm. A bactericidal effect has been attributed to the unsaturated glycerin ethers present in the rabbit Harderian gland (Kühnel 1971). Although the function of the Harderian gland itself is uncertain, it is closely related to the nature of its lipid secretions.

Acknowledgement. We express our deep appreciation to our colleages, Yamazaki, T., Otsuka, H., Ogawa, H., Yamakawa, T., Kawaguchi, A., Ohashi, K., Imamura, T., Otsuru, O., Kitamoto, T., Nishigai, M., Ikai, A., Kume, K. and Shimizu, T., who provided original data included in our original papers. We would like to thank Ohara, M. for critical comments on the manuscript.

References

Bareggi R, Crescenzi A, Narducci-Bareggi P (1979a) Lipids in the Harder's gland of certain rodents. I. Neutral lipids. Basic Appl Histochem 23:155–160

Bareggi R, Crescenzi A, Narducci-Bareggi P (1979b) Lipids in the Harder's gland of certain rodents. II. Phospholipids. Basic Appl Histochem 23:161–163

Bareggi R, Narducci-Bareggi P, Crescenzi A (1979c) Lipids in the Harder's gland of certain rodents. III. Glycolipids. Basic Appl Histochem 23:165–169

Bjöorkman N, Nicander L, Schantz B (1960) On the histology and ultrastructure of the Harderian gland in rabbits. Z Zellforsch Mikrosk Anat 52:93–104

Blank ML, Kasama K, Snyder F (1972) Isolation and identification of an alkyldiacylglycerol containing isovaleric acid. J Lipid Res 13:390–395

Bucana CD, Nadakavukaren MJ (1972) Fine structure of the hamster Harderian gland. Z Zellforsh Mikrosk Anat 129:178–187

Buckner JS, Kolattukudy PE, Rogers L (1978) Synthesis of multimethyl-branched fatty acids by avian and mammalian fatty acid synthetase and its regulation by malonyl-CoA decarboxylase in the uropygial gland. Arch Biochem Biophys 186:152–163

Cohn SA (1955) Histochemical observations on the Harderian gland of the albino mouse. J Histochem Cytochem 3:342–353

Deringer MK (1959) Occurrence of tumors, particularly mammary tumors, in agent-free strain C3HeB mice. J Natl Cancer Inst 22:995–1002

Deringer MK (1962) Development of tumors, especially mammary tumors, in agent-free strain DBA/2eB mice. J Natl Cancer Inst 28:203–210

Hais IM, Štrych A, Chmelař V (1968) Preliminary thin-layer chromatographic characterization of the rat Harderian gland lipid. J Chromatogr 35:179–191

Harder JJ (1694) Grandula nova lachrymalis una cum ductuexcretorio in cervis et damis. Acta Erudit Publ Lipsiae 49–53

Harlow HJ (1984) The influence of Harderian gland removal and fur lipid removal on heat loss and water flux to and from the skin of muskrats, *Ondotra zibethicus*. Physiol Zool 57:349–356

Harriman AE, Thiessen D (1983) Removal of Harderian exudates by sandbathing contributes to osmotic balance in Mongolian gerbils. Physiol Behav 31:317–323

Horie S, Suga T (1989) Participation of peroxisomes in lipid biosynthesis in the Harderian gland of guinea pig. Biochem J 262:677–680

Itoh K, Kasama K (1970) Transplantable lipid-producing tumor originating from the Harderian gland. Gann 61:271–273

Johnson DD, Rudeen PK, O'Steen WK (1979) Photically induced experimental exophthalmos: role of Harderian and pituitary glands. Invest Ophthalmol Visual Sci 18:1280–1285

Jost U (1974) 1-Alkyl-2,3-diacyl-*sn*-glycerol, the major lipid in the Harderian gland of rabbits. Hoppe-Seylers Z Physiol Chem 355:422–426

Jost U, Kühnel W, Schimassek H (1974) A morphologieal and biochemical analysis of the Harderian gland in the rabbit. Cytobiologie 8:440–456

Kahán IL, Juhász K, Mindszenti S (1967) Lipids of the Harderian gland of rabbits. Acta Biol Hung 18:295–301
Kasama K, Uezumi N, Itoh K (1970) Characterization and identification of glyceryl ether diesters in Harderian gland tumor of mice. Biochim Biophys Acta 202:56–66
Kasama K, Rainey WT, Snyder F (1973) Chemical identification and enzymatic synthesis of a newly discovered lipid class—hydroxyalkylglycerols. Arch Biochem Biophys 154:648–658
Kasama K, Blank ML, Snyder F (1989) Identification of 1-alkyl-2-acyl-3-(2′,3′-diacylglycerol) glycerols, a new type of lipid class, in Harderian gland tumors of mice. J Biol Chem 264:9453–9461
Kennedy GY (1970) Harderoporphyrin: a new porphyrin from the Harderian glands of the rat. Comp Biochem Physiol 36:21–36
Kitamoto T, Nishigai M, Ikai A, Ohashi K, Seyama Y (1985) The quaternary structure and activity of newly purified fatty acid synthetase from the Harderian gland of guinea pig. Biochim Biophys Acta 827:164–173
Kitamoto T, Nishigai M, Sasaki T, Ikai A (1988) Structure of fatty acid synthetase from the Harderian gland of guinea pig. J Mol Biol 203:183–195
Kolattukudy PE, Rogers LM, Balapangu A (1987) Synthesis of methyl-branched fatty acids from methylmalonyl-CoA by fatty acid synthase from both the liver and the Harderian gland of the guinea pig. Arch Biochem Biophys 255:205–209
Krieg T, Goerz G, Lissner R, Bolsen K, Ullrich V (1978) Drug monooxygenase activity in the Harderian gland. Biochem Pharmacol 27:575–577
Kühnel W (1971) Struktur und Cytochemie der Harderschen Drüse von Kaninchen. Z Zellforsch Mikrosk Anat 119:384–404
Kume K, Shimizu T, Seyama Y (1987) Characterization of *sn*-glycerol 3-phosphate acyltransferase from guinea pig Harderian gland microsomes. J Biochem 101:653–660
Kyushiki H, Ikai A (1989) Negative functional interference between two active centers is indicated in animal fatty acid synthetase. Biochem Biophys Res Commun 164:434–438
Lin WL, Nadakavukaren MJ (1981) Harderian gland lipids of male and female golden hamsters. Comp Biochem Physiol 70B:627–630
Lindsay JW (1976) Spontaneous occurrence of tumors in laboratory-reared *Arvicoline* rodents. Cancer Res 37:4092–4098
Madison RKM, Rabstein LS, Bryan WR (1968) Mortality rate and spontaneous lesions found in 2,928 untreated BALB/cC4 mice. J Natl Cancer Inst 40:683–685
Minucci S, Chieffi Baccari G, Di Matteo L, Chieffi G (1989) A sexual dimorphism of the Harderian gland of the toad, *Bufo viridis* Basic Appl Histochem 33:299–310
Minucci C, Chieffi Baccari G, Di Matteo L, Marmorino C,l d'Istria M, Chieffi G (1990) Influence of light and temperature on the secretory activity of the Harderian gland of the green frog, *Rana esculenta*. Comp Biochem Physiol 95A:249–252
Murawski U, Jost U (1974) Unsaturated wax esters in the Harderian gland of the rat. Chem Phys Lipids 13:155–158
Numa S, Yamashita S (1974) Regulation of lipogenesis in animal tissues. In: Horecker BL, Stadtman ER (eds) Current topics in cellular regulation, vol 8. Academic Press, New York, pp 197–246
Ohashi K, Otsuka H, Seyama Y (1985) Assay of fatty acid synthetase by mass fragmentography using [^{13}C]malonyl-CoA. J Biochem 97:867–875
Otsuka H, Otsuru O, Kasama T, Kawaguchi A, Yamazaki K, Seyama Y (1986) Stereochemistry of 2,3-alkanediols obtained from the Harderian gland of Mongolian gerbil (*Meriones unguiculatus*). J Biochem 99:1339–1344
Otsuru O, Otsuka H, Kasama T, Seyama Y, Sakai T, Yohro T (1983) The characterization of 2,3-alkanediol diacyl esters obtained from the Harderian glands of Mongolian gerbil (*Meriones unguiculatus*). J Biochem 94:2049–2054
Peters RL, Rabstein LS, Spahn GJ, Madison RM, Huebner RJ (1972) Incidence of spontaneous neoplasms in breeding and retired breeder BALB/cCr mice throughout the natural life span. Int J Cancer 10:273–282
Radominska-Pyrex A, Dabrowiecki Z, Horrocks LA (1979) Synthesis and content of ether-linked glycerophospholipids in the Harderian gland of rabbits. Biochim Biophys Acta 574:248–257
Rock CO, Snyder F (1974) Biosynthesis of 1-alkyl-*sn*-glycero-3-phosphate via adenosine triphosphate:1-alkyl-*sn*-glycerol phosphotransferase. J Biol Chem 249:5382–5387

Rock CO, Snyder F (1975a) Metabolic interrelations of hydroxy-substituted ether-linked glycerolipids in the pink portion of the rabbit Harderian gland. Biochim Biophys Acta 171:631–636
Rock CO, Snyder F (1975b) A short-chain acyl-CoA: 1-alkyl-2-acyl-*sn*-glycerol acyltransferase from a microsomal fraction of the rabbit Harderian gland. Biochim Biophys Acta 388:226–230
Rock CO, Fitzgerald V, Rainey Jr WT, Snyder F (1976) Mass spectral identification of 2-(*O*-acyl)hydroxy fatty acid esters in the white portion of the rabbit Harderian gland. Chem Phys Lipids 17:207–212
Rock CO, Fitzgerald V, Snyder F (1977a) Properties of dihydroxyacetone phosphate acyltransferase in the Harderian gland. J Biol Chem 252:6363–6366
Rock CO, Fitzgerald V, Snyder F (1977b) Activation of alkyldihydroxyacetone phosphate synthase by detergents. Arch Biochem Biophys 181:172–177
Sakai T (1981) The mammalian Harderian gland: morphology, biochemistry, function and phylogeny. Arch Histol Japn 44:299–333
Sakai T, Yohro T (1981) A histological study of the Harderian gland of Mongolian gerbils, *Meriones meridianus*. Anat Rec 200:259–270
Schlossman DM, Bell RM (1976) Triacylglycerol synthesis in isolated fat cells; Evidence that the *sn*-glycerol-3-phosphate and dihydroxyacetone phosphate acyltransferase activities are dual catalytic functions of a single microsomal enzyme. J Biol Chem 25:5738–5744
Seyama Y, Kawaguchi A, Okuda S, Yamakawa T (1978) New assay method for fatty acid synthetase with mass fragmentography. J Biochem 84:1309–1314
Seyama Y, Otsuka H, Kawaguchi A, Yamakawa T (1981) Fatty acid synthetase from the Harderian gland of guinea pig; biosynthesis of methyl-branched fatty acids. J Biochem 90:789–797
Seyama Y, Ohashi K, Imamura T, Kasama T, Otsuka H (1983) Branched chain fatty acids in phospholipids of guinea pig Harderian gland. J Biochem 94:1231–1239
Seyama Y, Ohashi K, Yasugi E, Otsuka H (1984) Fatty acid composition of cardiolipin from guinea pig Harderian gland. J Biochem 96:1639–1643
Smelser GK (1943) Water and fat content of orbital tissues of guinea pigs with experimental exoppthalmos produced by extracts of the anterior pituitary gland. Am J Physiol 140:308–315
Thiessen DD (1977) Thermoenergetic and the evolution of pheromone communication. In: Sprague JM, Epstein AN (eds) Progress in psychobiology and physiological psychology, vol 7, Academic Press, New York, pp 91–191
Thiessen DD (1988) Body temperature and grooming in the Mongolian gerbil. Ann NY Acad Sci 525:27–39
Thiessen DD, Pendergrass M (1985) Change of pelage lipids in the Mongolian gerbil, *Meriones unguiculatus*, as the result of autogrooming and sandliving. J Mammal 66:469–475
Thiessen DD, Graham M, Perkins J, Marcks S (1977) Temperature regulation and social grooming in the Mongolilan gerbil (*Meriones unguiculatus*). Behav Biol 19:279–288
Tvrzická E, Řezanka T, Krijt J, Janoušek V (1988) Identification of very-long-chain fatty acids in rat and mouse Harderian gland lipids by capillary gas chromatography-mass spectrometry. J Chromatog 431:231–238
Veleri V, Lopes RA, Migliorini RM, Camargo ACM (1974) Lipids in the Harderian gland of the guinea-pig (*Cavia porcelus*). Ann D'Histochim 18:301–310
Veselinovitch SD, Mihailovich N, Itze L (1970) Comparative studies on the kinetics of the neoplastic competence in mice. Cancer Res 30:2548–2551
Ward JJM, Goodman DG, Squire RA, Chu KC, Linhart MS (1979) Neoplastic and nonneoplastic lesions in aging (C57BL/6N × C3H/HeN) F1(B6C3F1) mice. J Natl Cancer Inst 63:849–854
Watanabe M (1980) An autoradiographic, biochemical, and morphological study of the Harderian gland of the mouse. J Morphol 163:349–365
Watanabe M, Yoshida Y, Watanabe M, Shimada M, Kurimoto K, Kihara T (1975) Effect of hydrofluoric acid on glucose metabolism of the mouse studied by whole-body autoradiography. Br J Ind Med 32:316–320
Yamaguchi K, Hayasaka S, Hara S, Shiono T, Mizuno K (1989) Lysosomal enzyme activities in ocular tissues and adnexa of rabbits. Ophthal Res 21:187–190
Yamazaki T, Seyama Y, Otsuka H, Ogawa H, Yamakawa T (1981) Identification of alkyldiacylglycerols containing saturated methyl-branched chains in the Harderian gland of guinea pig. J Biochem 89:683–691
Yasugi E, Kasama T, Seyama Y (1987) Branched long chain bases in cerebrosides of the guinea pig Harderian gland. J Biochem 102:1477–1482

Yasugi E, Kasama T, Seyamà Y (1988) Identification of 10-methylsphinganine in cerebrosides of the guinea pig Harderian gland. J Biochem 103:889–893
Yasugi E, Kasama T, Shibahara M, Seyama Y (1990) Composition of long-chain bases in sphingomyelin of the guinea pig Harderian gland. Biochem Cell Biol 68:154–160
Yokoyama Y, Kano K, Kaji K, Seyama Y (1989) Purification and characterization of a growth factor from guinea pig Harderian gland. J Biol Chem 264:17058–17063
Zackor B, Egge H, Murawski U (1976) 1-Alkyl-2,3-diacylglycerin in der Harderschen Drüse der Ratte. Hoppe-Seyler's Z Physiol Chem 357:288–289

Harderian Gland Indoles

Armando Menendez-Pelaez[1] and Gerald R. Buzzell[2]

[1]Departamento de Morfologia y Biologia Celular, Universidad de Oviedo, 33006 Oviedo, Principado de Asturias, Spain
[2]Department of Anatomy and Cell Biology, University of Alberta, Edmonton, Alberta T6G 2H7, Canada

1 Introduction

Indoles in Harderian glands have been the subjects of numerous studies. Melatonin was the first indole found in these glands (Bubenik et al. 1976a). Since that time, other indoles (e.g., N-acetylserotonin, 5-methoxytryptamine, etc.) have been found in the Harderian glands of a variety of species. The enzymatic machinery involved in the synthesis of these compounds has also been studied; Harderian gland N-acetyltransferase (NAT), hydroxyindole-*O*-methyltransferase (HIOMT), and monoamine oxidase (MAO) have been characterized.

This chapter describes the research done on indoles (particularly melatonin) and their synthesis in the Harderian glands of rodents and other animals in which they have been studied. The control of melatonin synthesis will be considered. Then appropriate comparisons will be made between indole synthesis in Harderian glands and that in the pineal gland. Finally, we will consider the question of whether indoles are secreted from the Harderian glands into the circulation to have effects elsewhere in the body.

2 Indole Synthesis and Metabolism

The biosynthesis and metabolism of indoles have been extensively studied in the pineal gland. Briefly, the essential amino acid, L-tryptophan, is taken up from the circulation and hydroxylated to 5-hydroxytryptophan. 5-Hydroxytryptophan can be either methylated (by HIOMT) producing 5-methoxytryptophan or decarboxylated to 5-hydroxytryptamine (serotonin). 5-Methoxytryptamine can be further metabolized along three different pathyways: methylation (by HIOMT) to 5-methoxytryptamine, N-acetylation (by NAT) to N-acetyl-5-hydroxytryptamine (N-acetylserotonin), or oxidation (by MAO) to 5-hydroxyindole acetaldehyde. 5-Hydroxyindole acetaldehyde may be the substrate of an aldehyde dehydrogenase (producing 5-hydroxyindole acetic acid) or of

Webb et al. (Eds).
Harderian Glands

an alcohol dehydrogenase (producing 5-hydroxytryptophol). These latter compounds are also the substrate of HIOMT, with 5-methoxyindole acetic acid and 5-methoxytryptophol being produced respectively. Finally, N-acetylserotonin (NAS) is methylated by HIOMT to melatonin.

In the pineal gland, these are normal pathways of indoleamine metabolism [see King and Steinlechner (1985) for an extensive review]. However, there is no evidence that these same metabolic pathways are present in the Harderian glands; this has not been investigated.

3 Indoles in the Rodent Harderian Gland

The presence of melatonin in the Harderian glands of rats was indirectly proposed by Vlahakes and Wurtman in 1972, who investigated the properties of the HIOMT present in this organ. Most of the published studies on Harderian gland indoles have been performed in rodents. Since the first direct identification of melatonin in the Harderian gland of rats by Bubenik et al. (1976a, b), this indole has been detected in Harderian glands of several other rodent species.

Immunocytochemical studies have revealed the existence of melatonin in the acinar secretory cells of rats, hamsters and mice (Bubenik et al. 1976a, b; Vivien-Roels et al 1981). In these studies, specific immunofluorescence was restricted to the cytoplasm surrounding the nucleus, and differences in the numbers of reacting cells from one alveolus to another were observed. In another immunocytochemical study (Bubenik et al. 1978), the amount of melatonin in the Harderian glands of rats was found to be higher during the dark phase than during the light phase. In this same study, melatonin injections resulted in increased levels of Harderian gland melatonin, suggesting the possible uptake of this indole from the circulation. Because of the extreme difficulty of melatonin immunocytochemistry, there are no electron microscope studies showing the cellular sites of melatonin synthesis and storage in the Harderian glands (or, indeed, in the pineal gland).

3.1 Melatonin

Melatonin and other indoles have been detected by radioimmunoassay in the Harderian gland of several rodents, though the functions of these chemicals have not been determined. In preliminary descriptive studies, both melatonin and N-acetylserotonin were detected in the rat Harderian gland (Pang et al. 1976, 1977). Several years later, melatonin levels were shown to exhibit a diurnal rhythm in the rat Harderian gland, with peak levels of immunoreactive melatonin detected 2 h after lights-on (Reiter et al. 1983). The peak levels of Harderian gland melatonin were augmented by pinealectomy. These authors suggested that something of pineal origin normally curtails melatonin levels within the Harderian glands.

As is shown in Tables 1 and 2, there have not been many studies focusing on indolamine metabolism of the rat Harderian gland. Most studies have been performed on the Syrian hamster (*Mesocricetus auratus*), though other rodents, including the mole rat, Richardson's ground squirrel, and Mongolian gerbil, have also been studied.

Melatonin in the Harderian glands of the Syrian hamster was first demonstrated by Pévet et al. (1980). In this paper, a circadian rhythm in the activity of HIOMT (involved in the synthesis of 5-methoxytryptophol and melatonin) was found in the Harderian glands of male hamsters. Melatonin was separated from 5-methoxytryptophol using bidimensional thin layer chromatography. Although the peaks of both indoles coincided late in the dark phase, the activity of HIOMT in 5-methoxytryptophol production was two times higher than for melatonin synthesis. Using the same procedure, Balemans et al. (1983) found seasonal variations in the HIOMT activity involved in the synthesis of different indoles. Melatonin was detectable in the Harderian glands during all seasons; however, 5-methoxytryptophan and 5-methoxytryptamine were absent during the summertime.

The first direct detection of melatonin by radioimmunoassay in the Harderian gland of hamsters was reported by Hoffman et al. (1985). Immunoreactive melatonin concentrations in the Harderian glands of female hamsters were 200 pg/mg protein during a 24-h period. However, 2 h after lights-on, the melatonin concentration dropped to 80pg/mg protein. Melatonin concentrations in glands of males were diurnally constant and low (20pg/mg protein). Castration of males resulted in a significant increase in melatonin concentration, while blinding had no effect. However, the combination of blinding and orchidectomy maintained male levels of melatonin concentrations in the Harderian glands. These combined effects of castration and blinding involved the pineal gland and, curiously, similar responses had been already observed for Harderian gland porphyrin and cholesterol (McMasters and Hoffman 1984). Thus, after this first evidence of a sexual dichotomy in the melatonin metabolism of the hamster Harderian gland, it seemed likely that light input and the pineal gland, as well as gonadal steroids, were involved in this phenomenon. Besides the observation of the striking sexual dimorphism, this paper suggested several intriguing questions. Is the melatonin found in the Harderian glands a consequence of de novo synthesis or uptake from the circulation? What is the reason for the dramatic drop in melatonin concentration, associated with light onset, observed in the female Harderian gland?

Two years later, following a study of melatonin (and the enzymes involved in its synthesis) in the Harderian glands of female Syrian hamsters (Menendez-Pelaez et al. 1987a), we were able to partially answer these questions. In this study, we confirmed some of the observations made by Hoffman and colleagues (1985). The melatonin concentrations dropped sharply after light

Table 1. Studies on indoles present in the Harderian glands

Indole	Species	Reference	Comments
Melatonin	Rat	Pang et al. 1976; 1977	
Melatonin	Rat	Reiter et al. 1983	Rhythm, effect of pinealectomy
Melatonin	Hamster	Pévet et al. 1980	
Melatonin	Hamster	Balemans et al. 1983	Seasonal variations
Melatonin	Hamster	Hoffman et al. 1985	Dimorphism, rhythm
Melatonin	Hamster	Menendez-Pelaez et al. 1987a	Synthesis, sexual differences
Melatonin	Hamster	Menendez-Pelaez et al. 1988a	Sexual differences, testosterone implants
Melatonin	Hamster	Menendez-Pelaez et al. 1988b	Short photoperiod, melatonin injections
Melatonin	Hamster	Hoffman et al. 1989a	Short photoperiod, effect of pinealectomy
Melatonin	Hamster	Hoffman et al. 1989b	Effects of melatonin infusion
Melatonin	Hamster	Buzzell et al. 1989b	Effects of cold and hypothyroidism
Melatonin	Hamster	Buzzell et al. 1990a	Short days and melatonin antagonist
Melatonin	Hamster	Buzzell et al. 1990c	Circadian rhythm
Melatonin	Mole-rat	Balemans et al. 1980	
Melatonin	Ground squirrel	Reiter et al. 1981	No rhythm
Melatonin	Mongolian gerbil	Buzzell et al. 1991	Gender, aging and castration
N-Acetylserotonin	Rat	Pang et al. 1977	
N-Acetylserotonin	Hamster	Menendez-Pelaez 1990	Secretion to circulation?
Serotonin	Hamster	Pévet et al. 1980	Rhythm
Serotonin	Hamster	Menendez-Pelaez et al. 1989a	Rhythm
Serotonin	Hamster	Hoffman et al. 1989a	Castration and pinealectomy
5-Methoxytryptamine	Rat	Prozialeck et al. 1978	Fluorometric determination
5-Methoxytryptophol	Hamster	Pévet et al. 1980	Rhythm
5-Methoxytryptamine	Hamster	Balemans et al. 1983	Seasonal changes
5-Methoxytryptophan	Hamster	Balemans et al. 1983	Seasonal changes
5-Methoxytryptophol	Hamster	Balemans et al. 1983	Seasonal changes
5-Methoxyindole acetic acid	Hamster	Balemans et al. 1983	Seasonal changes
5-Hydroxytryptophol	Hamster	Menendez-Pelaez et al. 1989a	Rhythm
5-Hydroxytryptophan	Hamster	Menendez-Pelaez et al. 1989a	Rhythm
5-Hydroxytryptophan	Hamster	Hoffman et al. 1989a	Sexual differences, pinealectomy
5-Hydroxyindole acetic acid	Hamster	Hoffman et al. 1989a	Sexual differences, pinealectomy

Table 2. Studies related to the enzymes involved in Harderian gland indole metabolism

Enzyme	Species	Reference	Comments
NAT	Rat	Brammer et al. 1978	Non-inducible enzyme
NAT	Rat	White et al. 1984	No rhythms
NAT	Rat	Osuna et al. 1990	Developmental study
NAT	Hamster	Menendez-Pelaez et al. 1987a	Rhythm in females
NAT	Hamster	Menendez-Pelaez et al. 1988a	Sexual dimorphism
NAT	Hamster	Menendez-Pelaez et al. 1988b	Differences between hamster strains
NAT	Hamster	Menendez-Pelaez et al. 1988c	Effects of short photoperiod
NAT	Hamster	Menendez-Pelaez et al. 1988d	Effects of pinealectomy and SCGx
NAT	Hamster	Menendez-Pelaez et al. 1989a	Rhythm in males
NAT	Hamster	Menendez-Pelaez et al. 1989b	Effects of lighter, β-adrenergic agonist
NAT	Hamster	Menendez-Pelaez et al. 1989c	Testosterone implants in females
NAT	Hamster	Menendez-Pelaez et al. 1990a	Effect of 5α-dihydrotestosterone
NAT	Hamster	Buzzell et al. 1989a	Castration and bromocriptine injections
NAT	Hamster	Buzzell et al. 1989b	Effects of hypothyroidism and cold
NAT	Hamster	Buzzell et al. 1990b	Castration and hypothyroidism
NAT	Hamster	Buzzell et al. 1990c	Circadian rhythm
HIOMT	Rat	Vlahakes and Wurtman 1972	Activation by divalent cations
HIOMT	Rat	Cardinali and Wurtman 1972	Biochemical properties
HIOMT	Rat	Cardinali et al. 1972	Effects of different wavelength
HIOMT	Rat	Osuna et al. 1990	Developmental study
HIOMT	Hamster	Pévet et al. 1980	Rhythms in male hamster
HIOMT	Hamster	Balemens et al. 1983	Seasonal changes
HIOMT	Hamster	Menendez-Pelaez et al. 1987a	Rhythms in female, drop after lights on
HIOMT	Hamster	Menendez-Pelaez et al. 1988a	Sexual dimorphism, effect of castration
HIOMT	Hamster	Menendez-Pelaez et al. 1988b	Differences in two hamster strains
HIOMT	Hamster	Menendez-Pelaez et al. 1989a	Very low activity in males, no rhythm
HIOMT	Hamster	Buzzell et al. 1990a	No effect of short photoperiod
HIOMT	Hamster	Buzzell et al. 1990b	Castration and hypothyroidsm
HIOMT	Hamster	Buzzell et al. 1990c	Lack of circadian rhythm
HIOMT	Mongolian gerbil	Buzzell et al. 1991	High activity
MAO	Rat	Menendez-Pelaez et al. 1988e	Detection by histoenzymology
MAO	Mouse	Menendez-Pelaez et al. 1988e	Detection by histoenzymology
MAO	Mongolian gerbil	Menendez-Pelaez et al. 1988e	Detection by histoenzymology
MAO	Hamster	Menendez-Pelaez et al. 1989d	Histoenzymology, sexual differences

onset, recovering to normal values at the middle of the light phase. This drop was accompanied by an equally dramatic decrease in HIOMT activity, while NAT activity exhibited an unexpected increase at the same time. These findings suggested that melatonin levels decreased because of the drop in the synthesis of this indoleamine due to the reduction in HIOMT activity.

In a separate experiment, we showed that melatonin is actually synthesized in the Harderian glands. Using 6-parachlorophenylalanine, a potent and specific blocker of melatonin synthesis in the hamster pineal gland, we reduced the melatonin concentrations in female hamster Harderian glands by 60% (Menendez-Pelaez et al. 1987a). This drug, which inhibits trytophan hydroxylase, did not affect NAT or HIOMT activity. The existence of melatonin synthesis in the Harderian glands of hamsters does not rule out the possibility of uptake from the circulation. However, Hoffman et al. (1989b) found that continuous infusion of melatonin from silastic implants did not modify melatonin levels in the Harderian glands of hamsters.

After these experiments, we concentrated our attention on the androgenic control of the sexual differences observed in melatonin content of the glands. We confirmed that melatonin concentrations in the Harderian glands from female hamsters are much higher than those from male hamsters (Menendez-Pelaez et al. 1988a). Castration of prepubertal males altered the levels of melatonin so that they were equivalent to those in females. However, orchidectomy after adulthood did not modify male melatonin concentrations. In this paper, we showed for the first time the existence of a marked sexual dimorphism in the activities of NAT and HIOMT in the Harderian glands of the Syrian hamster. Unlike the situation in the pineal gland, the melatonin content of the Harderian gland appears to run parallel to HIOMT instead of to NAT activity. Thus, Harderian gland HIOMT activity and melatonin levels are very low in male hamsters while NAT activity is high. The converse is true in females. In addition (and contrary to the melatonin content or the HIOMT activity), male NAT activity dropped to female levels when the animals, whether prepubertal or adult, were castrated. At this time, we concluded that in the male Harderian gland, HIOMT activity might be the rate-limiting enzyme for melatonin formation. We also found that ovariectomy did not affect melatonin metabolism in the hamster Harderian glands.

However, many intriguing questions are still open. What is the reason for the differential response to castration? How can we explain the inverse correlation between NAT and HIOMT activities? Is Harderian gland NAT involved in the production of other indoles? Besides the androgenic control, what other hormones affect indole metabolism in the Harderian glands? Currently, we can answer some of these questions, but most of them are still open to discussion.

In the next study (Menendez-Pelaez et al. 1988b) we compared the responses of indole metabolism in the Harderian glands to photoperiod and to melatonin injections in females of two strains of hamsters (outbred and inbred LHS/SsLak). Inbred LSH/SsLak hamsters respond to short-day photoperiods with gonadal regression more readily than do outbred hamsters. The purpose of this study was to determine if Harderian indoleamine metabolism is affected, directly or indirectly, by photoperiods. Surprisingly, the two strains of Syrian hamsters showed differences in basal NAT and HIOMT activities and melatonin content: inbred animals exhibited higher NAT and lower HIOMT activities and melatonin content than did outbred hamsters. In the inbred animals, HIOMT activity was not influenced by exposure to short photoperiods, but NAT activity was clearly depressed by short photoperiods in the inbred animals. In addition, melatonin concentrations increased after the exposure of inbred hamsters to short photoperiods, suggesting the possible uptake of melatonin from the circulation. Finally, we tried unsuccessfully to simulate short photoperiod exposure by injecting hamsters daily with melatonin in the afternoon. Although both short photoperiods and daily melatonin injections resulted in gonadal atrophy, the response of Harderian gland indole metabolism to these treatments was obviously different. From these results we concluded that the photoperiod might be involved in Harderian gland melatonin metabolism, independent of the gonadal status.

As we had obtained clearer and more conclusive results using male hamsters, we returned to the castration model. We compared the effects of castration under short and long days (Menendez-Pelaez et al. 1988c). Harderian gland NAT activity was clearly decreased by castration in both circumstances, but the castrated hamsters exposed to short days showed a significant decrease in NAT activity compared with the castrated animals under long-day conditions. Thus, light might directly affect NAT activity [as we had previously suggested due to the increase in NAT activity observed after the onset of light (Menendez-Pelaez et al. 1987a)]. Further evidence for this effect was found when we showed that extending the dark phase for 1 h prevented the rise in NAT activity after lights-on in the Harderian glands of female hamsters (Menendez-Pelaez et al. 1989b).

Contrary to the situation in the pineal gland, NAT activity and melatonin content in the hamster Harderian gland were independent of the sympathetic neural input to the gland (the superior cervical ganglion) (Menendez-Pelaez et al. 1988d). Indeed, several years ago, Brammer et al (1978) had suggested that the NAT detected in the rat Harderian gland was not inducible by β-agonists. We have confirmed this lack of response to adrenergic agonists and also demonstrated that the hamster Harderian gland does not respond to cholinergic input (Menendez-Pelaez et al. 1989b).

Nevertheless, we observed that pinealectomy was able to modify, although in different ways, NAT activity and melatonin concentration in the Harderian glands (Menendez-Pelaez et al. 1988d). In pinealectomized female hamsters, the NAT rise after light onset was delayed. On the other hand, melatonin levels in these animals decreased rapidly after the onset of light (as usual), but the recovery to normal values was faster than in control animals. In other studies, pinealectomy has been shown to be ineffective in changing melatonin levels in the Harderian glands of male hamsters (Hoffman et al. 1989a). However, Panke et al. (1979) earlier demonstrated that the bilateral ablation of the hamster Harderian gland resulted in alterations of the timing of the pineal melatonin peak, which suggests a possible feed-back between the pineal and the Harderian glands.

No rhythm of melatonin concentrations in male hamster Harderian glands was detected by Hoffman et al. (1985) or Memendez-Pelaez et al. (1989a). However, more recently, we detected significant circadian rhythms in Harderian gland melatonin in male hamsters kept in long-day photoperiods and in those kept in decreasing autumnal photoperiods (Buzzell et al. 1990c). The reason for these conflicting results is not apparent. HIOMT activity was very low in male hamster Harderian glands and no circadian rhythm was detected (Menendez-Pelaėz et al. 1989a; Buzzell et al. 1990c). NAT activity was lower in Harderian glands of male hamsters kept in short photoperiods than in those kept in long days, and a weak circadian rhythm was seen only in the long-day animals (Buzzell et al.1990c).

Male or female hamsters kept in short days have reduced melatonin concentration in their Harderian glands; melatonin levels, NAT, and HIOMT activity and their responses to short days were not substantially altered by treatment with a putative melatonin antagonist N-(2,4-dinitrophenyl)-5-methoxytryptamine (Buzzell et al. 1990a).

Sexual differences in Harderian gland indole metabolism of the Syrian hamster are regulated by hormones, especially androgens (Menendez-Pelaez et al. 1989e). High-affinity specific intracellular androgen receptors have been demonstrated in the hamster Harderian glands; long-term castration did not affect androgen receptor concentration or sedimentation coefficient (Vilchis et al. 1987). In addition, Hoffman and colleagues (1989a) found that exposure of castrated males to short photoperiods prevented the castration-associated rise in melatonin concentration, suggesting a possible mediation by the pineal gland or by hormones influenced by pineal melatonin. In our opinion, these results suggest that other hormones, possibly from the pituitary and/or thyroid gland, also play an important role in the regulation of indoleamine metabolism. A possible mechanism for this is that orchidectomy elevates levels of luteinizing hormone (LH) and LH interferes with Harderian gland indole metabolism; castrated animals exposed to short photoperiods have reduced levels of LH (Turek et al. 1975). In addition, we have recently given evidence that thyrotropin also mediates Harderian gland NAT activity (Buzzell and Menendez-Pelaez, Chap. 16, this vol; Buzzell et al. 1990b).

It seems clear that sex steroids, while undoubtedly influencing Harderian indole metabolism, are not the only hormones involved. Recently it has been proposed that melatonin and porphyrin synthesis are differentially controlled in the Harderian glands of hamsters, and that gonadal steroids are not necessarily involved in the regulation of sexual "type" (Hoffman et al. 1989b). A link between the thyroid and the Harderian glands has been proposed (Buzzell and Menendez-Pelaez, chap.16, this vol.; Hoffman et al. 1990). Finally, although immunoreactive melatonin in the Harderian glands was unaffected by the state of thyroid secretion, melatonin content declined after 3 h of exposure to 4 °C, suggesting that Harderian gland melatonin might be involved in thermoregulation (Buzzell et al. 1989b).

In the last few years we have accumulated a substantial amount of information regarding melatonin metabolism in rodent Harderian glands. We have investigated the existence of melatonin and the enzymes involved in its synthesis in gerbils (Buzzell et al. 1991, degus, and rats Buzzell and Menendez-Pelaez, unpublished data). In spite of all these studies, we know very little about the physiological role of the melatonin present in the Harderian glands. The dichotomy between the regulation of NAT and HIOMT activities and the melatonin content suggests that perhaps other indoles are produced by these enzymes and should be examined in future studies.

3.2 Other Indoles

Harderian gland indoles other than melatonin have received little attention. 5-Methoxyindoles produced by incubation of tissue extracts from hamster Harderian glands were investigated by Pévet et al. (1980) and Balemans et al. (1983). In these studies, production of 5-methoxytryptophol increased after lights on. Production of some of the indoles studied exhibited seasonal differences (Table 1), but the rhythms in these substances were not very pronounced. While some indoles were produced through the year, others, such as 5-methoxytryptamine, were absent during the spring. More studies are needed in order to confirm and understand the cyclical variations observed.

Serotonin and its precursor, 5-hydroxytryptophan, occur in the Harderian glands of female hamsters in much higher concentrations than in those of males. Orchidectomy led to female levels of 5-hydroxytryptophan, but not of serotonin. Pinealectomy raised 5-hydroxytryptophan levels in male hamsters. In these experiments, the pooled data suggested that levels of these indoles [measured by high performance liquid chromatography (HPLC)] reflect the concentration of melatonin. However, the large variability in the data precluded definitive statements regarding temporal changes (Hoffman et al. 1989a).

We have also studied some aspects of Harderian indole metabolism by HPLC. Thus, male hamsters showed day:night rhythms in serotonin, 5-hydroxytryptophol, and 5-hydroxyindole acetic acid concentrations (Menendez-Pelaez et al. 1989a). Serotonin concentrations during the night were much higher than during the light phase (which is opposite to the situation in the pineal gland). Serotonin and 5-hydroxytryptophol levels decreased dramatically 1 h after the onset of light. In another study, we observed a strong correlation between concentrations of serotonin and N-acetylserotonin (NAS) in the Harderian glands from female hamsters studied during the estrous cycle (Menendez-Pelaez et al. unpublished observations). NAS was also detected in rat Harderian glands by Pang et al. (1977).

4 Indoles in the Harderian Glands of Nonrodent Species

There have been very few studies on indole metabolism in the Harderian glands of nonrodent species. In reptiles, melatonin was detected immunocytochemically in secretory cells of the Harderian gland from a tortoise (*Testudo hermanni*), a lizard (*Lacerta agilis*), and a snake (*Natrix tesselata*) (Vivien-Roels et al. 1981). In birds, melatonin was found in Harderian gland of pigeons (*Columba livia*) (Vakkuri et al. 1985a, b) and chickens (*Gallus domesticus*) (Cogburn et al. 1987).

In the pigeon, Harderian gland melatonin showed a diurnal variation, with low levels at midday and high levels at midnight (Vakkuri et al. 1985a). Pinealectomized pigeons showed a significant increase in Harderian gland melatonin concentrations (Vakkuri et al. 1985b). According to these authors (who developed their own radioimmunoassay), the Harderian glands might be the main organs responsible for the circulating melatonin observed after pinealectomy.

In young chickens, the melatonin rhythm observed in the Harderian glands was very modest, with the concentrations found during the scotophase only 51% higher than during the photophase (Cogburn et al. 1987). Pinealectomy did not affect this rhythm. Based on the fact that both the Harderian and pineal glands of chickens contain large populations of immunologically competent B-lymphocytes, the authors speculated that the rhythmic melatonin synthesis in the pineal and Harderian glands might influence these resident lymphocytes.

Contrary to the situation in pigeons, the Harderian glands of chickens do not contribute to the plasma melatonin rhythm under normal circumstances (Cogburn et al. 1987).

To our knowledge, there are no publications regarding the indole metabolism of Harderian glands from other mammals but rodents. Nevertheless, Weaker (1981) described as unpublished observations the presence of melatonin in the Harderian gland of the nine-banded armadillo (*Dasypus novemcinctus*). Curiously, the armadillo is one of the few animals with no known pineal gland (Harlow et al. 1980). Finally, we have found immunoreactive melatonin in the Harderian glands of rabbits (Menendez-Pelaez et al. unpublished data) and western spotted skunks (Buzzell et al. unpublished data). In conclusion, the existence of melatonin in vertebrate Harderian glands seems to be a common phenomenon; however, its physiological role is unknown.

5 Are Harderian Gland Indoles Secreted to Circulation?

Although the possible functions of this intriguing orbital structure will be addressed in other chapters of this book, we should reflect upon the physiological meaning of indoles in the Harderian glands. Thermoregulation has been suggested as a major function of the Harderian glands, at least in rodents (Harlow 1984; Thiessen 1988). In this context, the observed Harderian-thyroid interactions do not seem to be casual (Hoffman et al. 1989c, 1990; Buzzell and Menendez-Pelaez, Chapt. 16, this vol; Buzzell et al. 1990b). If we admit the possibility of Harderian-thyroid interactions, how do indoles from the Harderian glands affect other tissues and organs? Are Harderian gland indoleamines secreted into the circulation to influence the thyroid (or other structures)?

Morphological evidence suggests a possible endocrine function of the Harderian glands; this is based on the type of blood supply and special morphological features observed in the basal pole of the secretory cells (Weaker 1981; Menendez-Pelaez et al. 1987b, 1990b). The best experimental approach to study the possible endocrine secretion of Harderian gland indoles is the bilateral ablation of the Harderian glands. However, this surgical procedure is complicated, may damage the optic nerve, and produces a moderate mortality, at least in Syrian hamsters.

We have previously mentioned contradictory results of pinealectomy on Harderian gland melatonin rhythms found in pigeons and chickens; it is also not clear what the contribution of the Harderian glands to the serum melatonin rhythm is in rodents. It had been known for years that pinealectomy abolishes the nocturnal melatonin peak in serum in almost all animals (Reiter 1989). However, in rats, pinealectomy does not suppress basal (daytime) melatonin levels in serum (Yu et al. 1981). We recently described the combined effects of Harderectomy and pinealectomy on serum melatonin levels in male Syrian hamsters (Menendez-Pelaez et al. 1990c). Surprisingly, neither pinealectomy, Harderectomy, nor a combination of both surgeries was able to modify daytime basal melatonin levels (Fig. 1). Apparently, the daytime serum melatonin originated from another source.

If melatonin is not secreted from the Harderian glands, what is its physiological role? Without denying its possible synthesis, are the Harderian glands mainly sponges taking melatonin up from the circulation? The answers to these questions are still not known.

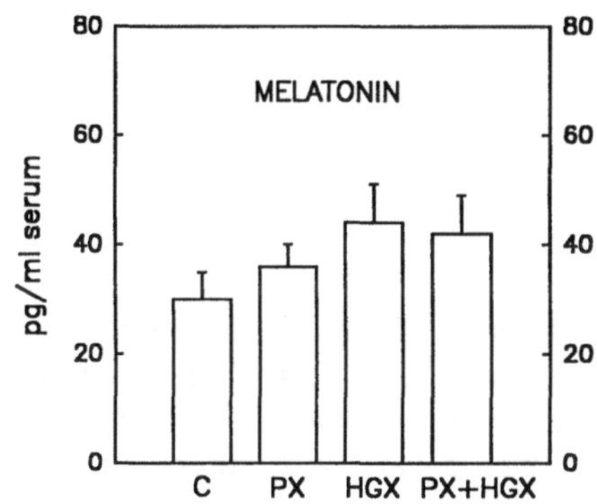

Fig. 1. Serum concentrations of melatonin in male Syrian hamsters sacrificed in the middle of the light phase. Neither pinealectomy (*PX*), Harderectomy (*HGX*), nor the combination of both surgeries (*PX + HGX*) modified the basal levels, in comparison with control animals (*C*). (Menendez-Pelaez et al. 1990c)

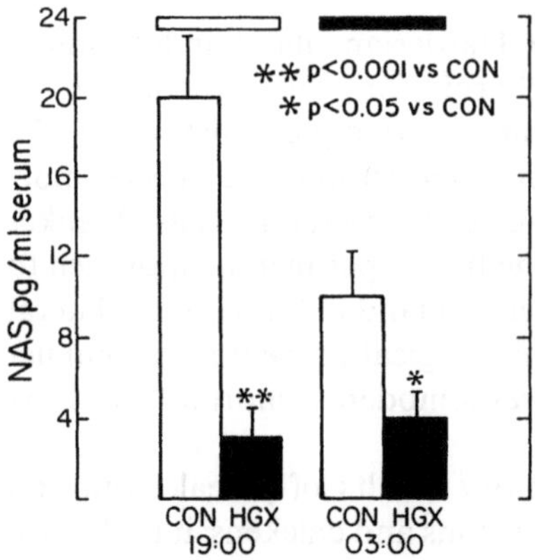

Fig. 2. Serum levels of N-acetylserotonin (*NAS*) in female Syrian hamsters late in the afternoon (19.00 h) and in the middle of the dark phase (03.00 h). Harderectomy (*HGX*) significantly reduced NAS in serum, compared with control animals (*CON*). (Menendez-Pelaez 1990)

N-Acetylserotonin (NAS) has been found in the serum of rats and other rodents (Pang et al. 1977; Brown et al. 1987). This indole has been shown to exert some influence on endocrine tissues (Porter et al. 1971; Lewinski et al. 1989). Interestingly, hamster serum NAS exhibits a rhythm that does not correlate with the NAS rhythm in the pineal (Sato et al. 1989). Since this serum rhythm persists after pinealectomy (Yu et al. 1981), it has been proposed that serum NAS originates from a pool distinct from that in the pineal (Brown et al. 1987). To test this, we performed an experiment where female Syrian hamsters were bilaterally Harderectomized and the serum concentrations of NAS were measured by HPLC at two time points, NAS levels from Harderectomized animals were significantly reduced below those of controls (Menendez-Pelaez 1990, Fig. 2). Thus, the possibility of an endocrine secretion of NAS from the Harderian glands should be taken in account.

6 Conclusions

The specific and precise functions of the Harderian glands are unknown. The Harderian glands of most of the vertebrates which have been studied contain indoleamines. The regulation of these indoleamines in rodents is substantially different from that in the pineal gland. However, a variety of other mammals should be studied in order to increase our very partial knowledge.

The Harderian glands may influence the thyroid, pituitary, pineal, and/or other endocrine organs. The possibility of an indole endocrine secretion is certainly open. Perhaps indoles secreted by the Harderian gland are involved in the modulation of the gonadal or thyroid function mediated by the pineal gland. From an exocrine or an endocrine point of view, the possible role of the Harderian glands in thermoregulation is an enormous open field, where Harderianologists might concentrate their efforts in future years.

Acknowledgments. We would like to express our gratitude to Dr. R.J. Reiter for his constant support and friendship. Also, the help and encouragement of all the members of the San Antonio Pineal Lab are truly appreciated. Dr. R.A. Hoffman provided us many of the references regarding this manuscript.

References

Balemans MGM, Pévet P, Legerstee WC, Nevo E (1980) Preliminary investigations on melatonin and 5-methoxytryptophol synthesis in the pineal, retina, and Harderian gland of the mole rat and in the pineal of the mouse "eyeless". J Neural Transm 49:247–255

Balemans MGM, Pévet P, van Benthem J, Haldar-Misra C, Smith I, Hendriks H (1983) Day/night rhythmicity in the methylating capacities for different 5-hydroxyindoles in the pineal,the retina and the Harderian gland of the golden hamster (*Mesocricetus auratus*) during the annual seasons. J Neural Transm 56:53–72.

Brammer GL, Yuwiler A, Wetterberg L (1978) N-Acetyltransferase activity of the rat Harderian gland. Biochem Biophys Acta 526:93–99

Brown GM, Ho Ak, Chik CL (1987) Effects of feeding on pineal indoleamines. In: Reiter RJ, Fraschini F (eds) Advances in pineal research, vol II. John Libbey, London, pp 67–81

Bubenik GA, Brown GM, Grota LJ (1976a) Immunohistochemical localization of melatonin in the rat Harderian gland. J. Histochem Cytochem 24:1173–1177

Bubenik GA, Brown GM, Grota LJ (1976b) Differential localization of N-acetylated indolealkylamines in CNS and the Harderian gland using immunohistology. Brain Res 118:417–427

Bubenik GA, Purtill RA, Brown GM, Grota LJ (1978) Melatonin in the retina and the Harderian gland. Ontogeny, diurnal variations and melatonin treatment. Exp Eye Res 27:323–333

Buzzell GR, Menendez-Pelaez A, Porkka-Heiskanen T, Pangerl A, Pangerl B, Vaughan MK, Reiter RJ (1989a) Bromocriptine prevents the castration-induced rise in porphyrin concentration in the Harderian glands of the male Syrian hamster,*Mesocricetus auratus*. J Exp Zool 249:172–176

Buzzell GR, Chen ZL, Vaughan MK, Reiter RJ (1989b) Effects of inhibition of thyroid function and of cold on melatonin synthesis and porphyrin content in the Harderian glands of male Syrian hamsters, *Mesocricetus auratus*. Comp Biochem Physiol 94A:427–429

Buzzell GR, Menendez-Pelaez A, Troiani ME, McNeill ME, Reiter RJ (1990a) Effects of short-day photoperiods and of N-(2,4-dinitrophenyl)-5-methoxytryptamine, a putative melatonin antagonist, on melatonin synthesis in the Harderian gland of the Syrian hamster, *Mesocricetus auratus*. J Pineal Res 8:229–237

Buzzell GR, Menendez-Pelaez A, Hoffman RA, Vaughan MK, Reiter RJ (1990b) N-Acetyltransferase activity in the Harderian glands of the Syrian hamster, *Mesocricetus auratus*, is regulated by androgens and by hormones of the pituitary-thyroid axis. J Endocrinol 127:59–67

Buzzell GR, Pangerl A, Pangerl B, Menendez-Pelaez A, Vaughan MK, Little JC, Hill SM, Vaughan GM, Reiter RJ (1990c) Melatonin and porphyrin in the Harderian glands of the Syrian hamster: circadian patterns and response to autumnal conditions. Int J Biochem 22:1465–1469

Buzzell GR, Menendez-Pelaez A, Blank JL, Nonaka KO, Reiter RJ (1991) Effects of gender, age, and castration on porphyrin concentration and melatonin synthesis in the Harderian glands of the Mongolian gerbil, *Meriones unguiculatus*. Can J Zool 69:151–155

Cardinali DP, Wurtman RJ (1972) Hydroxyindole-O-methyl transferases in rat pineal, retina and Harderian gland. Endocrinology 91:247–252

Cardinali DP, Larin F, Wurtman RJ (1972) Action spectra for effects of light on hydroxyindole-O-methyl transferases in rat pineal, retina and Harderian gland. Endocrinology 91:877–886

Cogburn LA, Wilson-Placentra S, Letcher LR (1987) Influence of pinealectomy on plasma and extrapineal melatonin rhythms in young chickens (*Gallus domesticus*). Gen Comp Endocrinol 68:343–356

Harlow HJ (1984) The influence of Harderian gland removal and fur lipid removal on heat loss and water flux to and from the skin of muskrats, *Ondatra zibethicus*.Physiol Zool 57:349–356

Harlow H, Phillips J, Ralph C (1980) Circadian rhythms of an animal without pineal complex, the nine-banded armadillo. Anat Rec 196:73A (Abstr)

Hoffman RA, Johnson LB, Reiter RJ (1985) Harderian glands of golden hamsters: Temporal and sexual differences in immunoreactive melatonin. J Pineal Res 2:161–168

Hoffman RA, Johnson LB, Reiter RJ (1989a) Regulation of melatonin in the Harderian glands of golden hamsters. J Pineal Res 6:63–71

Hoffman RA, Vaughan MK, Reiter RJ (1989b) Photic input and pineal-Harderian-thyroid interactions in golden hamsters. In. Reiter RJ, Pang SF (eds) Advances in pineal research, vol III. John Libbey, London, pp 225–230

Hoffman RA, Johnson LB, Vaughan MK, Reiter RJ (1989c) Interaction of Harderian glands, illumination, and temperature on thyroid hormones in golden hamsters. J Comp Physiol B 158:697–702

Hoffman RA, Habeeb P, Buzzell GR (1990) Further studies on the regulation of the Harderian glands of golden hamsters by the thyroid gland. J. Comp Physiol B 160:269–275

King TS, Steinlechner S (1985) Pineal indoalkylamine synthesis and metabolism: Kinetic considerations. In: Reiter RJ (ed) Pineal research reviews, vol III. Alan R Liss, New York, pp 69–113

Lewinski A, Sewerynek E, Zerek-Melen G, Kunert-Radek J, Pawlikowski M, Karasek E (1989) Influence of melatonin and N-acetylserotonin on cyclic AMP concentration in the rat thyroid lobes incubated in vitro. J. Pineal Res 7:55–61

McMasters KM, Hoffman RA (1984) Harderian gland: regulation of sexual "type" by gonads and pineal gland. Biol Reprod 31:579–585

Menendez-Pelaez A (1990) Melatonin and other indoles in the rodent Harderian gland: regulation and physiological significance. In: Reiter RJ, Lukaszyk A (eds) Advances in pineal research, vol. IV. John Libbey, London pp 75–80

Menendez-Pelaez A, Howes KA, Gonzalez-Brito A, Reiter RJ (1987a) N-Acetyltransferase activity, hydroxyindole-O-methyltransferase activity, and melatonin levels in the Harderian glands of the female Syrian hamster: changes during the light:dark cycle and the effect of 6-chlorophenylalanine administration. Biochem Biophys Res Commun 145:1231–1238

Menendez-Pelaez A, Lopez JM, Alvarez-Uria M, Howes KA, Reiter RJ (1987b) Cambios ultrastructurales en la glandula Harderiana del hamster (*Mesocricetus auratus*) tras gonadectomia bilateral. Rev Biol Univ Oviedo 5:57–64

Menendez-Pelaez A, Reiter RJ, Guerrero JM, Puig-Domingo M, Howes KA (1988a) Sexual dimorphism in N-acetyltransferase activity, hydroxyindole-O-methyltransferase activity and melatonin content in the Harderian glands of Syrian hamsters: changes following gonadectomy. Proc Soc Exp Biol Med 187:287–291

Menendez-Pelaez A, Santana C, Howes KA, Sabry I, Reiter RJ (1988b) Effects of photoperiod or exogenous melatonin administration on the activity of N-acetyltransferase and hydroxyindole-O-methyltransferase and the melatonin content of the Harderian gland of two strains of female Syrian hamsters. J Pineal Res 5:293–300

Menendez-Pelaez A, Reiter RJ, Howes KA, Puig-Domingo M, Vaughan MK, Troiani ME, Little JC (1988c) Harderian gland N-acetyltransferase activity in the male Syrian hamster:effects of gonadectomy, short photoperiod exposure or subcutaneous melatonin implants. Endocr Res 14:121–130

Menendez-Pelaez A, Reiter RJ, Guerrero JM, Santana C, Howes KA, Gonzalez-Brito A (1988d) N-Acetyltransferase activity and melatonin concentrations in the Harderian glands of female Syrian hamsters: alterations following either pinealectomy or bilateral superior cervical ganglionectomy. Biochem Arch 4:231–238

Menendez-Pelaez A, Coalla C, Martinez-Telleria A, Vega JA, Reiter RJ (1988e) Estudio histoquimico de la activadad monoamino oxidasa en la glandula Harderiana de la rata, hamster, raton y gerbo. Rev Biol Univ Oviedo 6:13–19

Menendez-Pelaez A, Reiter RJ, Gonzalez-Brito A, Howes KA (1989a) N-Acetyltransferase activity and indole contents of the male Syrian hamster Harderian gland: changes during the light:dark cycle. Arch biochem Biophys 268:20–25

Menendez-Pelaez A, Buzzell GR, Delgado MJ, Gonzalez-Brito A, Reiter RJ (1989b) N-Acetyltransferase activity in the Syrian hamster Harderian gland is independent of the nonadrenergic and cholinergic input to the gland. Neurosci Res Commun 5:35–42
Menendez-Pelaez A, Marrufo B, Reiter RJ, Santana C, Guerrero JM (1989c) Testosterone increases N-acetyltransferase activity in both male and female Syrian hamster Harderian glands. Biochem Int 19:115–121
Menendez-Pelaez A, Martinez-Telleria A, Vega JA, Coalla C, Reiter RJ (1989d) Histochemical detection of monoamine oxidase and alcohol dehydrogenase activities in the Syrian hamster Harderian glands: existence of a sexual dimorphism. Histochem J 21;125–131
Menendez-Pelaez A, Reiter RJ, Gonzalez-Brito A, Santana C, Guerrero JM (1989e) Regulation of N-acetyltransferase in the Syrian hamster Harderian glands. In: Reiter RJ, Pang SF (eds) Advances in pineal research, vol III. John Libbey, London, pp 77–81
Menendez-Pelaez A, Buzzell GR, Gonzalez-Brito A, Reiter RJ (1990a) The androgenic control of Syrian hamster Harderian gland N-acetyltransferase is mediated by 5α-dihydrotestosterone. J Cell Biochem 42:95–100
Menendez-Pelaez A, Tolivia D, Rodriguez-Colunga MJ, Reiter RJ (1990b) Ultrastructure of the blood vessels in the Syrian hamster Harderian gland:existence of sinusoids. J Morphol 204:1–7
Menendez-Pelaez A, Vaughan MK, Tolivia D, Rodriguez C, Reiter RJ (1990c) Pineal-Harderian gland interactions: morphological and physiological evidences for an endocrine function of the Syrian hamster Harderian gland. In Reiter RJ, Fraschini F (eds) Role of melatonin and pineal peptides in neuroimmunomodulation, Plenum, pp 185–190
Osuna C, Rubio A, Goberna R, Guerrero JM (1990) Ontogeny of type II thyroxine 5′-deiodinase, N-acetyltransferase and hydroxyindole-O-methyltransferase activities in the rat Harderian gland. Life Sci 46:1945–1952
Pang SF, Brown GM, Grota LJ, Rodman RL (1976) Radioimmunoassay of melatonin in pineal glands, Harderian glands, retinas and sera of rats and chickens. Fed Proc Fed Am Soc Exp Biol 35:691
Pang SF, Brown GM, Grota LJ, Chambers JW, Rodman RL (1977) Determination of N-acetylserotonin and melatonin activities in the pineal gland, retina, Harderian gland, brain and serum of rats and chickens. Neuroendocrinology 23:1–13
Panke ES, Reiter RJ, Rollag MD (1979) Effect of removal of the Harderian glands on pineal melatonin concentrations in the Syrian hamster, *Mesocritus auratus*. Experientia 35:1405–1406
Pévet P, Balemans MGM, Legerstee WC, Vivien-Roels B (1980) Circadian rhythmicity of the activity of hydroxyindole-O-methyltransferase (HIOMT) in the formation of melatonin and 5-methoxytryptophol in the pineal, retina, and Harderian gland of the golden hamster. J. Neural Transm 49:229–245
Porter JC, Mical RS, Cramer OM (1971) Effects of serotonin and other indoles on the release of LH, FSH, and prolactin. Gynecol Invest 2:13–22
Prozialeck WC, Boehme DH, Vogel WH (1978) The fluorometric determination of 5-methoxytryptamine in mammalian tissues and fluids. J Neurochem 30:1471–1477
Reiter RJ (1989)Melatonin: Its sources, its message and the interpretation of the message. In: Reiter RJ, Pang SF (eds) Advances in pineal research vol III. John Libbey, London, pp 165–173
Reiter RJ, Richardson BA, Hurlbut EC (1981) Pineal, retinal and Harderian gland melatonin in a diurnal species, the Richardson's ground squirrel (*Spermophilus richardsonii*). Neurosci Lett 22:285–288
Reiter RJ, Richardson BA, Matthews SA, Lane SJ, Ferguson BN (1983) Rhythms in immunoreactive melatonin in the retina and the Harderian glands of rats: persistence after pinealectomy. Life Sci 32:1229–1236
Sato T, Attanasio A, Wake K, Gupta D, (1989) Rhythm development in pineal and circulating serotinin, N-acetylserotonin and melatonin in the Syrian hamster. J Pineal Res 7:45–54
Thiessen DD (1988) Body temperature and grooming in the Mongolian gerbil. Ann NY Acad Sci 525:27–39
Turek FW, Elliot JA, Alvis JD, Menaker M (1975) The interaction of castration and photoperiod in the regulation of hypophyseal and serum gonadotropin levels in male golden hamsters. Endocrinology 96:854–859
Vakkuri O, Rintamäki H, Leppäluoto J (1985a) Presence of immunoreactive melatonin in different tissue of the pigeon (*Columbia livia*). Gen Comp Endocrinol 58:69–75

Vakkuri O, Rintamäki H, Leppäluoto J (1985b) Plasma and tissue concentration of melatonin after midnight light exposure and pinealectomy in the pigeon. J Endocrinol 105:263–268
Vilchis F, Hernandez A, Perez AE, Perez-Palacios G (1987) Hormone regulation of the rodent Harderian gland: binding properties of the androgen receptor in the male golden hamster. J Endocrinol 112:3–8
Vivien-Roels B, Pévet P, Dubois MP, Arendt J, Brown GM (1981) Immunohistochemical evidence for the presence of melatonin in the pineal gland, the retina and the Harderian gland. Cell Tissue Res 217:105–115
Vlahakes GJ, Wurtman RJ (1972)A Mg^{2+} dependent hydroxyindole-O-methyltransferase in rat Harderian gland. Biochem Biophys Acta 261:194–197
Weaker FJ (1981) Light microscopic and ultrastructural features of the Harderian gland of the nine-banded armadillo. J. Anat 133:49–65
White BH, Mosher K, Binkley S (1984) Daily profiles of N-acetyltransferase measured at a single time in rat pineal glands, retinas, and Harderian glands. J Pineal Res 1:129–137
Yu HS, Pang SF, Tang PL, Brown GM (1981) Persistence of circadian rhythms of melatonin and N-acetylserotinin in the serum of rats after pinealectomy. Neuroendocrinology 32:262–265

Harderian Gland Peptides

SUSAN M. WEBB, MANUEL PUIG-DOMINGO, MERCÈ VIADER, EUGENIA MATO, MIGUEL-ANGEL PEINADO

Laboratory of Experimental Endocrinology, Dept. of Endocrinology and Nutrition, Hospital de la Santa Creu i Sant Pau, Autonomous University, Padre Claret 167, 08025-Barcelona, Spain

1 Introduction

The description of various peptides in the Harderian glands of different rodents, the rabbit, the cat, and the chicken is recent, dating back only to the early 1980s. Since then, 11 different peptides have been reported in this gland and various hypotheses have been advanced for their function. However, their physiological role in this gland has not yet been defined. This chapter reviews the available information on these peptide substances (Table 1).

2 Somatostatin

This tetradecapeptide with a molecular weight of 1600 was first described in the Harderian gland by Puig-Domingo et al. (1988a, b) using radioimmunoassay. In the first of these reports, the androgenic control of immunoreactive somatostatin (IRS) was described in the Syrian hamster. As also occurs with melatonin and porphyrins, sexual differences were observed in IRS concentrations, with values being higher in females (10 ± 2.1 ng/mg protein) than in males (2.6 ± 0.4 ng/mg protein). Furthermore, one month after gonadectomy, males had IRS values in the female range (12.4 ± 2.3 ng/mg protein); this rise was prevented when castrated males received a testosterone implant. After 2 months, gonadectomized males had lower Harderian gland IRS concentrations than did control males (1 ± 0.2 ng/mg protein). In female Syrian hamsters, gonadectomy did not alter IRS concentration. However, in normal females, significant differences were observed during the estrus cycle with a nadir detected at diestrus-1 (approximately 1 ng/mg protein) and peak values were found at the time of ovulation (around 13 ng/mg protein) when the animals were sacrificed at 13.00 h. This was not the case in female rats, where no changes were observed throughout the estrus cycle (mean values oscillated between 29.5 and 31.7 pg/mg protein) (Peinado et al. 1990). Changes in other peptide concentrations throughout the estrus cycle have not yet been reported.

Webb et al. (Eds).
Harderian Glands

Table 1. Peptides described in the Harderian gland

1981	Johansson and Lundberg	VIP
1984	Butler et al.	VIP, Substance P
1987	Tsukahara and Jacobowitz	VIP, NT, CCK, NPY, CGRP
1987	Ghaemmaghami et al.	AVP
1988	Gauquelin et al.	AVP, OT
1988a	Puig-Domingo et al.	IRS
1988b	Puig-Domingo et al.	IRS
1989	Walcott et al.	VIP, Substance P
1989	Yokoyama et al.	HGDGF
1990	Peinado et al.	IRS
1990	Duntas et al. (unpublished data)	TRH

Ovariectomy had no effect on Harderian gland IRS, but a possible relationship between the pituitary and the Harderian gland deserves further study (Puig-Domingo et al. 1988a). In hypophysectomized rats, the Harderian gland is known to undergo marked atrophy within 3 weeks (Joó and Kahán 1975; Kok-van Alphen and Noach 1978), but the factors responsible for this atrophy have not been elucidated.

In a second paper, Puig-Domingo et al. (1988b) described the presence of IRS by radioimmunoassay in the male rat Harderian gland (10.8 ± 3.5 ng/mg protein). Hypophysectomy did not change the IRS concentration over the relatively short 2-week study period, but injections of growth hormone for 4 days increased Harderian gland IRS six-to sevenfold in hypophysectomized animals. After the administration of the beta-adrenergic agonist isoproterenol, Harderian gland IRS dropped from 10.6 ± 3.2 to 2.3 ± 0.9 ng/mg protein; this was prevented by a subsequent injection of the calcium channel blocker, verapamil (9.6 ± 1.7 ng/mg protein), but not by the injection of nifedipine, another calcium channel blocker. The decrease in IRS after isoproterenol treatment suggested a calcium-mediated secretory depletion of the peptide which could, therefore, be a lumone factor being secreted into the lumen of the acini of the Harderian gland. Our preliminary immunohistochemical results on rat Harderian gland would support this hypothesis; positive staining with a somatostatin antibody was observed in the apical part of some of the alveolar cells and in isolated lumina of the gland.

Peinado et al. (1990) have shown diurnal and infradian rhythms of radioimmunoassayable IRS in rat Harderian gland. Lower values in males were again observed (10.3 ± 0.8 pg IRS/mg protein versus 30.6 ± 3.5 pg/mg protein in females). A diurnal rhythm with an acrophase at 02.00 h during darkness was observed in females but not in males (Fig. 1 top panel) when the experiment was performed in November. A similar study carried out with females in May again showed a significant rhythm and peak value, although the rise in IRS occured slightly earlier, at the beginning of the dark period at 20.00 and 24.00 h (Fig. 1, bottom panel); this lack of seasonal variation in the Harderian gland IRS

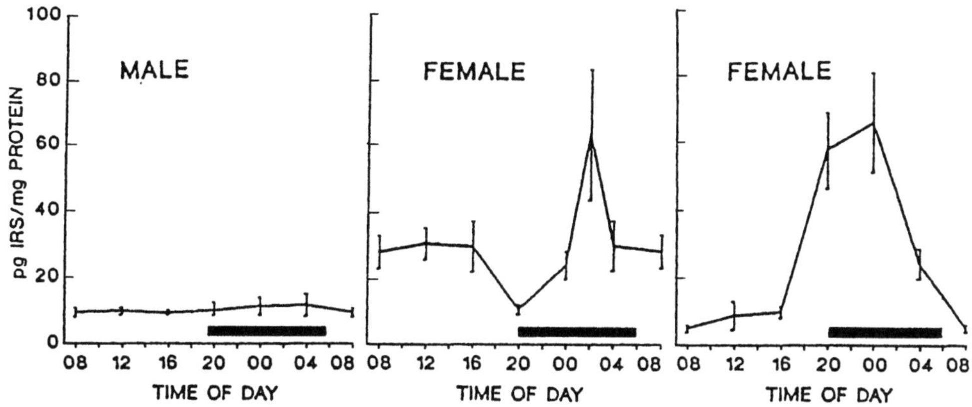

Fig. 1. Harderian gland immunoreactive somatostatin (*IRS*) throughout a 24-h -period in male and female rats, sacrificed in November. The *black bar* depicts period of darkness. A significant rhythm ($P < 0.01$) was observed only in females. *Right* A 24-h -profile of IRS in female rats sacrificed in May, again showing a significant diurnal rhythm ($P < 0.001$). (Peinado et al. 1990)

rhythm contrasted with findings in pineal and retina, where lower values were detected in spring. Even though no changes were observed throughout the estrus cycle, it is worth recalling that all these animals were sacrificed at a single timepoint at the beginning of the dark period; thus, diurnal changes, as demonstrated in the 24-h experiments, were not evaluated.

The differences in rat Harderian gland IRS concentration measured in the United States (ng/mg protein) and Spain (pg/mg protein) by the same individuals is probably related to the standard and separation technique since the antibody, tracer, and other reagents were similar. This fact exemplifies the difficulty in comparing results from different laboratories as far as absolute concentrations are concerned.

Finally, the injection of a long-acting somatostatin analog (SMS 201-995) at a dose of 1μg/100 g b.w. at 2200 h to female rats did not produce significant changes 3 h later in Harderian gland or pineal melatonin content, but did increase retinal melatonin (Webb et al. 1989).

3 Vasoactive Intestinal Polypeptide (VIP)

Of the four reports on VIP in the Harderian gland, three used immunohistochemical localization of the peptide in cats[1] of both sexes (Johansson and Lundberg 1981), in male rats (Tsukahara and Jacobowitz 1987), and chickens

[1]But, according to Sakai (Chap. 2, this Vol.), cats possess nictitans glands but not Harderian glands.

(Walcott et al. 1989), while the other measured VIP by radioimmunoassay in rabbits of both sexes (Butler et al. 1984). In this latter study, VIP was measured in substantial amounts (20.2 ± 4.9 pmol/g wet weight) in the Harderian gland of these rabbits, but it virtually disappeared (0.6 ± 0.3 pmol/g wet weight) 7 days after parasympathetic denervation by diathermic damage to the pterygopalatine ganglion, and thus VIIth nerve degeneration. These findings led the authors to suggest that this VIP was of neural origin, contained only within unmyelinated parasympathetic fibers of facial nerve origin where VIP-ergic fibers would co-exist with cholinergic ones; they also pointed out an almost certain physiological function for VIP in both the lacrimal and Harderian gland as a vasodilator. Furthermore, in the isolated rat lacrimal gland, VIP has been reported to stimulate protein secretion (Dartt et al. 1983), and thus could conceivably have a similar function in the Harderian gland, although species variations must, of course, be considered.

An extensive study performed in six cats of both sexes (Johansson and Lundberg 1981) demonstrated VIP-like immunoreactivity in nerve endings of the Harderian gland as well as in the submandibular salivary and lacrimal glands by the peroxidase-antiperoxidase method and routine electron microscopy. These findings were in agreement with previous findings at the light microscopic level. Both nerve fibers and varicosities stained positively for VIP and both could be seen close to the secretory acini and further away from ducts and blood vessels of the glands. Sympathectomy did not alter the number or distribution of these VIP-positive fibers. These authors were among the first who supported the concept of co-localization of acetylcholine and VIP in the same nerve fibers, and suggested a role for VIP in evoking vasodilation and exocrine secretion of these glands. In fact, they hypothesized that this vasodilation following parasympathetic stimulation had two components: one atropine-sensitive, mainly cholinergic, which occurred at low frequency stimulation, and one atropine-resistant, of VIP-ergic nature, which took place at a higher frequency stimulation. There are data suggesting that both substances must act in combination to produce an adequate vasodilation, and that both may also be important at other levels such as on electrolyte and water secretion into the glandular ducts.

Concordant findings have been described in the chicken Harderian gland by Walcott et al. (1989) using immunocytochemistry. The innervation of the chicken and the pigeon HG differs from that of the mammalian lacrimal gland by the absence of nerve fibers surrounding cortical secretory epithelial acini and the high density of fibers in the medullary interstitial region which contains lymphoid cells of the immune system. These authors demonstrated positive VIP staining both for nerve fibers within the Harderian gland and for neuronal somata in the adjacent pterygopalatine ganglion complex. An extensive acetylcholinesterase (AChE) network was also demonstrated in both locations, as well as catecholamine-positive fibers among the interstitial cells. VIP, AChE and catecholamine staining within the Harderian gland was markedly reduced after surgical excision of the pterygopalatine ganglion, while superior cervical ganglionectomy (SCGX) or severance of the radix autonomica resulted in loss of catecholamine-staining in the pterygopalatine ganglion and the Harderian gland, but did not affect AChE or VIP staining. Thus, these results also support the concept that VIP is a co-transmitter for the cholinergic neurons of the pterygopalatine ganglion, and may mediate lacrimal secretion. As mentioned above, VIP in salivary glands has been suggested to potentiate secretion induced by acetylcholine and can directly stimulate lacrimal protein secretion.

Co-localization of VIP and AChE has also been described in the rat HG by immunocytochemistry (Tsukahara and Jacobowitz 1989). Numerous VIP-positive nerve fibers were observed in close proximity to the glandular acini and within connective tissues near the basement membrane of the acini and enveloping blood vessels, often exhibiting concomitant AChE activity also, especially in proximity to the acini and blood vessels. This again raises the possibility that VIP and acetylcholine play cooperative roles in processes of vasodilation and glandular secretion.

4 Substance P

Substance P has been reported to exist in the Harderian gland of chickens (Walcott et al. 1989) and rabbits (Butler et al. 1984). In the latter, the peptide concentration (0.6 ± 0.1 pmol/g wet weight) was considerably smaller than that of VIP, but similarly fell after disruption of the VIIth nerve tract by damage to the pterygopalatine ganglion (0.2 ± 0.0; $p < 0.0125$), indicating some additional injury to sensory fibers of probable trigeminal origin. In chickens, Substance P-immunoreactive fibers were observed throughout the Harderian gland, being particularly dense and varicose among the interstitial cells. As was the case with VIP, the pterygopalatine ganglion also contained Substance P- as well as catecholamine- and AChE-positive fibers. In some areas of this ganglion, the somata appeared surrounded by Substance P varicosities. However, the origin of these fibers was unclear since no Substance P cell bodies were found in the pterygopalatine ganglion, and its ablation failed to eliminate these positive fibers from the Harderian gland. Even though Substance P has been found in the trigeminal ganglion, ablation of branches of the trigeminal nerve had no significant effect on the presence of Substance P fibers in the gland. In rats, Substance P has been found in salivary glands, where it can stimulate secretion and depolarize postganglionic neurons in some autonomic ganglia. Thus, it is conceivable that in certain species, Substance P may play an effector role in exocrine secretion and probably also in the Harderian gland. Furthermore, Substance P is known to induce an enhancement of macrophage and T-cell function. It is, therefore, possible that, together with the autonomic nervous system, Substance P fibers may play a role in the production and release of immunoglobulins from plasms cells in the chicken Harderian gland.

5 Arginine Vasopressin and Oxytocin

Arginine vasopressin (AVP) and oxytocin (OT) have been reported in Harderian glands of Wistar rats (Gauquelin et al. 1988; Ghaemmaghami et al. 1987). Using HPLC coupled with specific radioimmunoassays, these authors demonstrated day/night rhythms of AVP and OT in the Harderian gland of males sacrificed in February, with peak values at 22.00 and 02.00 h (around 200 and 600 fmol/mg protein) and nadirs at 04.00 h (near the lower limit of detection and 200 fmol/mg protein). In female rats which underwent either treadmill running or swimming, a significant rise in Harderian gland AVP concentration was seen only after running (approximately 0.55 compared to 0.38 in controls and 0.35 pmol/mg protein in the swimming group). No parallelism was seen in the variations of AVP concentration in the neurohypophysis, retina, different brain regions, or the Harderian gland. Curiously, the pineal gland did resemble the Harderian gland in having higher values of AVP after running; this led the authors to suggest that the increase could be related to a lesser thermal loss during running than that during swimming. Alternatively, osmotic balance could have been modified, since Harderian gland exudate spread over the coat contributes to water balance in gerbils (Harriman and Thiessen 1983). However, no data are available on the effect of water deprivation on Harderian gland AVP. Numerical data were not offered by these authors nor did they specify the time of day or month the exercising animals were sacrificed. Thus, possible sex differences in the Harderian gland concentrations of AVP cannot be determined.

6 Neurotensin (NT), Cholecystokinin (CCK), Neuropeptide Y (NPY) and Calcitonin-Gene-Related-Peptide (CGRP)

Apart from immunohistochemically localizing VIP in male Sprague-Dawley rats, Tsukahara and Jacobowitz (1987) also studied the distribution of these other peptides. CCK-positive fibers were found to exhibit a density and distribution in close apposition to the acini, similar to those observed for VIP. The other three peptides were only observed in sparse numbers of varicose nerve fibers in the interlobular space. The origin and function of these peptides has not been studied.

7 Harderian Gland-Derived Growth Factor (HGDGF)

Yokoyama et al. (1989) have recently described an extract from adult guinea-pigs which possessed an important stimulatory action on the growth of Harderian gland in culture. Based on its physicochemical and biological properties, it appears to be a previously undescribed polypeptidic growth factor with a molecular weight of 13 000. The authors suggest that HGDGF may regulate or somehow stabilize the distinctive metabolic nature of the HG, or once secreted to the surface of the eye, is used to repair injured parts of the cornea.

8 Thyrotropin-Releasing Hormone (TRH)

Very preliminary measurements carried out by Dr. L. Duntas of the University of Ulm in Germany on Harderian gland tissue supplied to him by us have been successful in detecting TRH in some samples. In four male Syrian hamsters sacrificed in October, TRH concentration was measurable in the Harderian gland of all four (between 6.5 and 18.5 fmol/mg protein). In 40 female rats sacrificed at 2100 h in May throughout the estrous cycle, only seven were above the limit of detection (15 fmol/mg protein); six of these seven were animals in estrus ($n = 18$), with values oscillating between 15 and 57 fmol/mg protein. The remaining animal (26 fmol/mg protein) corresponded to one of the 16 females in diestrus. These samples were obtained from the same animals in which IRS was measured by Peinado et al. (1990). In view of these initial results, further measurements of HG-TRH are presently being planned.

9 Summary

Immunoreactive somatostatin (IRS), vasoactive intestinal polypeptide (VIP), Substance P, arginine vasopressin (AVP), oxytocin (OT), neurotensin (NT), cholecystokinin (CCK), neuropeptide Y (NPY), calcitonin gene-related peptide (CGRP), and thyrotropin-releasing hormone (TRH), as well as the Harderian gland-derived growth factor (HGDGF) have been described in the Harderian gland of different rodents, rabbits, cats, and chickens. IRS is higher in female rats and Syrian hamsters than in males, and in the female hamster is highest at ovulation. In rats, hypophysectomy had no effect, but prior in vivo growth hormone treatment did increase HG-IRS, while neuropharmacological studies suggest secretion of IRS into the lumen of the acini, partially by a calcium-mediated mechanism. Thus, IRS could be a so-called lumone factor. In female rats, a circadian rhythm with an acrophase during the dark period was observed.

Co-existence of VIP and acetylcholine-esterase (AChE)-positive fibers has been reported in parasympathetic cholinergic fibers originating in the pterygopalatine ganglion complex of the VIIth nerve in rabbits, cats, chickens, and rats, and is believed to mediate secretion and vasodilation of the HG. Substance P, described in chickens, also coexists with AChE but its origin is unclear since superior cervical ganglionectomy (SCGX), pterygopalatine ganglionectomy (PPGX), and Vth nerve damage hardly influenced the positive reaction in these fibers. It may regulate immune mechanisms in the avian Harderian gland as well as influencing gland secretion. In rabbits, PPGX did reduce Substance P.

AVP and OT exhibit day/night rhythms with acrophases at 22.00 and 02.00 h in male rats. In female rats, AVP rose after treadmill running but not after swimming, similar to AVP behavior in the pineal gland.

Finally, the purification and characterization of an apparently new type of polypeptidic growth factor in the Harderian gland of guinea pigs (HGDGF) may open new horizons in elucidating the physiological role of these enigmatic glands.

References

Butler JM, Ruskell GL, Cole DF, Unger WG, Zhang SQ, Blank MA, McGregor GP, Bloom SR (1984) Effects of VIIth (facial) nerve degeneration on vasoactive intestinal polypeptide and substance P levels in ocular and orbital tissues of the rabbit. Exp Eye Res 39:523–532

Dartt DA, Baker AK, Vaillant C (1983) Effect of vasoactive intestinal polypeptide on protein secretion from rat lacrimal gland. Invest Ophtalmol Visual Sci 24 (Suppl):155

Ebling FJ, Ebling E, Randall V, Skinner J (1975) The synergistic action of alpha-melanocyte-stimulating hormone and testosterone on the sebaceous, prostate, preputial, Harderian and lachrymal glands, seminal vesicles and brown adipose tissue in the hypophysectomized-castrated rat. J Endocrinol 66:407–412

Gauquelin G, Gharib C, Ghaemmaghami F, Allevard AM, Cherbal F, Geelen G, Bouzeghrane F, Legros JJ (1988) A day/night rhythm of vasopressin and oxytocin in rat retina, pineal and Harderian gland. Peptides 9:289–293

Ghaemmaghami F, Gauquelin G, Gharib C, Yoccoz D, Desplanches D, Favier R, Allevard AM (1987) Effects of treadmill running and swimming on plasma and brain vasopressin levels in rats. Eur J Appl Physiol 56:1–6

Harriman AE, Thiessen D (1983) Removal of Harderian exudates by sandbathing contributes to osmotic balance in Mongolian gerbils. Physiol Behav 31:317–323

Johansson O, Lundberg JM (1981) Ultrastructural localization of VIP-like immunoreactivity in large dense-core vesicles and "cholinergic-type" nerve terminals in cat exocrine glands. Neuroscience 6:847–862

Joó I, Kahán A (1975) The porphyrin content of Harderian glands in rats and the melatonin-melanocyte stimulating hormone system. Endokrinologie 65:308–312

Kok-van Alphen CC, Noach EL (1978) Hormonal influences on ocular refraction in the rat. Nature 271:62–64

Peinado MA, Fajardo N, Hernández G, Puig-Domingo M, Viader M, Reiter RJ, Webb SM (1990) Immunoreactive somatostatin diurnal rhythms in rat pineal, retina and Harderian gland: Effects of sex, season, continuous darkness and estrous cycle. J Neural Transm 81:63–72

Puig-Domingo M, Guerrero JM, Reiter RJ, Peinado MA, Menéndez-Peláez A, Santana C, Webb SM (1988a) Androgenic control of immunoreactive somatostatin in the Harderian gland of the Syrian hamster. J Reprod Fertil 82:753–759

Puig-Domingo M, Guerrero JM, Reiter RJ, Peinado MA, Sabry I, Viader M, Webb SM (1988b) Identification of immunoreactive somatostatin in the rat Harderian gland: regulation of its content by growth hormone, beta-adrenergic agonists and calcium channel blockers. Peptides 9:571–574

Tsukahara S, Jacobowitz DM (1987) Peptidergic innervation of the rat Harderian gland. Histochemistry 87:233–236

Walcott B, Sibony PA, Keyser KT (1989) Neuropeptides and the innervation of the avian lacrimal gland. Invest Ophtalmol Visual Sci 30:1666–1674

Webb SM, Peinado MA, Puig-Domingo M, Viader M, Hernández G, Reiter RJ (1989) Somatostatin and melatonin in the rat pineal gland, retina and Harderian gland. In: Reiter RJ, Pang SF (eds) Advances in pineal research, vol 3. John Libbey, London, pp 151–154

Yokoyama Y, Kano K, Kaji K, Seyama Y (1989) Purification and characterization of a growth factor from guinea pig Harderian gland. J Biol Chem 264:17058–17063

Adrenergic Activity in the Male Rat Harderian Gland

PEDRO ABREU, GUADALBERTO HERNÁNDEZ, NATALIA FAJARDO and RAFAEL ALONSO

Laboratorio de Neuroendocrinología, Departamento de Fisiología, Universidad de La Laguna, Tenerife, Spain

1 Introduction

The rodent Harderian glands (HG) are large tubuloalveolar glands which occupy a considerable portion of the posterior half of the orbital cavity, and their secretions are vented by means of merocrine mechanisms (Sakai 1981). Although the function of this gland is poorly understood, its lipid-containing secretions may lubricate the nictitating membrane in those animals with a third eyelid (Davis 1929).

Previous investigations of the mammalian HG have demonstrated two major cell types: secretory cell types I and II, and myoepithelial cells. The activity of these cells is regulated by both cholinergic and adrenergic nerves. The myoepithelial cells seem to be cholinergically innervated and their contraction squeezes the secretory products from the secretory cells. Furthermore, this secretory process seems to be indirectly under the dual control of both cholinergic and adrenergic fibers, which innervate the blood vessels of the gland, thus affecting the nutritional state of the secretory cells (Bucana and Nadakavukaren 1972; Norvell and Clabough 1972; Huhtala et al. 1977). Recently, the possibility of co-innervation between cholinergic and peptidergic nerves has been suggested (Uddman et al. 1980; Tsukahara and Jacobowitz 1987).

Among other compounds of interest, the HG of a number of species contain porphyrins (Buzzel et al. 1989; Van Jaarsveld et al. 1989), indoleamines such as melatonin and the synthesizing enzymes N-acetyltransferase (NAT) and hydroxyindole-O-methyltransferase (HIOMT) (Hoffman et al. 1989a; Menéndez-Pelaez et al. 1989a). In addition, several authors have reported the presence of peptides such as somatostatin, VIP, neurotensin, and neuropeptide Y (Tsukahara and Jacobowitz 1987; Puig-Domingo et al. 1988a).

In Syrian hamsters (*Mesocricetus auratus*), these glands exhibit an interesting sexual dimorphism which involves both morphological and biochemical characteristics (Menéndez-Pelaez et al. 1988a; Buzzel et al. 1989; Pangerl et al. 1989a, b). The Harderian gland NAT activity shows a marked increase (especially in females) shortly after lights-on. Several authors have reported the ineffectiveness of superior cervical ganglionectomy (SCGx) and adrenergic

Webb et al. (Eds).
Harderian Glands

stimulation to modify the NAT activity in male hamster HG; however, the adrenergic denervation in female animals is able to reduce the enzyme activity. Male castrated hamsters exposed to long days showed higher NAT activity than those of castrated animals exposed to short photoperiods, indicating that light may have an influence on HG NAT activity independent of gonadal status (Menéndez-Pelaez et al. 1988b, 1989b). In the same way, short photoperiods (which induce gonadal regression) significantly decreased NAT activity in male and female Syrian hamsters HG (Menéndez-Pelaez et al. 1988b, c). On the other hand, Hoffman et al. (1985, 1989a) suggested that immunoreactive melatonin concentrations in the HG of hamsters is controlled by testosterone or its derivatives. Even though a number of studies support the possible androgenic control of several compounds in the HG (Puig-Domingo et al. 1988b; Hoffman et al. 1989b), to our knowledge no data have been reported concerned with the HG adrenergic activity and its relationship to changes in gonadal status in the male rat.

This work was undertaken in order to investigate the tyrosine hydroxylase (TH) activity and norepinephrine (NE) content in the male rat HG during a 24-h period, and its possible modifications by means of gonadal manipulations conditioned by different profiles of serum testosterone levels.

2 Experimental Procedures

2.1 Animals

Adult male Sprague-Dawley rats (150–200 g) were housed in a 12:12 light/dark cycle, under controlled lighting of 60–80 $\mu W/cm^2$ (lights on 23.00 h; lights off 11.00 h) and temperature (22 ± 2 °C) with ad libitum access to rat chow and tap water. Bilateral castration was performed under ether anesthesia 10 days before sacrifice. Animals were killed by decapitation at 60 days, and trunk blood collected for hormone assays. Those animals sacrificed during the dark phase were under dim red light. Harderian glands were quickly removed and frozen on solid CO_2. The glands were homogenized in 2 ml of 0.25 M sucrose, aliquoted and stored together with serum samples at − 80 °C until assayed.

Experiment 1. The animals received the following treatment: (1) Control animals. (2) Bilateral castration + vehicle injections (propylene glycol). (3) Bilateral castration + two subcutaneous silastic implants (60 mm) of testosterone, 10 days prior to sacrifice. Animals were sacrificed at 00.00, 04.00, 08.00, 12.00, 16.00 and 20.00 h (eight animals per hour) (Fig. 1).

Experiment 2. In this experiment, castrated animals were subcutaneously injected with testosterone propionate (TP) (40 μg/100 g body weight) at 11.00,

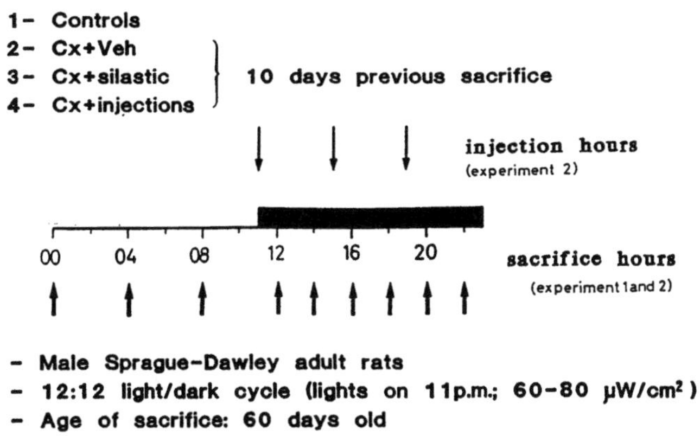

Fig. 1. Experimental design to elucidate the role of circulating gonadal hormones on adrenergic activity in the HG male rat. The *upper arrows* represent the subcutaneous injection times of TP (40 µg/100 g body weight, *experiment 2*) and *bottom arrows* indicate the sacrifice hours

15.00 and 19.00 h, during the 10 days previous to sacrifice. Animals were sacrificed at 00.00, 04.00, 08.00, 12.00, 14.00, 16.00, 18.10, 20.00 and 22.00 h (eight animals per hour) (Fig. 1). This injection schedule was chosen in order to simulate a pulsatile serum testosterone pattern (Södersten et al. 1983).

2.2 Tyrosine Hydroxylase Activity

Harderian gland aliquots were thawed at room termperature and TH activity was determined by high performance liquid chromatography (HPLC) with amperometric detection (ED) of enzymatically formed L-dihydroxy-phenylalanine (L-Dopa) (Abreu et al. 1987). The incubation mixture consisted of the following components: 10 µl of 1.0 M sodium acetate buffer (pH 6.0), 20 µl of 1.5 mM L-tyrosine, 10 µl of 10 mM 6-methyl-5, 6, 7, 8 tetrahydropterin, 10 µl of 10 mM ferrous ammonium sulfate, 50 µl of homogenate, and 20 µl of water. After 15 min of incubation at 37 °C, the reaction was stopped by the addition of 500 µl of 0.5 M perchloric acid containing 300 pmol of alpha-methyl-Dopa (as internal standard). After 10 min, the mixture was centrifuged at 2500 *g* for 10 min; then 20 µl of 0.2 M EDTA and 400 µl of 2.5 M Tris buffer were added. Dopa and alpha-methyl Dopa were adsorbed onto alumina from the clear supernatant, and eluted in 70 µl of 0.2 M perchloric acid. Aliquots of 20 µl were injected into HPLC-ED system. The results are expressed in pmol Dopa/min/mg protein.

2.3 Catecholamine Concentration

Harderian gland aliquots were thawed at room temperature and NE content was determined by HPLC-ED. To 200 µl of homogenate, 500 µl 0.5 M perchloric acid were added, containing 5 ng dihydroxybenzylamine (DHBA) as internal standard in an ice bath for 10 min. After centrifugation at 23000 *g* for 10 min, the supernatant was adsorbed onto alumina previously adjusted to pH 8.5 with 400 μl of 2.5 M Tris buffer, and then eluted in 70 μl 0.2 M perchloric acid. Aliquots of 20 μl were injected into HPLC-ED system. The results are expressed in ng NE/mg protein.

Protein content was measured using Bradford's method (Bradford 1976).

2.4 Chromatographic Conditions

The samples were injected onto a C18 Novapak column (particle size 5 μm, Millipore Iberica, S.A., Barcelona, Spain). The mobile phase consisted of 0.1 M monosodium phosphate buffer containing 0.1 mM EDTA, 5 mM sodium heptanosulphonate, and 0.42 M acetic acid, pH 3.1, at a flow rate of 0.9 ml/min. The detector potential was set at 0.70 V versus Ag/AgCl reference electrode (Water 460, Millipore Iberica, S.A., Barcelona, Spain).

2.5 Hormonal Determination

Testosterone serum levels were determinated using a commercial kit, (Inmunotech S.A.) and are expressed as ng/ml.

2.6 Statistical Analysis

Data are expressed as means ± SEM and were analyzed using two-way ANOVA followed by Tukey test.

3 Experimental Findings

Harderian gland NE content remained stable through the 24-h period. Neither castration nor testosterone implants were able to induce any significant changes of this variable, although a slight elevation in two light phase hours (00.00, 04.00 h) were observed (Fig. 2). Testosterone propionate injections likewise did not modify the HG NE content (Fig. 3).

Harderian gland TH activity exhibited changes over the 24-h period in the control group, with the acrophase at 08.00 h, 3 h before dark onset (Fig. 4, left panel, Co). Light/dark variations observed in this group ($p < 0.03$) (Fig. 4, right panel, Co) were abolished by castration (Fig. 4 right panel, Cx). Testosterone replacement with silastic implants produced an increase in TH activity during the light phase (00.00, 04.00 h) (Fig. 4, left panel, Cx + Sil), and restored the light/dark variations observed in control animals ($p < 0.05$) (Fig.4, right panel, Cx + Sil).

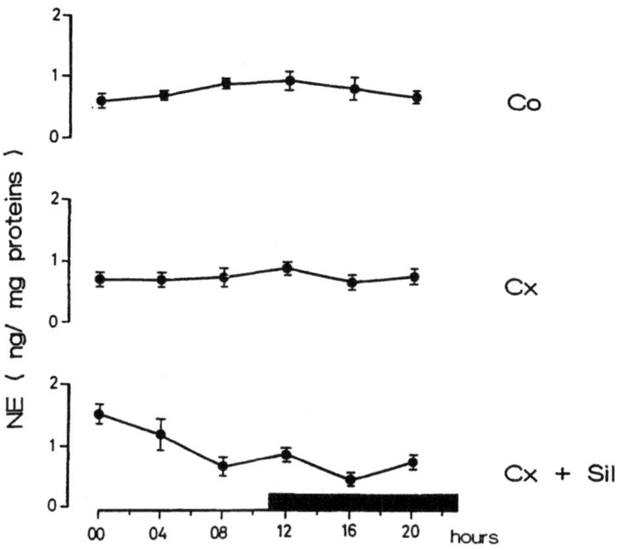

Fig. 2. Norepinephrine HG content in control animals (*Co*), castrated (*Cx*), and castrated plus silastic implant (*Cx + Sil*). *Black horizontal bar* represents the dark phase of the 12:12 light/dark cycle (light off 11.00 h). Data are expressed as mean ± SEM of eight animals per hour

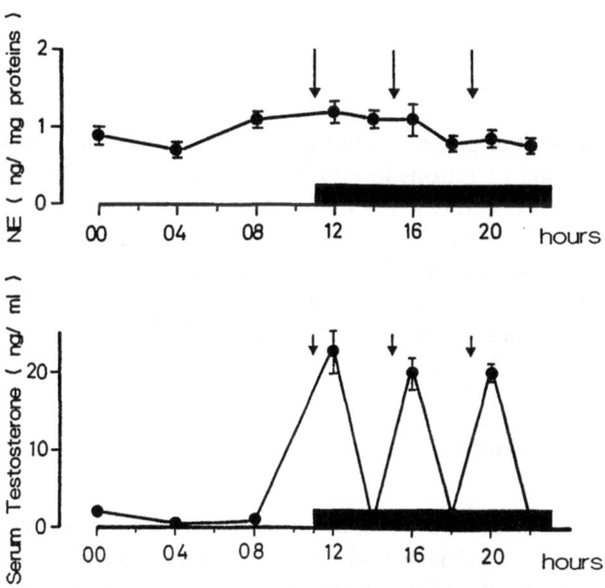

Fig. 3. Serum testosterone (*bottom panel*) and norepinephrine content (*upper panel*) in testosterone propionate-injected animals. *Black horizontal bar* represents the dark phase of the 12:12 light/dark cycle (light off 11.00 h). The *arrows* represent the injection hours. Data are expresed as mean ± SEM of eight animals per hour

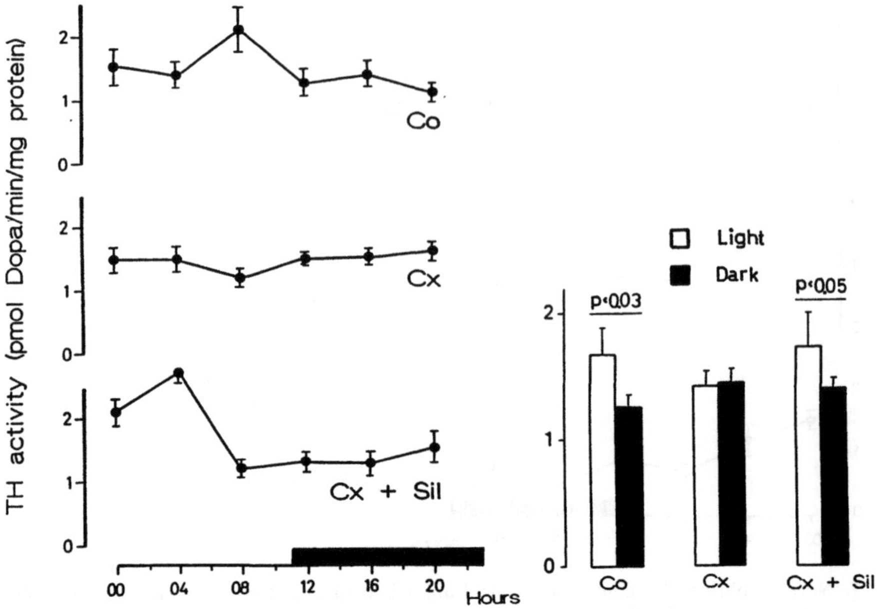

Fig. 4. Tyrosine hydroxylase HG activity in control animals (*Co*), castrated (*Cx*), and castrated plus silastic implant (*Cx + Sil*). *Black vertical and horizontal bars* represent the dark phase of the 12:12 light/dark cycle (light off 11.00 h). The *arrows* represent the injection hours. Data are expresed as mean ± SEM of eight animals per hour (*left panel*), or mean ± SEM of 24 animals, regrouping them light vs. dark (*right panel*)

The pulsatile serum testosterone pattern, conditioned by TP injection, was apparently able to induce parallel variations in HG TH activity (Fig. 5). During the period of TP injection, HG TH activity was higher at hours coinciding with peak serum testosterone levels (12.00, 16.00 and 20.00 h), as compared with those coinciding with low serum testosterone levels (14.00, 18.00, and 22.00 h). Furthermore, a positive correlation between mean serum testosterone and HG TH activity during darkness in TP-injected animals was observed (Fig. 6).

4 Commentary and Conclusions

Even though day/night variations of several biochemical parameters in the rat HG are well documented (Menéndez-Pelaez et al. 1988c, d, 1989a; Van Jaarsveld et al. 1989), to our knowledge, no data have been reported on the adrenergic activity of the rodent HG. Therefore, to assess whether the HG adrenergic activity could be affected either by light/dark cycle or the gonadal status, adult male rats with four different profiles of serum testosterone were studied.

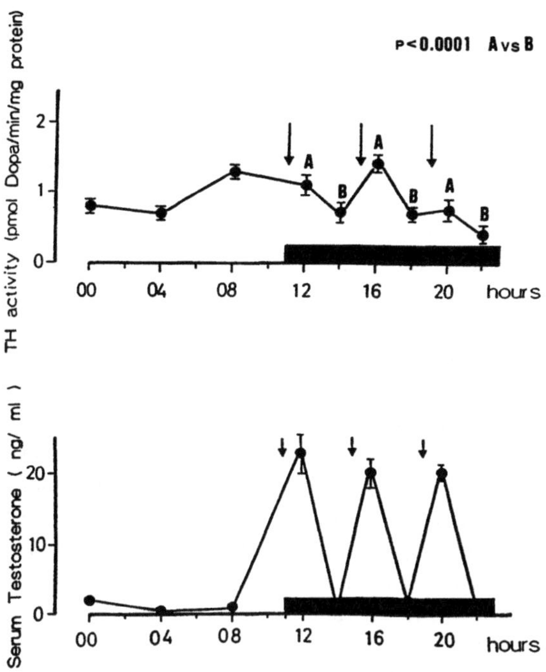

Fig. 5. Serum testosterone (*bottom panel*) and tyrosine hydroxylase HG activity (*upper panel*) in testosterone propionate-injected animals. *Black horizontal bar* represents the dark phase of the 12:12 light/dark cycle (light off 11.00 h). The *arrows* represent the injection hours. Data are expressed as mean ± SEM of eight animals per hour. *A* vs. *B* points ($P < 0.0001$)

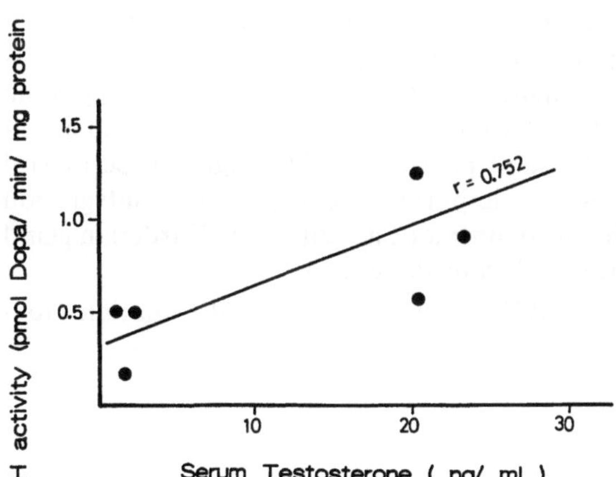

Fig. 6. Relationship between serum testosterone levels and TH activity in the HG of castrated plus testosterone propionate-injected animals during darkness. The correlation coefficient was $r = 0.752$

Harderian gland TH activity was higher during light hours in control animals, with no apparent changes of NE content. Castration abolished day/night variations of TH activity, whereas testosterone replacement by means of silastic implants was able to prevent such effects. On the other hand, daily administration of testosterone was apparently able to stimulate HG TH activity, with a peak coinciding with the rise of serum testosterone levels. Since circulating testosterone displays episodic variations in the normal male rat (Sördersten et al. 1983), similar to those induced in the present study, these data might suggest a functional role of androgens in the control of adrenergic inputs to the rat HG. This was further indicated by the positive relationship between serum testosterone levels and Harderian gland TH activity (Fig. 6).

In other central and peripheral regions with adrenergic innervation, castration has been reported to reduce TH activity, an effect which was prevented by testosterone administration (Abreu et al. 1988; Hernández et al. 1991). These reports, as well as the results presented here, could be partially in agreement with a general effect of castration reducing the number of neurons (Wright and Smolen 1985) and the levels of TH activity (Hamill et al. 1984) in peripheral sympathetic ganglia. Alternatively, other studies were interpreted as demonstrating the presence of a specific high-affinity intracellular androgen receptor in the male hamster HG (Vilchis et al. 1986).

Unexpectedly, neither gonadectomy nor testosterone treatment caused any apparent variation on HG NE content. Since NE is the neurotransmitter in the adrenergic nerve fibers arising from the cervical sympathetic ganglia, and coupled to blood vessels that supply the HG (Huhtala et al. 1977), the observed alteration of TH activity should have caused a similar change in NE level.These data are in agreement with the view that TH activity is a good index of adrenergic activity, while the single determination of NE content can be influenced by modification of neurotransmitter synthesis rate, release, or degradation (Westerink 1985).

In summary, the results reported here are consistent with one castration-testosterone replacement-induced change in HG adrenergic activity, suggesting that circulating levels of gonadal and/or pituitary hormones are able to participate in the control of Harderian gland physiology. Although the peripheral sympathetic ganglia might constitute important neuroendocrine integration centers (Cardinali et al. 1987), the site of hormone action on the Harderian gland regulation system cannot be identified from these data.

Acknowledgements. Supported by DGICYT PM88-0206, and Canarian Government 70/31.07.89 grants.

References

Abreu P, Santana C, Hernandez G, Calzadilla CH, Alonso R (1987) Day–night rhythm of rat pineal tyrosine hydroxylase activity as determined by HPLC with amperometric detection. J Neurochem 48:665–668

Abreu P, Hernandez G, Hernandez CH, Alonso R (1988) Reproductive hormones control striatal tyrosine hydroxylase activity in the male rat. Neurosci Lett 95:213–217

Bradford MM (1976) A rapid and sensitive method for the quantitation of quantities of protein utilizing the principle of protein-dye binding. Anal Biochem 72:248–254

Bucana CD, Nadakavukaren MJ (1972) Innervation of the hamster Harderian gland. Science 175:205–206

Buzzell GR, Menéndez-Pelaez A, Porkka-Heiskanen T, Pangerl A, Pangerl B, Vaughan MK, Reiter RJ (1989) Bromocriptine prevents the castration-induced rise in porphyrin concentration in the Harderian glands of the male Syrian hamster, *Mesocricetus auratus*. J Exp Zool 249:172–176

Cardinali DP, Romeo HE, Vacas MI (1987) Neuroendocrine projections of superior cervical ganglia. In: Reiter RJ, Fraschini F (eds) Advances in pineal research, vol 2. John Libbey, London, pp 35–49

Davis FA (1929) The anatomy and histology of the eye and orbit of the rabbit. Trans Am Ophthalmol Soc 27:401–441

Hamil RW, Earley CJ, Guernseyl LA (1984) Hormonal regulation of adult sympathetic neurons: the effects of castration on tyrosine hydroxylase activity. Brain Res 299:331–337

Hernández G, Abreu P, Alonso R, Santana C, Moujir F, Calzadilla CH (1991) Castration reduces the nocturnal rise of pineal melatonin levels in the male rat by impairing its noradrenergic input. J Neuroendocrinol (in press)

Hoffman RA, Johnson LB, Reiter RJ (1985) Harderian glands of golden hamsters: temporal and sexual differences in immunoreactive melatonin. J Pineal Res 2:161–168

Hoffman RA, Johnson LB, Reiter RJ (1989a) Regulation of melatonin in the Harderian glands of golden hamsters. J Pineal Res 6:63–71

Hoffman RA, Vaughan MK, Reiter RJ (1989b) Photic input and pineal–Harderian–thyroid gland interactions in golden hamsters. In: Reiter RJ, Pang SF (eds) Advances in pineal research, vol 3. John Libbey, London, pp 225–230

Huhtala A, Huikuri KT, Palkama A, Tervo T (1977) Innervation of the rat Harderian gland by adrenergic and cholinergic nerve fibers. Anat Rec 188:263–272

Menéndez-Pelaez A, Reiter RJ, Guerrero JM, Puig-Domingo M, Howes KA (1988a) Sexual dimorphism in N-acetyltransferase activity, hydroxyindole-O-methyltransferase activity and melatonin content in the Harderian gland of Syrian hamsters: changes following gonadectomy. Proc Soc Exp Biol Med 187:287–291

Menéndez-Pelaez A, Reiter RJ, Howes KA, Puig-Domingo M, Vaughan MK, Troiani ME, Little JC (1988b) Harderian gland N-acetyltransferase activity in the male Syrian hamster: effects of gonadectomy, short photoperiod exposure, or subcutaneous melatonin implants. Endocr Res 14:121–130

Menéndez-Pelaez A, Santana C, Howes KA, Sabry I, Reiter RJ (1988c) Effects of photoperiod or exogenous melatonin administration on the activity of N-acetyltransferase and hydroxyindole-O-methyltransferase and the melatonin content of the Harderian gland of two strains of female Syrian hamster. J Pineal Res 5:293–300

Menéndez-Pelaez A, Reiter RJ, Guerrero JM, Santana C, Howes KA, González-Brito A (1988d) N-acetyltransferase activity and melatonin concentrations in the Harderian glands of female Syrian hamsters: alterations following either pinealectomy or bilateral superior cervical ganglionectomy. Biochem Arch 4:231–238

Menéndez-Pelaez A, Reiter RJ, González-Brito A, Howes KA, (1989a) N-acetyltransferase activity and indole contents of the male Syrian hamster Harderian gland: changes during the light:dark cycle. Arch Biochem Biophys 268:20–25

Menéndez-Pelaez A, Reiter RJ, González-Brito A, Santana C, Guerrero JM (1989b) Regulation of N-acetyltransferase in the Syrian hamster Harderian glands. In: Reiter RJ, Pang SF (eds) Advances in pineal research, vol 3. John Libbey, London, pp 77–81

Norvell JE, Clabough JW (1972) Adrenergic and cholinergic innervation of the hamster Harderian gland. Science 178:1102–1103

Pangerl A, Pangerl B, Buzzell GR, Jones DJ, Reiter RJ (1989a) Characterization of B-adrenoceptors in the Syrian hamster Harderian gland: sexual differences and effects of either castration or superior cervical ganglionectomy. J Neurosci Res 22:456–460

Pangerl A, Pangerl B, Jones DJ, Reiter RJ (1989b) B-adrenoceptors in the extraorbital lacrimal gland of the Syrian hamster. Characterization with I-iodopindolol and evidence of sexual dimorphism. J Neural Transm 77:153–162

Puig-Domingo M, Guerrero JM, Reiter RJ, Peinado MA, Sabry I, Viader M, Webb SM (1988a) Identification of immunoreactive somatostatin in the rat Harderian gland: regulation of its content by growth hormone, beta-adrenergic agonists and calcium channel blockers. Peptides 9:571–574

Puig-Domingo M, Guerrero JM, Reiter RJ, Peinado MA, Menéndez-Pelaez A, Santana C, Webb SM (1988b) Androgenic control of immunoreactive somatostatin in the Harderian gland of the Syrian hamster. J Reprod Fertil 82:753–759

Sakai T (1981) The mammalian Harderian gland: morophology, biochemistry, function and phylogeny. Arch Histol Jpn 44:299–333

Södersten P, Eneroth P, Pettersson A (1983) Episodic secretion of luteinizing hormone and androgen in male rats. J Endocrinol 97:145–153

Tsukahara S, Jacobowitz DM (1987) Peptidergic innervation of the rat Harderian gland. Histochemistry 87:233–236

Uddman R, Alumets J, Ehinger B, Hakanson R, Lorén I, Sundler F (1980) Vasoactive intestinal peptide nerves in ocular and orbital structures of the cat. Invest Ophthalmol Visual Sci 19:878–885

van Jaarsveld A, Mhatre MC, Reiter RJ (1989) Porphyrine levels in the Harderian glands of female and male Syrian hamsters: early changes following either gonadectomy or light deprivation and lack of a circadian rhythm. Biomed Res 10:1–8

Vilchis F, Hernández A, Pérez AE, Pérez-Palacios G (1987) Hormone regulation of the rodent Harderian gland: binding properties of the androgen receptor in the male golden hamster. J Endocrinol 112:3–8

Westerink BHC (1985) Sequence and significance of dopamine metabolism in the rat brain. Neurochem Int 7:221–227

Wright LL, Smolen AJ (1985) Effects of neonatal castration or treatment with dihydrotestosterone on number of neurons in the rat superior cervical sympathetic ganglion. Dev Brain Res 20:314–316

The Interrelationship Between the Harderian Glands and the Neuroendocrine-Thyroid Axis in Rodents

GERALD R. BUZZELL[1] and ARMANDO MENENDEZ-PELAEZ[2]

[1]Department of Anatomy and Cell Biology; University of Alberta, Edmonton, Alberta T6G 2H7, Canada
[2]Departamento de Morfologia y Biologia Celular, Universidad de Oviedo, 33006 Oviedo, Spain

1 Introduction

It was noted some time ago that Harderian glands are usually enlarged in guinea pigs with exophthalmos resembling that seen in human hyperthyroidism (Graves' disease). For this reason, a Harderian gland-thyroid gland connection was thought likely. Studies have confirmed that thyroid hormones and thyroid-stimulating hormone from the pituitary influence Harderian gland morphology and biochemistry. In addition, the Harderian glands have been shown to influence thyroid hormone levels.

In this review, we will describe the contributions of studies involving the Harderian glands during investigations of Graves' ophthalmopathy. We will discuss the effects of hormones of the neuroendocrine-thyroid axis on the Harderian glands and the effects of the Harderian glands on the neuroendocrine-thyroid axis. We will then make some generalizations on the interactions of these two systems in the normal physiology of rodents.

2 Graves' Ophthalmopathy and the Harderian Gland

Graves' disease is a common thyroid disorder, characterized by high levels of circulating thyroid hormones with resultant overactivity, weight loss, and weakness. The disease is often accompanied by exophthalmos (proptosis), which results from swelling of retrobulbar tissues, especially the extraocular eye muscles. This puts pressure on the contents of the orbit. The eyeballs protrude, contributing to the characteristic exophthalmic stare. In extreme cases, there may be compression of the optic nerve, with loss of vision.

Guinea pigs or mice injected with thyrotropin (TSH) or with pituitary extracts develop a proptosis resembling human exophthalmos (Fig. 1). This has been a commonly used model for the experimental study of exophthalmos

Webb et al. (Eds).
Harderian Glands

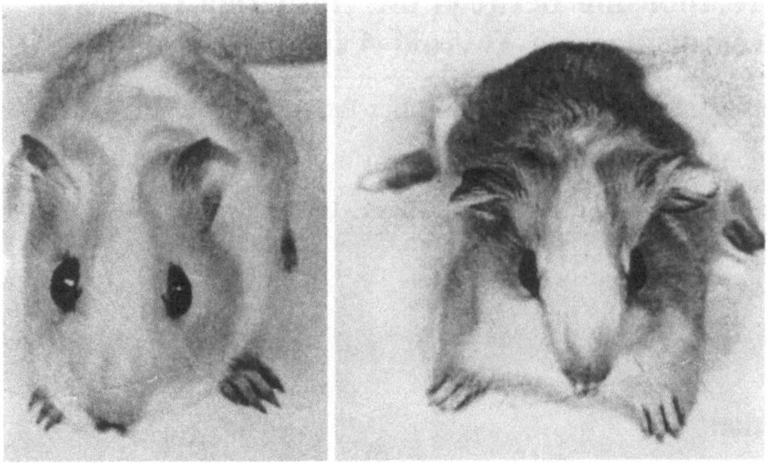

Fig. 1. Guinea pig siblings. The one on the left received TSH daily for 10 days and shows exophthalmos. The one on the right is an untreated control. (Asboe-Hansen and Iversen 1951)

(Smelser 1938, 1962; Wegelius et al. 1959), although its use has been sporadic since the mid-1930s.

In early studies, researchers produced exophthalmos by injecting thyroidectomized guinea pigs with crude pituitary extracts containing thyrotropic activity (Smelser 1936–38, 1943b,c; Paulson 1937). The exophthalmos resulting from this treatment is a long-term condition. Pochin (1945) and Ludwig et al. (1950) found that exophthalmos also results from injection of TSH preparations into intact guinea pigs (Fig. 1). Exophthalmos in intact guinea pigs tends to be transient, becoming evident a few days after injections begin and subsiding a week or two later. Most studies have used this latter model. Because proptosis is difficult to quantitate, researchers usually measured the increased ability of TSH-stimulated Harderian gland tissues to take up $^{35}SO_4$ as an indication of the degree of exophthalmos produced (Lamberg et al. 1956; Wegelius et al. 1959; Wegelius and Lamberg 1960; Brunish et al. 1962; Hadad 1963; Reichlin and Hadad 1963).

Exophthalmos in guinea pigs stimulated with TSH normally results from an increase in the size of the Harderian glands and other retrobulbar tissues (Smelser 1936), due to edema of the gland, its sheath, and other orbital tissues (Smelser 1937; Pochin 1945; Smelser and Ozanics 1959). There is an accumulation of hyaluronic acid (Ludwig et al. 1950; Asboe-Hansen and Iversen 1951; Sisson and Miles 1967; Winand and Kohn 1973) and sulfated glycosaminoglycans (Brunish 1962; Singh and McKenzie 1969b, 1970; Winand and Kohn 1973) in the tissues.

It is interesting, however, that the presence of a Harderian gland (or even an eye) is not necessary for the development of exophthalmos (or swelling of orbital tissues) in thyroidectomized guinea pigs treated with TSH (Smelser 1943a,c).

Swelling of the Harderian glands and other retrobulbar tissues as a response to pituitary extracts is probably stimulated by TSH, with some contributions from corticosteroids stimulated by corticotropin (Smelser and Ozanics 1954, 1955). However, the whole TSH molecule is not needed for this effect. Wegelius et al. (1959) and Winand and Kohn (1973) found that a pepsin digest of TSH with little to no thyrotropic activity also has ophthalmotropic activity, increasing the uptake of $^{35}SO_4$ into Harderian gland tissue. This digest of TSH appears to be the TSH β-chain and the amino terminus of the α-chain (Kohn and Winand 1971).

In a series of experiments, it was found that [^{3}H]TSH, [^{125}I]TSH, and the [^{3}H]-exophthalmos-producing factor derived from TSH by partial pepsin digestion bind to membrane preparations of guinea pig Harderian glands (Winand and Kohn 1972; Bolonkin et al. 1975) and stimulate adenylate cyclase activity in these membranes (Winand and Kohn 1975). Both TSH and the exophthalmos-producing factor stimulate the accumulation of hyaluronic acid and sulfated glycosaminoglycans in the Harderian glands (Winand and Kohn 1973).

In contrast, Teng et al. (1975), Davies et al. (1978), and Kendall-Taylor et al. (1984) failed to detect binding sites for TSH on membranes prepared from guinea pig Harderian glands and other orbital tissues.

Rodent exophthalmos may be influenced by antibodies stimulating the Harderian glands. Exophthalmos was produced in guinea pigs by immunization with emulsified Harderian glands (Pisarev et al. 1968) and $^{35}SO_4$ uptake, water content, and glycosaminoglycan content were increased in Harderian glands of mice immunized against Harderian glands (Singh and McKenzie 1971). These effects mimic those of TSH stimulation; the possibility must therefore be kept in mind that TSH or an exophthalmos-producing factor in some way stimulates an autoimmune response of the animal to its own Harderian gland tissue.

Contradictory results have been reported from workers determining whether serum from exophthalmic patients could stimulate swelling or uptake of $^{35}SO_4$ by the Harderian glands. Singh and McKenzie (1969a, b) found that there was no correlation between serum levels of long-acting thyroid stimulator [i.e., the antibody to the TSH receptor (Rees Smith et al. 1988)] and either exophthalmos-producing properties or ^{35}S uptake by mouse Harderian glands. Banovac et al. (1988) found that, although an immunoglobulin from human serum bound to guinea pig Harderian gland membranes and to membranes from human eye muscle, there was little correlation between binding of this immunoglobulin and whether the serum was from exophthalmic patients or not.

On the other hand, Hadad (1963) noted that serum from patients with progressive endocrine exophthalmos caused increased uptake of $^{35}SO_4$ by guinea pig Harderian glands. Serum from patients with inactive exophthalmos had little to no effect. Dandona and ElKabir (1970) found that either serum or immunoglobulin from patients with active exophthalmos had a tendency to increase the weights of Harderian glands in guinea pigs, but that no effect was evident with serum from normal subjects, from a patient in remission, or from patients with high levels of the long-acting thyroid stimulator but no exophthalmos. The authors suggested that there is a factor in the immunoglobulin fraction of serum from exophthalmic patients (distinct from the long-acting thyroid stimulator) which stimulates an increase in Harderian gland weight. Kendall-Taylor et al. (1984) also found an immunoglobulin from exophthalmic patients which bound to guinea pig Harderian gland or to porcine eye muscles.

In this context, it is interesting that γ-globulin from patients with active exophthalmos enhanced the binding of TSH and of the exophthalmos-producing substance derived from TSH to guinea pig Harderian gland membranes, regardless of whether the serum from which the γ-globulin was prepared tested positive for the long-acting thyroid stimulator (Winand and Kohn 1972; Bolonkin et al. 1975).

In summary, the guinea pig model of exophthalmos has been used sporadically, by well-respected scientists with interests in the pathogenesis of thyroid disease, for more than 50 years. However, its usefulness as a model for human Graves' ophthalmopathy is limited. Graves' disease is usually a condition of hyperthyroidism, whereas thyroidectomized guinea pigs display exophthalmos if injected with TSH. Graves' exophthalmos is due to swelling of extraocular muscles whereas guinea pig exophthalmos is usually due to swelling of the Harderian gland. Humans do not have Harderian glands. Although one might not agree with Weetman and McGregor (1984), who stated categorically that "no animal model of Graves' disease is currently available . . .", one must admit, with Jacobson and Gorman (1984) that "This is a very imperfect model for Graves' ophthalmopathy . . ."

3 Effects of the Neuroendocrine-Thyroid Axis on Harderian Glands

3.1 Weights, Morphology, Composition

In rats, thyroid hormones appear to be important for the normal development and maintenance of the Harderian glands. Injection of newborn rats with thyroxine led to slower Harderian gland growth than that seen in saline-treated controls (Wetterberg et al. 1970b). In adult male rats, thyroidectomy led to reduced Harderian gland weights, but to relatively little change in glandular weights relative to the body weight. Thyroxine treatment of thyroidectomized rats prevented this decrease in Harderian gland weight (Boas and Bates 1954; Boas and Scow 1954).

Hypophysectomy in rats led to a considerable decrease in both the absolute and the relative weights of the Harderian glands and to a drop in the fat content of these glands (Boas and Bates 1954; Ebling et al. 1975a, b). This decrease in weight and fat content was prevented by treatment of hypophysectomized rats with pituitary extracts containing thyrotropic activity, with partially purified TSH, or with fairly high dosages of thyroxine (Boas and Bates 1954).

A pituitary extract, from which most TSH, corticotropin, and growth hormone had been removed, caused a relative enlargement of the Harderian glands of castrated hypophysectomized male rats over the size of glands in castrated hypophysectomized control rats. However, partially purified TSH had no effect on Harderian gland weight in these animals (Lorincz and Lancaster 1957). Thus, in rats there is little evidence that TSH has any effect on Harderian gland weight distinct from its effect on hormones from the thyroid gland.

In guinea pigs, there is no good evidence that thyroid hormones directly influence Harderian gland size or composition. In published studies, thyroidectomy either had no effect (Smelser 1937) or a moderately stimulatory effect (Smelser 1943b) on Harderian gland weights. Thyroxine injections in thyroidectomized guinea pigs led to reduced body weights and reduced weights of the Harderian glands. The relative Harderian gland weights were similar to those of thyroidectomized controls.

Injection of guinea pigs with TSH led to a decrease in Harderian gland weight during the first 2-24 h, followed by a rise in weight during the next 3 days (Lamberg et al. 1956). Long-term treatment of guinea pigs with pituitary extracts containing thyrotropic, gonadotropic, and somatotropic activity induced an increase in Harderian gland absolute and relative weights, in intact and (especially) thyroidectomized guinea pigs (Smelser 1936, 1937, 1938, 1943b). The cause of this increase in weight was probably edema (Smelser 1943b; Pochin 1945); it was accompanied by a fall in the percentage of the tissue consisting of fat (Smelser 1943b). The increase in weight following injection of pituitary extracts was proportional to the enlargement of the thyroid glands caused by this treatment, which suggests (but does not prove) that TSH is the pituitary factor having this effect (Pochin 1945).

Thus, in guinea pigs, TSH may stimulate Harderian gland edema, resulting in an increase in size. This is probably the biological basis of this model of exophthalmos.

Hypothyroidism in female Syrian hamsters led to Harderian glands which were smaller than those of euthyroid animals, but these treatments usually had no effect on relative Harderian gland weights. In intact or castrated male hamsters, hypothyroidism led to reduced relative and/or absolute Harderian gland weights (Hoffman 1971; Hoffman et al. 1989b, 1990). Cellular and tissue damage and increased mitotic index were evident in hypothyroid hamsters of either sex (Hoffman 1971; Hoffman et al. 1989b, 1990).

In female or male hamsters, hyperthyroidism led to increased Harderian gland weight, but no change in relative gland weight, compared with those of euthyroid controls (Hoffman 1971; Hoffman et al. 1989b, 1990). Treatment of female hamsters with thyroid hormones generally led to the appearance of type II cells (which are usually only found in Harderian glands of male hamsters).

Similar treatment of intact or castrated male hamsters had no effect on Harderian gland histology (Hoffman et al. 1989b, 1990).

Thus, in Syrian hamsters, thyroid hormones appear to maintain the size and structural integrity of the Harderian glands. It seems likely that this effect is due to thyroid hormones and not to TSH because, although hypophysectomy reduced Harderian gland weights in both male and female hamsters, this effect was at least partially prevented by treatment with triiodothyronine (T_3) (Hoffman et al. 1990).

3.2 Porphyrins

A high porphyrin content (of unknown function) is one of the most striking features of the Harderian gland of most rodents. Thyroid hormones are important for the control of porphyrin levels, at least in the two species which have been examined.

Wetterberg et al. (1970b) noted that injection of thyroxine into newborn rats led to an early appearance of Harderian gland porphyrins, but that the attainment of maximal porphyrin content or concentration was delayed. This implies that thyroxine has an influence on early porphyrin synthesis in the rat; whether this is independent of other effects of this treatment in neonatal rats (e.g., premature eye opening) has not been addressed.

In adult male rats, treatment with T_3 led to reduced Harderian gland porphyrin concentrations (Buzzell, unpublished data).

All other published studies on influences of the thyroid gland on porphyrins in the Harderian glands have been done on Syrian hamsters. In order to interpret the results of studies on hamster Harderian gland porphyrins, it is first necessary to consider the biological meaning of variations in porphyrin levels. First, and most obvious, are the differences between intact males and females, with male porphyrin concentrations being in the ng/mg range and those of females in the μg/mg range. Hormonal treatment of males may do one of four things: it may decrease porphyrin concentrations, leave them unchanged, increase them 2–20-fold, or increase them 500–1000-fold. Decreased porphyrin levels in males are seldom seen; this is probably because Harderian gland porphyrin levels in males are near the lower limits of detection with the assays used.

A 2–20-fold increase in porphyrin concentrations is usually statistically significant but pales beside the 500–1000-fold increase in porphyrin concentrations routinely seen after a few weeks of castration. A 2–20-fold increase is usually seen in situations where androgen levels are low, but something else is preventing the in porphyrin levels which should accompany low androgens. It is as though a very powerful influence (low androgens) trying to increase porphyrin concentrations is competing against a weaker, but still powerful, influence trying to keep porphyrin concentrations low. The weaker influence is blunting the effect of the stronger influence!

For this reason, a 2–20-fold rise in male Harderian gland porphyrin levels is probably of little biological significance, regardless of the statistics. And, because of the fact that porphyrin concentrations in males are near the lower limits of measurement, it is not surprising if this magnitude of porphyrin increase in males is matched by no change or even by a significant decrease in porphyrin concentrations in female hamsters subjected to similar treatments. Such a situation can be seen, for instance, in Figs. 2 and 4.

Injection of T_3 or thyroxine into intact male hamsters did not change Harderian gland porphyrin concentrations (Buzzell, unpublished data; Hoffman et al. 1989b). This treatment generally led to a decrease in Harderian gland porphyrin concentrations in females; T_3 was more potent in causing this effect than was thyroxine (Hoffman 1971; Hoffman et al. 1989b, 1990). Injection of T_3 into castrated male hamsters prevented the castration-induced rise in Harderian gland porphyrins (Hoffman et al. 1990) (Fig. 2).

Antithyroid treatments (thiourea, methimazole, $KClO_4$) in intact male hamsters led to either no change in Harderian gland porphyrin concentrations (Buzzell, unpublished data; Buzzell et al. 1989a) or to a small significant rise in porphyrin concentrations (Hoffman et al. 1989b). In females, these treatments led to no change (Buzzell, unpublished data; Hoffman et al. 1990) or to a significant decrease in porphyrin concentrations (Hoffman 1971; Hoffman et al. 1989b). In castrated males, these treatments prevented the castration-induced rise in porphyrin concentrations (Buzzell unpublished data; Hoffman et al. 1990) (Fig. 3).

Thus, both hyperthyroid and hypothyroid conditions tend to reduce porphyrin concentrations. Since thyroid hormones regulate TSH levels by negative feedback mechanisms, the explanation of this contradiction might be found at the level of the pituitary gland.

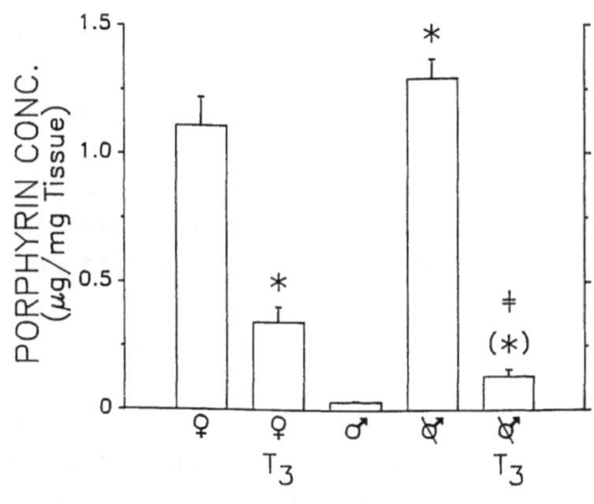

Fig. 2. Effects of daily T_3 injections on Harderian gland porphyrin concentrations in female and male Syrian hamsters. Significance: *, $P < 0.05$ vs. untreated controls; (*), $P < 0.05$ vs. untreated controls, but effect blunted relative to castrates; ǂ, $P < 0.05$ vs. castrated males. (After Hoffman et al. 1990)

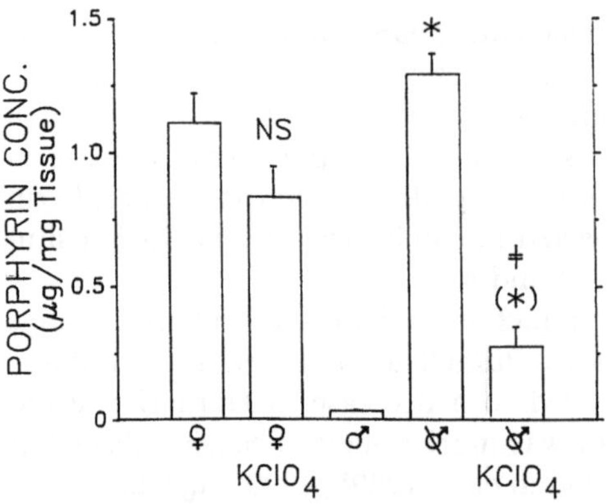

Fig. 3. Effects of $KClO_4$-induced hypothyroidism on Harderian gland porphyrin concentrations in female and male Syrian hamsters. Significance: *, $P < 0.05$ vs. untreated controls; (*), $P < 0.05$ vs. untreated controls, but effect blunted relative to castrates; ⧺, $P < 0.05$ vs. castrated males. (After Hoffman et al. 1990)

All hormones of the pituitary gland are eliminated by hypophysectomy. This treatment led to a small significant rise in Harderian gland porphyrin concentrations in male hamsters and to a significant decrease in porphyrin concentrations in females (Buzzell et al., unpublished data; Hoffman 1971; Hoffman et al. 1990). Replacement of thyroid hormones in hypophysectomized hamsters reduced the increase in porphyrin levels in males and led to a further decrease in porphyrin levels in females (Hoffman et al. 1990) (Fig. 4). Thus, thyroid hormones stimulate a decrease in Harderian gland porphyrin levels, even in the absence of TSH from the pituitary.

Because TSH does not seem to be important in the thyroid control of porphyrin levels in Harderian glands, the effects of antithyroid treatments is probably best explained by direct or indirect influences of the treatments on other pituitary hormones which affect Harderian gland porphyrin concentrations. For instance, low growth hormone levels are a common consequence of hypothyroidism (Oppenheimer et al. 1987).

Thus, hyperthyroidism generally leads to reduced Harderian gland porphyrin concentrations in rats or hamsters. Conversely, hypothyroidism does not stimulate increased porphyrin concentrations. For this reason, it seems unlikely that thyroid hormones directly influence Harderian gland porphyrin synthesis. Because of the well-documented stimulatory effects of thyroid hormones on metabolism in general, it seems likely that they modulate Harderian gland porphyrin concentrations by their effects on metabolism. We propose that there is a certain basal level of porphyrin metabolism which occurs in euthyroid or even hypothyroid animals. Metabolism increases in the hyperthyroid state with the result that porphyrin concentrations drop in females or that increased porphyrin concentrations due to orchidectomy are prevented by the removal of newly produced porphyrins.

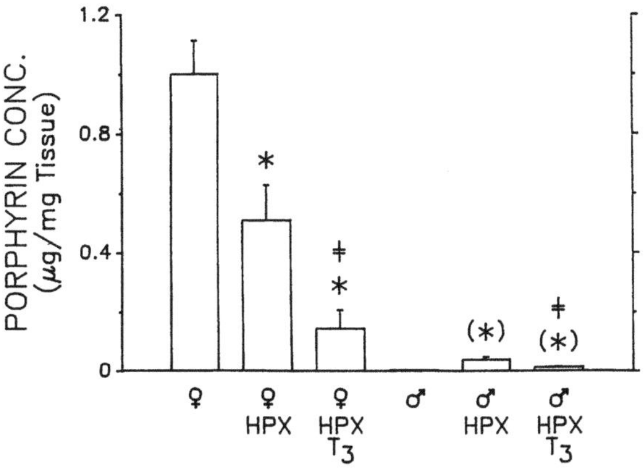

Fig. 4. Effects of hypophysectomy, with or without replacement of thyroid hormones by daily T_3 injections, on Harderian gland porphyrin concentrations in female and male Syrian hamsters. Significance: *, $P < 0.05$ vs. untreated controls; (*), $P < 0.05$ vs. untreated controls, but effect blunted relative to castrated males; ǂ, $P < 0.05$ vs. hypophysectomized animals of the same sex. (After Hoffman et al. 1990)

3.3 Indoles

Serotonin, melatonin, and several other hydroxy- and methoxyindoles are present within the rodent Harderian gland; also present are the enzymes N-acetyltransferase (NAT) and hydroxyindole-*O*-methyltransferase (HIOMT) which, in the pineal gland, convert serotonin to melatonin. In Syrian hamsters, the activity of Harderian gland NAT is particularly responsive to the endocrine state. In males, NAT activity is 3-5 times greater than that in females. Orchidectomy reduces NAT activity to female levels and implantation of testosterone pellets into females leads to male levels of NAT activity (Buzzell et al. 1989b; Marrufo et al. 1989; Menendez-Pelaez et al. 1988a, 1989). This androgenic control of Harderian gland NAT activity occurs in Syrian hamsters but not in other species which we have examined (Buzzell et al. 1991; Menendez-Pelaez and Buzzell, unpublished data).

Intact or castrated male hamsters kept in short-day photoperiods show a reduction in Harderian gland NAT activity (Fig. 5) (Buzzell et al. 1990b,c; Menendez-Pelaez et al. 1988b). This suggests that another hormone (besides androgens) controlled by the photoperiod influences Harderian gland NAT activity. Short photoperiods have a number of endocrine consequences, including reducing levels of circulating thyrotropin releasing hormone, which thus reduces TSH and thyroid hormone levels (Vriend 1983).

Reducing thyroid hormone levels (and increasing TSH levels) with methimazole led to an increase in Harderian gland NAT activity in both intact and castrated male and in female hamsters (Fig. 6). Increasing thyroid hormone levels (and decreasing TSH levels) with T_3 injections led to reduced Harderian

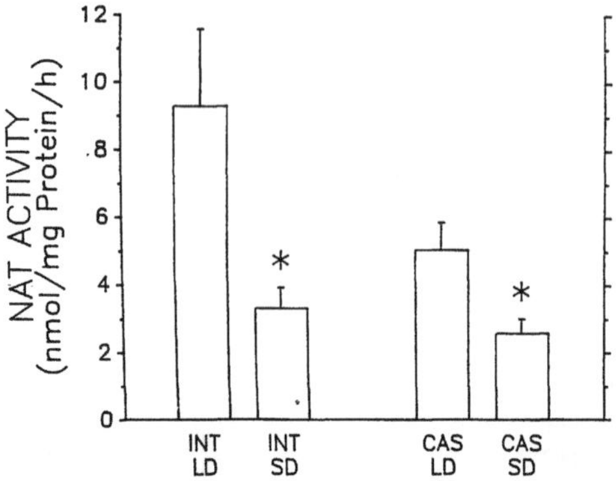

Fig. 5. Effects of photoperiods on Harderian gland N-acetyltransferase (*NAT*) activity in intact (*INT*) and castrated (*CAS*) male Syrian hamsters. Animals were kept in long days (*LD*) (14 h L: 10 h D) or exposed to naturally decreasing autumnal photoperiods (*SD*) (10 h L: 14 h D at the time of sacrifice) for 2 months. Significance: *, $P < 0.05$ vs. appropriate LD control. (After Menendez-Pelaez et al. 1988b)

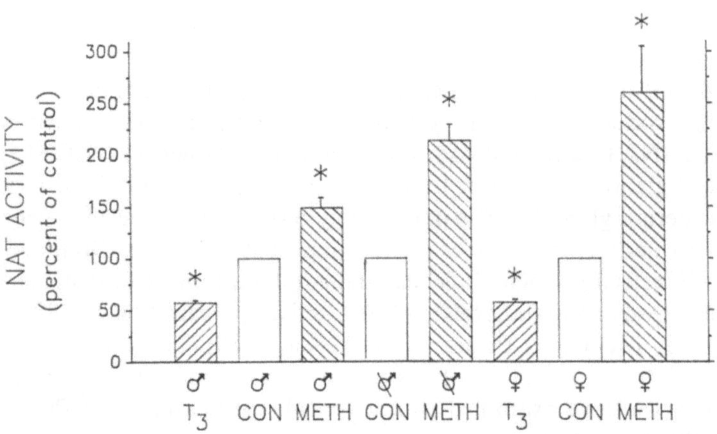

Fig. 6. Effects of hypothyroidism induced by methimazole (*METH*) and hyperthyroidism by T_3 injections on Harderian gland N-acetyltransferase (*NAT*) activity in Syrian hamsters. Data are expressed as percent of control (*CON*) values, as this graph represent the results of three experiments. Significance: *, $P < 0.05$ vs. appropriate control. (After Buzzell et al. 1990a)

gland NAT activity in both male and female hamsters (Fig. 6) (Buzzell et al. 1989a, 1990a).

Thus, the photoperiod data indicate that NAT activity decreases with decreasing TSH and decreasing thyroid hormone levels. Hyperthyroid and hypothyroid experiments indicate that NAT activity decreases with decreasing TSH and increasing thyroid hormone levels whereas it increases with increasing

Table 1. Effects, in Syrian hamsters, of short photoperiods, hypothyroidism, and hyperthyroidism on circulating levels of thyrotropin (TSH) and thyroid hormones and on Harderian gland N-acetyltransferase (NAT) activity. Because NAT activity is correlated with TSH and not with thyroid hormone levels, we suggest that TSH stimulates NAT activity in the Harderian gland

	Endocrine consequences		
Treatment	TSH	Thyroid hormones	Harderian NAT
Short days	Decrease	Decrease	Decrease[a]
Hypothyroid	Increase	Decrease	Increase[b]
Hyperthyroid	Decrease	Increase	Decrease[b]

[a]Menendez-Pelaez et al. 1988b; Buzzell et al. 1990b,c.
[b]Buzzell et al. 1990a.

TSH and decreasing thyroid hormone levels (Table 1). These data suggest that NAT activity in the hamster Harderian gland is stimulated by TSH.

Neither HIOMT activity nor melatonin concentration in the Harderian glands is influenced by hypothyroid or hyperthyroid conditions (Buzzell, unpublished data; Buzzell et al. 1989a).

4 Response of Harderian Glands to Thyrotropin and Thyroid Hormones

Thyrotropin is a glycoprotein hormone and thyrotropin receptors are located on the plasmalemma of target tissues (Rees Smith et al. 1988). In three different studies, investigators have looked for thyrotropin binding in Harderian gland membranes, with contradictory results. On the one hand, Winand and Kohn (1972) found that purified bovine TSH or a partial pepsin digest of TSH (with ophthalmogenic, but not thyrotropic, activity), specifically bound to guinea pig Harderian gland membranes. On the other hand, Teng et al. (1975) and Kendall-Taylor et al. (1984) found no thyrotropin binding on similar guinea pig Harderian gland membrane preparations.

The reason for this discrepancy is not apparent. All three studies used purified bovine TSH. The fact that Winand and Kohn found that thyrotropic activity was not necessary for binding suggests that these receptors, if real, might not be the same as those in other target tissues. If this is the case, the preparations of Teng and Kendall-Taylor might not have recognized these binding sites. In addition, TSH receptors are usually few in number and difficult to study without special techniques (Rees Smith et al. 1988).

At any rate, it is apparent that more studies, preferably molecular, are necessary before the question of the presence or absence of TSH receptors is resolved. These receptors should also be sought in hamster Harderian gland tissue, as it seems likely that indole metabolism in this gland responds to TSH (see above) (Buzzell et al. 1990a).

Thyroxine and T_3 are iodinated amino acids and bind to nuclear receptors (Oppenheimer et al. 1987). No one has looked for these receptors in Harderian glands. However, 5′-deiodinase enzymes (5′D), which convert thyroxine to T_3 (the active form of the thyroid hormone), are present in the Harderian glands of male rats and Syrian hamsters (Delgado et al. 1988; Guerrero et al. 1987, 1989; Puig-Domingo et al. 1989; Osuna et al. 1990). This is the type II isoenzyme, as its activity is strongly inhibited by iopanoic acid (Guerrero et al. 1987). A diurnal rhythm of 5′D activity was detected in rats (Guerrero et al. 1987) and in hamsters in one study (Guerrero et al. 1989) but not in another (Delgado et al. 1988). Hamster Harderian gland 5′D activity appears to be stimulated by a β-adrenergic mechanism (Delgado et al. 1988; Osuna et al. 1990); it is not influenced in hamsters by melatonin, a hormone which stimulates 5′D activity in brown adipose tissue (Puig-Domingo et al. 1989).

5 Effects of Harderian Glands on Thyroid Hormones

Several studies have suggested that removal of the Harderian glands influences the secretory patterns of endocrine glands (Wetterberg et al. 1970a,c; Panke et al. 1979; Hoffman et al. 1989a; Porkka-Heiskanen and Buzzell, unpublished data). Removal of the Harderian glands in Syrian hamsters kept in low photic intensities or moderately low temperatures influenced circulating levels of thyroid hormones (Hoffman et al. 1989a). This suggests that the Harderian glands form part of the response of these animals to potentially stressful environmental conditions. The mechanism by which the Harderian glands influence endocrine glands is unknown. The presence in the Harderian glands of melatonin, a hormone which modulates other hormones [including thyroid hormones (Vriend 1983)], suggests that this influence is endocrine in nature.

6 Conclusions

Interrelationships between the Harderian glands and the neuroendocrine-thyroid axis have been postulated for some time. The TSH-injected guinea pig (with enlarged Harderian glands) remains one of the only animal models of human exophthalmos available and continues to be used to study Graves' disease, despite the fact that humans have no Harderian glands. There are species differences in the responses of the Harderian glands to stimulation by hormones of the neuroendocrine-thyroid axis. In rats and hamsters, thyroid hormones appear to maintain the integrity of these glands whereas in guinea pigs, TSH is more important. There is evidence for both TSH receptors and 5′-deiodinase activity in Harderian glands, although some of this evidence is disputed and should be reexamined.

In Syrian hamsters, porphyrin concentrations are depressed in hyperthyroid animals, probably as a consequence of the increased metabolism which generally accompanies this condition. The biological role of Harderian gland porphyrins remains unestablished, so the significance of thyroidal modulation of porphyrin levels must await further studies.

Also in hamsters, Harderian gland NAT activity is modulated by the neuroendocrine-thyroid axis, probably by TSH. Again, the significance of this must await studies to establish the role of Harderian gland indoles and indole metabolism in the biology of the animal.

Thyroid hormones regulate metabolic rates and this is important for thermoregulation, especially in low temperatures. Evidence that lipid secretions from the Harderian glands are important to gerbils during cold weather (Thiessen and Kittrell 1980) suggests that the neuroendocrine-thyroid-Harderian gland axis is relevant to the survival of the animal during these adverse conditions. The possible roles of these hormones during lipid synthesis and secretion has not been examined. It seems likely that these studies would be rewarding.

The fact that almost all studies on the relationships of the Harderian glands and the neuroendocrine-thyroid axis have been done on only three species means that important insights into the significance of these relationships have almost certainly been missed. We would encourage researchers to look at these relationships in a variety of rodents in relation to a variety of glandular characteristics. Studies of the Harderian glands relative to the neuroendocrine-thyroid axis, thermoregulation, hibernation, estivation, etc. would almost certainly be rewarding.

Acknowledgements. It is a pleasure to acknowledge Drs. RJ Reiter and RA Hoffman. Both of us became interested in Harderian glands while on study leave in Dr. Reiter's lab and his encouragement and friendship remains an inspiration to us. Dr. Hoffman did some critical experiments for us and gave us access to data before publication.

References

Asboe-Hansen G, Iversen K (1951) Influence of thyrotrophic hormone on connective tissue. Pathogenic significance of mucopolysaccharides in experimental exophthalmos. Acta Endocrinol 8:90–96

Banovac K, Levis S, Byrne SF (1988) A study of the binding of Graves' immunoglobulins to orbital antigens. J Endocrinol Invest 11:483–488

Boas NF, Bates RW (1954) Role of the thyroid and anterior pituitary glands in the maintenance of the Harderian glands of the rat. Endocrinology 55:601–612

Boas NF, Scow RB (1954) Apparent exophthalmos in the rat following cortisone treatment or thyroidectomy. Endocrinology 55:148–155

Bolonkin D, Tate RL, Luber JH, Kohn LD, Winand RJ (1975) Experimental exophthalmos. Binding of thyrotropin and an exophthalmogenic factor derived from thyrotropin to retro-orbital tissue plasma membranes. J Biol Chem 250:6516–6521

Brunish R (1962) The nature of the lacrimal mucoprotein stimulated by exophthalmos-producing substance. J Biol Chem 237:2816–2819

Brunish R, Hayashi K, Jayashi J (1962) Exophthalmos-producing substance: mode of action. Exp Eye Res 1:221–228

Buzzell GR, Chen ZL, Vaughan MK, Reiter RJ (1989a) Effects of inhibition of thyroid function and of cold on melatonin synthesis and porphyrin content in the Harderian glands of male Syrian hamsters, *Mesocricetus auratus*. Comp Biochem Physiol 94A:427–429

Buzzell GR, Menendez-Pelaez A, Porkka-Heiskanen T, Pangerl A, Pangerl B, Vaughan MK, Reiter RJ (1989b) Bromocriptine prevents the castration-induced rise in porphyrin concentration in the Harderian glands of the male Syrian hamster, *Mesocricetus auratus*. J Exp Zool 249:172–176

Buzzell GR, Menendez-Pelaez A, Hoffman RA, Vaughan MK, Reiter RJ (1990a) N-Acetyltransferase activity in the Harderian glands of the Syrian hamster, *Mesocricetus auratus*, is regulated by androgens and by hormones of the pituitary-thyroid axis. J Endocrinol 127:59–67

Buzzell GR, Menendez-Pelaez A, Troiani ME, McNeill ME, Reiter RJ (1990b) Effects of short-day photoperiods and of N-(2,4-dinitrophenyl)-5-methoxytryptamine, a putative melatonin antagonist, on melatonin synthesis in the Harderian gland of the Syrian hamster, *Mesocricetus auratus*. J Pineal Res 8:229–235

Buzzell GR, Pangerl A, Pangerl B, Menendez-Pelaez A, Vaughan MK, Little JC, Hill SM, Vaughan GM, Reiter RJ (1990c) Melatonin and porphyrin in the Harderian glands of the Syrian hamster: circadian patterns and response to autumnal conditions. Int J Biochem 22:1465–1469

Buzzell GR, Menendez-Pelaez A, Blank JL, Nonaka KO, Reiter RJ (1991) Effects of gender, age, and castration on porphyrin concentration and melatonin synthesis in the Harderian glands of the Mongolian gerbil, *Meriones unguiculatus*. Can J Zool 69:151–155

Dandona P, ElKabir DJ (1970) The effect of the injections of exophthalmic sera on the Harderian gland of the guinea-pig. Clin Sci 38:2P

Davies TF, Teng CS, McLachlan SM, Rees Smith B, Hall R (1978) Thyrotropin receptors in adipose tissue, retro-orbital tissue and lymphocytes. Mol Cell Endocrinol 9:303–310

Delgado MJ, Guerrero JM, Santana C, Menendez-Pelaez A, Gonzalez-Brito A, Chen ZL, Reiter RJ (1988) Thyroxine 5′-deiodinase activity in brown adipose tissue, Harderian gland, pineal gland and pituitary gland of the male Syrian hamster (*Mesocricetus auratus*). Neuroendocrinol Lett 10:363–368

Ebling FJ, Ebling E, Randall V, Skinner J (1975a) The effects of hypophysectomy and of bovine growth hormone on the responses to testosterone of prostate, preputial, Harderian and lachrymal glands and of brown adipose tissue in the rat. J Endocrinol 66:401–406

Ebling FJ, Ebling E, Randall V, Skinner J (1975b) The synergistic action of α-melanocyte-stimulating hormone and testosterone on the sebaceous, prostate, preputial, Harderian and lachrymal glands, seminal vesicles and brown adipose tissue in the hypophysectomized-castrated rat. J Endocrinol 66:407–412

Guerrero JM, Puig-Domingo M, Vaughan GM, Reiter RJ (1987) Characterization of type-II thyroxine 5′-deiodinase activity in rat Harderian gland. Life Sci 41:1179–1185

Guerrero JM, Gonzalez-Brito A, Santana C, Reiter RJ (1989) Nocturnal increase of type II thyroxine 5′-deiodinase activity in the Syrian hamster Harderian gland is abolished by light exposure and induced by isoproterenol. Proc Soc Exp Biol Med 190:186–189

Hadad HM (1963) Stimulation of radiosulfate incorporation into guinea pig Harderian gland by thyroid-stimulating hormone (TSH) and exophthalmic serum. J Clin Invest 41:1363

Hoffman RA (1971) Influence of some endocrine glands, hormones and blinding on the histology and porphyrins of the Harderian glands of golden hamsters. Am J Anat 132:463–478

Hoffman RA, Johnson LB, Vaughan MK, Reiter RJ (1989a) Interaction of Harderian glands, illumination, and temperature on thyroid hormones in golden hamsters. J Comp Physiol B 158:697–702

Hoffman RA, Wertz P, Habeeb P (1989b) Harderian glands of golden hamsters: morphological and biochemical responses to thyroid hormones. J Comp Physiol B 159:293–299
Hoffman RA, Habeeb P, Buzzell GR (1990) Further studies on the regulation of the Harderian glands of golden hamsters by the thyroid gland. J Comp Physiol B 160:269–275
Jacobson DH, Gorman CA (1984) Endocrine ophthalmopathy: current ideas concerning etiology, pathogenesis, and treatment. Endocrinol Rev 5:200–220
Kendall-Taylor P, Atkinson S, Holcombe M (1984) A specific IgG in Graves' ophthalmopathy and its relation to retro-orbital and thyroid autoimmunity. Br Med J 288:1183–1186
Kohn LD, Winand RJ (1971) Relationship of thyrotropin to exophthalmos-producing substance. Formation of an exophthalmos-producing substance by pepsin digestion of pituitary glycoproteins containing both thyrotropic and exophthalmogenic activity. J Biol Chem 246:6570–6575
Lamberg B-A, Wegelius O, Harjanne A (1956) The uptake of radioactive sulphate by the retrobulbar connective tissue and Harder's gland in guinea pigs treated with thyrotrophin. Acta Endocrinol 22:407–414
Lorincz AL, Lancaster G (1957) Anterior pituitary preparation with tropic activity for sebaceous, preputial, and Harderian glands. Science 126: 124–125
Ludwig AW, Boas NF, Soffer LJ (1950) Role of mucopolysaccharides in pathogenesis of experimental exophthalmos. Proc Soc Exp Biol Med 73:137–141
Marrufo B, Menendez-Pelaez A, Buzzell GR, Gonzalez-Brito A, Reiter RJ (1989) 5α-Dihydrotestosterone administration converts indolamine metabolism and porphyrin content of the female Syrian hamster Harderian gland to the male type. Proc Soc Exp Biol Med 192:192–195
Menendez-Pelaez A, Reiter RJ, Guerrero JM, Puig-Domingo M, Howes KA (1988a) Sexual dimorphism in N-acetyltransferase activity, hydroxyindole-O-methyltransferase activity, and melatonin content in the Harderian gland of Syrian hamsters: changes following gonadectomy. Proc Soc Exp Biol Med 187:287–291
Menendez-Pelaez A, Reiter RJ, Howes KA, Puig-Domingo M, Vaughan MK, Troiani ME, Little JC (1988b) Harderian gland N-acetyltransferase activity in the male Syrian hamster: effect of gonadectomy, short photoperiod exposure, or subcutaneous melatonin implants. Endocrinol Res 14:121–130
Menendez-Pelaez A, Marrufo B, Reiter RJ, Santana C, Guerrero JM (1989) Testosterone increases N-acetyltransferase activity in both male and female Syrian hamster Harderian glands. Biochem Int 19:115–121
Oppenheimer JH, Schwartz HL, Mariash CN, Kinlaw WB, Wong NCW, Freake HC (1987) Advances in our understanding of thyroid hormone action at the cellular level. Endocrinol Rev 8:288–308
Osuna C, Rubio A, Goberna R, Guerrero JM (1990) Ontogeny of type-II thyroxine 5′-deiodinase, N-acetyltransferase, and hydroxyindole-O-methyltransferase activities in the rat Harderian gland. Life Sci 46:1945–1951
Panke ES, Reiter RJ, Rollag MD (1979) Effect of removal of the Harderian glands on pineal melatonin concentrations in the Syrian hamster. Experientia 35:1405
Paulson DL (1937) Experimental exophthalmos in the guinea pig. Proc Soc Exp Biol Med 36:604–605
Pisarev MA, Altschuler N, Davison TA (1968) Immune exophthalmos due to Harderian gland antigen. Endocrinology 83:903–906
Pochin EE (1945) Exophthalmos in guinea pigs injected with pituitary extracts. Clin Sci 5:75–91
Puig-Domingo M, Guerrero JM, Menendez-Pelaez A, Reiter RJ (1989) Melatonin specifically stimulates type-II thyroxine 5′-deiodination in brown adipose tissue of Syrian hamsters. J Endocrinol 122:553–556
Rees Smith B, McLachlan SM, Furmaniak J (1988) Autoantibodies to the thyrotropin receptor. Endocrinol Rev 9:106–121
Reichlin S, Hadad HM (1963) Influence of thyrotrophin on radiosulfate uptake by orbital and other tissues of the guinea pig. J Lab Clin Med 61:44–50
Singh SP, McKenzie JM (1969a) Effects of thyrotropin on ^{35}S-sulfate uptake by mouse Harderian gland. Endocrinology 85:952–955
Singh SP, McKenzie JM (1969b) Relationship between exophthalmos, LATS and ^{35}S uptake by mouse Harderian gland. Clin Res 17:591
Singh SP, McKenzie JM (1970) The influence of thyrotropin on Harderian gland glycosaminoglycans of the mouse. Proc Soc Exp Biol Med 135:478–482

Singh SP, McKenzie JM (1971) Studies of homologous immunization with mouse Harderian gland. Metabolism 20:293–298
Sisson JC, Miles M (1967) Acid mucopolysaccharide alterations in experimental exophthalmos. Endocrinology 80:931–937
Smelser GK (1936) Experimental production of exophthalmos resembling that found in Graves disease. Proc Soc Exp Biol Med 35:128–130
Smelser GK (1937) A comparative study of experimental and clinical exophthalmos. Am J Opthalmol 20:1189–1203
Smelser GK (1938) Treatment of experimentally produced exophthalmos with thyroxin and other iodine compounds. Am J Ophthalmol 21:1208–1218
Smelser GK (1943a) Reaction of orbital tissues in experimental exophthalmos following removal of Harder's gland. Anat Rec 85:245–259
Smelser GK (1943b) Changes induced in the Harderian gland of the guinea pig by the injection of hypophyseal extracts. Anat Rec 86:41–57
Smelser GK (1943c) A study of retrobulbar tissues in experimental exophthalmos in guinea pigs with reference to primary and secondary modifications. Am J Anat 72:149–169
Smelser GK (1962) Experimental studies on exophthalmos. Am J Ophthalmol 54:929–951
Smelser GK, Ozanics V (1954) Studies on the nature of exophthalmos-producing principle in pituitary extracts. Am J Ophthalmol 38:107–122
Smelser GK, Ozanics V (1955) Further studies on the nature of the exophthalmos-producing principles in pituitary extracts. Am J Ophthalmol 39:146–155
Smelser GK, Ozanics V (1959) Hydrophilia of the orbital connective tissue in experimental exophthalmos. Am J Ophthalmol 47:380–387
Teng CS, Rees Smith B, Anderson J, Hall R (1975) Comparison of thyrotrophin receptors in membranes prepared from fat and thyroid tissue. Biochem Biophys Res Commun 66:836–841
Thiessen DD, Kittrell EM (1980) The Harderian gland and thermoregulation in the gerbil. Physiol Behav 24:417–424
Vriend J (1983) Evidence for pineal gland modulation of the neuroendocrine-thyroid axis. Neuroendocrinology 36:68–78
Weetman AP, McGregor AM (1984) Autoimmune thyroid disease: developments in our understanding. Endocrinol Rev 5:309–355
Wegelius O, Naumann J, Brunish R (1959) Uptake of ^{35}S-labelled sulfate in the Harderian and the ventral lacrymal gland of the guinea pig during stimulation with ophthalmotrophic pituitary agents. A new assay method for ophthalmotrophic activity in thyrotrophic preparations. Acta Endocrinol 30:53–60
Wegelius O, Lamberg B-A (1960) ^{35}S-Autoradiography of the retrobulbar connective tissue and Harder's gland in thyrotrophin-stimulated guinea-pigs. Acta Endocrinol 33:473–476
Wetterberg L, Geller G, Yuwiler A (1970a) Harderian gland: An extraretinal photoreceptor influencing the pineal gland in neonatal rats? Science 167:884–885
Wetterberg L, Yuwiler A, Geller E, Schapiro S (1970b) Harderian gland: development and influence of early hormonal treatment on porphyrin content. Science 168:996–998
Wetterberg L, Yuwiler A, Ulrich R, Geller E, Wallace R (1970c) Harderian gland: influence on pineal hydroxyindole-O-methyltransferase activity in neonatal rats. Science 170:194–196
Winand RJ, Kohn LD (1972) The binding of [^{3}H]thyrotropin and an ^{3}H-labeled exophthalmogenic factor by plasma membranes of retro-orbital tissue. Proc Natl Acad Sci USA 69:1711–1715
Winand RJ, Kohn LD (1973) Retrobulbar modifications in experimental exophthalmos: the effects of thyrotropin and an exophthalmos-producing substance derived from thyrotropin on the $^{35}SO_4$ incorporation and glycosaminoglycan content of Harderian glands. Endocrinology 93:670–680
Winand RJ, Kohn LD (1975) Stimulation of adenylate cyclase activity in retro-orbital tissue membranes by thyrotropin and an exophthalmogenic factor derived from thyrotropin. J Biol Chem 250:6522–6526

Type II 5′-Deiodinase in the Harderian Gland

M.K. Vaughan[1] and J.M. Guerrero[2]

[1]Department of Cellular and Structural Biology, University of Texas Health Science Center at San Antonio, San Antonio, TX 78284, USA
[2]Departamento de Bioquimica, Facultad de Medicina, Sevilla, Spain

1 Introduction

The major secretory products of the thyroid gland are the hormones thyroxine (T_4) and to a much lesser extent, triiodothyronine (T_3). The intrathyroidal ratio of T4 to T3 is about 10:1 to 15:1 (Mazzaferri 1980). Thyroxine is taken up from the circulation by tissues of the body where monodeiodination by a deiodinase enzyme results in the formation of reverse T_3 (rT_3) or the more biologically active T_3. The T_3 which is secreted by the peripheral tissues accounts for approximately 80% of the extrathyroidal T_3 pool with the thyroid itself contributing only about 20% in normal individuals (Mazzaferri 1980). Reverse T_3 and the other diiodotyrosine (T_2) metabolites have little or no biological activity. The structure and possible breakdown products of T_4 are shown in Fig. 1.

2 Characteristics of the Type II 5′-Deiodinase Enzyme

There are at least two isozymes of 5′-deiodinase (5′D); for the current paper Type I and Type II 5′D are of interest. The Type I isozyme has a wide body distribution, has a low affinity for T_4 and is decreased when the animal becomes hypothyroid. The Type II form of the enzyme has a limited distribution and can be found in the central nervous system (Crantz and Larsen 1980), anterior pituitary (Kaplan 1980), brown adipose tissue (Leonard et al. 1982), and pineal gland (Tanaka et al. 1986; Guerrero et al. 1988a). The Type II enzyme has a high affinity for T_4. The Harderian gland also contains the Type II 5′-D. The most important regulatory mechanism for this enzyme is the thyroid status, exhibiting an important increase in its activity during hypothyroidism (Dratman et al. 1983; Silva and Leonard 1985). The enzyme is believed to play a major role in maintaining intracellular levels of T_3 and could serve as a defense against thyroid hormone deficiency, especially in tissues depending on thyroid hormones.

Webb et al. (Eds).
Harderian Glands

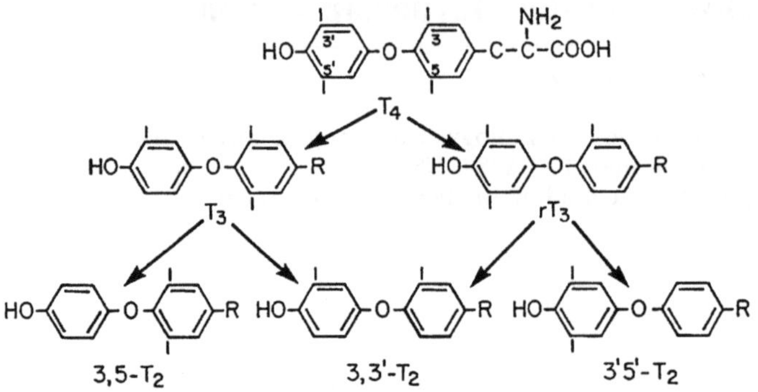

Fig. 1. Thyroid hormones and their metabolites. Thyroxine (T_4), triiodothyronine (T_3), reverse triiodothyronine (rT_3), diiodotyrosines (T_2)

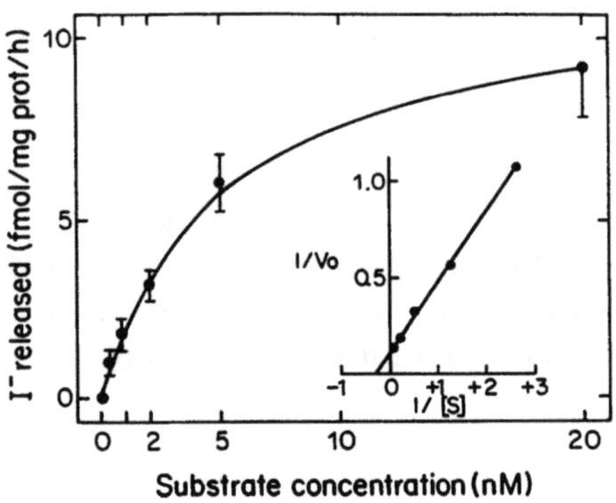

Fig. 2. Effect of substrate concentration on ^{125}I release in rat Harderian homogenates. Incubates were performed in the presence of 20 mM dithiothreitol (DTT) at 37 °C for 60 min. The insert is the Lineweaver-Burk plot of the same data. (Guerrero et al. 1987)

The presence of 5′-D activity in the Harderian gland was first described in the rat (Guerrero et al. 1987). In this rodent, Harderian gland 5′-D activity was pH-dependent, with an optimum at pH 7.0, and related to temperature, between 4–37 °C. The enzyme activity was augmented by dithiothreitol in a dose-dependent manner and inhibited moderately by propylthiouracil and completely by iopanoic acid (IA). Hypothyroidism induced by thyroidectomy increased enzymatic activity (up to 30-fold). In addition, Harderian gland 5′-D activity was also related to the T_4 concentration (K_m = 3.3 nM; V_{max} = 10 fmol ^{125}I released/mg protein, Fig. 2). Based on the characteristics of the enzyme, it is virtually certain that the specific enzyme in the rat Harderian gland is the type II enzyme.

The enzyme has been found in both the rat and Syrian hamster Harderian gland (Delgado et al. 1988; Guerrero et al. 1989) but studies with Swiss mice have shown very low activity of the enzyme (Rubio et al. 1991). All these studies used exclusively male animals and it can only be assumed that female rat and hamster Harderian glands would show similar levels of 5′-D activity.

3 Photoperiod and Type II 5′-Deiodinase Activity

Several regulatory mechanisms have been described which may determine Harderian gland 5′-D activity. In addition to hypothyroidism, which clearly increases the enzyme activity, environmental lighting conditions also appear to regulate the enzyme. Thus, a conspicuous 24-h rhythm in 5′-D activity exists in both rat and hamster Harderian gland (Guerrero et al. 1987, 1989). When animals are kept under a light:dark (LD) 14:10h photoperiodic regimen, the late light phase is associated with a low 5′-D activity (Figs. 3 and 4); with darkness onset, 5′-D activity rises to reach a peak 7 h later, in the second half of the dark period. The rat Harderian gland rhythm in 5′-D activity differs in at least one important way from that in pineal gland. In the rat pineal, high deiodinase activity is reached 2-3h earlier (Guerrero et al. 1988a, b).

Alterations in the amount of light to which the animal is exposed have dramatic effects on the circadian rhythms in 5′-D activity. In the Syrian hamster, a nocturnal increase in Harderian gland 5′-D activity is prevented if hamsters are exposed to light at night (Fig. 4). The results suggest that, at least in the hamster, lighting environment regulates the activity of the enzyme in the Harderian gland as occurs with 5′-D activity in rat pineal gland (Guerrero et al. 1988a, c). In another experiment (Vaughan et al. 1991), chronic exposure of male rats to a 10:14 LD photoperiod altered the rhythm in 5′-D activity such that the acrophase occurred during the late afternoon and the minima was during the night (Fig. 5). Mechanisms involved in the alteration by light of the nocturnal

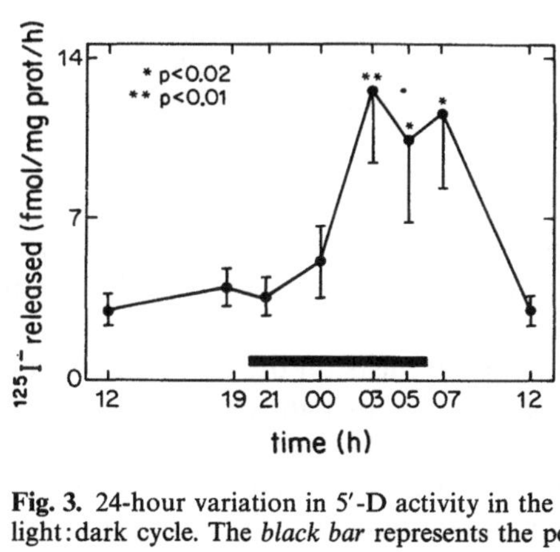

Fig. 3. 24-hour variation in 5′-D activity in the Harderian glands of male rats kept under a 14:10 light:dark cycle. The *black bar* represents the period of darkness. (Guerrero et al. 1987)

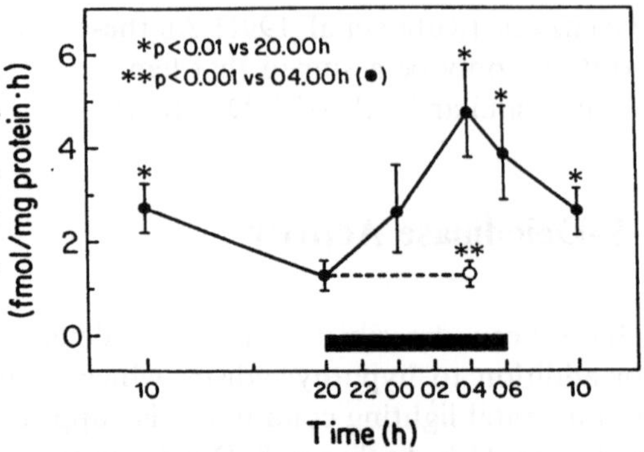

Fig. 4. Nyctohemeral profile of 5′-D activity in Syrian hamster Harderian glands (*filled circle*). One group of animals (*open circle*) was maintained under light conditions rather than entering darkness at 20.00 h (from Guerrero et al. 1989)

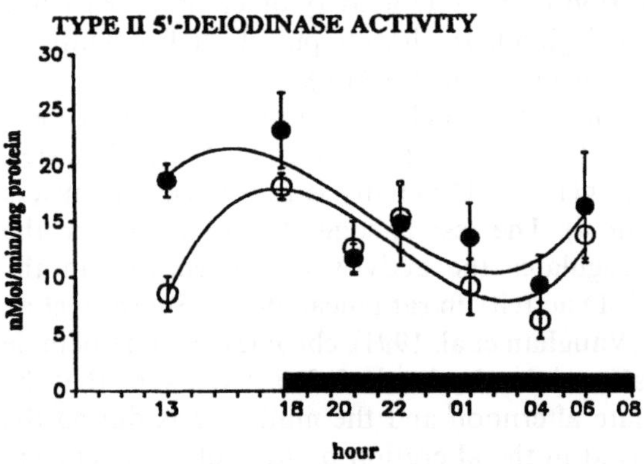

Fig. 5. Type II 5′-deiodinase activity in male Sprague-Dawley (*open circles*) and Fischer-344 rats (*dark circles*) rats exposed to a short photoperiod (10L:14D; lights on 06.00 h) for 8 weeks. Twenty-four-h mean for Fischer-344 rats (15.77) was significantly different ($p < 0.05$) from Sprague-Dawley mean (12.28). Amplitude of the 5′D rhythm in Fischer-344 (4.45) was significantly different ($p < 0.05$) from the amplitude of the 5′D rhythm in Sprague-Dawley rats (2.65). *Dark bar* indicates the length of the dark period. (Vaughan et al. 1991)

5′-D activity increase in the Harderian gland remain to be investigated. In rat, changes in the levels of the enzyme after continuous light exposure have not been examined.

On the other hand, the effect of adrenergic agonists on the Harderian gland 5′-D activity have been studied. Thus, in the Syrian hamster, the nocturnal increase of the enzyme activity in the Harderian gland of light-exposed hamsters can be induced by injecting isoproterenol, a β-adrenergic agonist (Guerrero et al. 1989). Isoproterenol was found to be a potent activator of the enzyme, and when the drug was injected at 2-h invervals, it was able to produce an increase in activity similar to that seen during the normal dark phase. Similarly, rat Harderian glands also respond to isoproterenol injections but only in euthyroid animals. Isoproterenol was not able to activate Harderian gland 5′-D activity when the enzyme had been previously activated by hypothyroidism (Osuna et al. 1990). The results suggest that the adrenergic innervation of the Harderian gland may mediate the nyctohemeral rise in 5′-D activity. However, since the noradrenergic fibres innervating the Harderian gland end primarily on the blood vessels within the organ (rather than on the secretory cells; Norvell and Clabough 1972), it would seem that 5′-D activity in the Harderian gland is regulated differently than in the pineal gland where postganglionic sympathetic nerves terminate directly on the secretory cells of the gland. At least one other explanation has been suggested for the effect of isoproterenol on 5′-D activity in the Harderian gland (Guerrero et al. 1989). Since chronic melatonin treatment is shown to increase 5′-D activity in brown adipose tissue (Puig-Domingo et al. 1988), isoproterenol might indirectly activate Harderian gland 5′-D activity after stimulating melatonin production. This is consistent with the suppression of 5′-D activity at night in light-exposed hamsters; light is known to suppress melatonin production in the hamster pineal gland. However, melatonin injections fail to increase Harderian gland 5′-D activity (unpublished data). Further studies are required to elucidate whether the noradrenergic input regulates Harderian gland 5′-D activity either directly or indirectly through other messengers.

The ontogeny and regulation by isoproterenol of 5′-D activity has also been studied in the rat Harderian gland during development (Osuna et al. 1990). Results showed that the enzyme exhibits higher activities in the early stages of development than in adult rats. The results are similar to those obtained in the rat brain where there is a complex pattern of maturation of the iodothyronine-deiodinating enzymes (Kaplan and Yaskoski 1981). In the rat brain, a peak in 5′-D activity has been described at 2weeks, while tyrosyl ring deiodination exhibits a progressive decline from birth. During the same period, most of the structures in the rat central nervous system undergo major developmental changes. In rat Harderian gland, the higher 5′-D activity during the first 2weeks of age could also indicate that thyroid hormones play a role in the maturation of the gland. However, another possibility is that the increases in the enzyme activity could be only a response to low serum T_3 during the first days of life (Kaplan and Yaskoski 1982). Osuna et al. (1990) also showed the regulation by isoproterenol of 5′-D activity during development. During the first 2weeks of life, when basal 5′-D activity exhibits high values, 5′-D activity in Harderian glands is insensitive to the action of isoproterenol. In the following weeks, when basal 5′-D activity gradually decreases, isoproterenol is able to stimulate the enzyme, reaching the maximal values exhibited at the first week. There are at least two explanations for these findings; perhaps β-adrenergic receptors may not be present in the rat Harderian gland until 2weeks of age, although this possibility should be confirmed. Another explanation is that the enzyme, during the first 2weeks of life, is already activated and isoproterenol can not induce additional stimulation.

A major question remaining to be answered relates to the functional significance of 5′-D activity in the Harderian glands. Although it can be assumed that the enzyme is involved in the maturation of the gland during development, and/or the synthesis of porphyrins, further research is necessary.

References

Crantz FR, Larsen PR (1980) Rapid thyroxine to 3, 5, 3′triiodothyronine conversion and nuclear 3, 5, 3′, triiodothyronine binding in rat cerebral cortex and cerebellum. J Clin Invest 69:935–938

Delgado MJ, Guerrero JM, Santana C, Menendez-Pelaez A, Gonzalez-Brito A, Chen ZL, Reiter RJ (1988) Thyroxine 5′-deiodinase activity in brown adipose tissue, Harderian gland, pineal gland and pituitary gland of the male Syrian hamster (*Mesocricetus auratus*). Neuroendocrinol Lett 10:363–368

Dratman MB, Crutchfield FL, Gordon JT, Jennings AS (1983) Iodothyronine homeostasis in rat brain during hypo- and hyperthyroidism. Am J Physiol 245:E184–E193

Guerrero JM, Puig-Domingo M, Vaughan GM, Reiter RJ (1987) Characterization of type II thyroxine 5′-deiodinase activity in rat Harderian gland. Life Sci 41:1179–1185

Guerrero JM, Puig-Domingo M, Reiter RJ (1988a) Thyroxine 5′-deiodinase activity in pinal gland and frontal cortex: nighttime increase and the effect of either continuous light exposure or superior cervical ganglionectomy. Endocrinology 122:236–241

Guerrero JM, Puig-Domingo M, Santana C, Menendez-Pelaez A, Reiter RJ (1988b) Inhibition of pineal type-II 5′-deiodinase does not affect the nocturnal increase of N-acetyltransferase activity and melatonin content in either euthyroid or thyroidectomized rats. J Pineal Res 5:513–520

Guerrero JM, Puig-Domingo M, Santana C, Menendez-Pelaez A, Gonzalez-Brito A, Reiter RJ (1988c) Differential responses of rat pineal thyroxine type-II 5′-deiodinase and N-acetyltransferase activities to either light exposure, isoproterenol, phenylephrine or propranolol. Cell Mol Neurobiol 8:447–458

Guerrero JM, Gonzalez-Brito A, Santana C, Reiter RJ (1989) Nocturnal increase of type-II thyroxine 5′-deiodinase activity in the Syrian hamster Harderian gland is abolished by light exposure and induced by isoproterenol. Proc Soc Exp Biol Med 190:186–189

Kaplan MM (1980) Thyroxine 5′-monodeiodination in rat anterior pituitary gland homogenate. Endocrinology 106:567–576

Kaplan MM, Yaskoski KA (1981) Maturational patterns of iodothyronine phenolic and tyrosyl ring deiodinase activities in rat cerebrum, cerebellum and hypothalamus. J Clin Invest 67:1208–1214

Kaplan MM, Yaskoski KA (1982) Effects of congenital hypothyroidism and partial and complete food deprivation on phenolic and tryosyl ring iodothyronine deiodination in rat brain. Endocrinology 110:761–767

Kaplan MM, Pan C, Gordon PR, Lee J, Gilchrest BA (1988) Human epidermal keratinocytes in culture convert thyroxine to 3, 5. 3′-triiodothyronine by type II iodothyronine deiodination: A novel endocrine function of the skin. J Clin Endocrinol Metab 66:815–822

Leonard JL, Mellen SA, Larsen PR (1982) Thyroxine 5′-deiodinase activity in brown adipose tissue. Endocrinology 112:1153–1155

Mazzaferri EL (1980) Endocrinology, 2nd edn. Medical Examination Publishing, Bern

Norvell JE, Clabough JW (1972) Adrenergic and cholinergic innervation of the hamster Harderian gland. Science 178:1102–1103

Osuna C, Rubio A, Goberna R, Guerrero JM (1990) Ontogeny of type II thyroxine 5′-deiodinase, N-acetyltransferase, and hydroxyindole-O-methyltransferase activities in the rat Harderian gland. Life Sci 46:1945–1951

Puig-Domingo M, Guerrero JM, Reiter RJ, Tannenbaum MJ, Hurlbut EC, Gonzalez-Brito A, Santana C (1988) Thyroxine 5′-deiodination in brown adipose tissue and pineal gland: implications for thermogenic regulation and role of melatonin. Endocrinology 123:677–680
Reiter RJ, Guerrero JM, Puig-Domingo M, Santana C (1988) Type II iodothyronine deiodination: nyctohemeral rhythms in the cerebral cortex, the pineal gland, and the Harderian glands. In: Rietveld W (ed) Advances in the biosciences vol 73 Pergamon London, pp 123–136
Rubio A, Osuna C, Lopez-Gonzalez MA, Reiter RJ, Guerrero JM (1991) Nyctohemeral rhythmicity of type II thyroxine 5′-deiodinase activity in the pineal gland but not in the Harderian gland of the Swiss mouse. Biosci Rep in press
Silva JE, Leonard JL (1985) Regulation of rat cerebrocortical and adenohypophyseal type II 5′-deiodinase by thyroxine, triiodothyronine and reverse triiodothyronine. Endocrinology 116:1627–1635
Tanaka K, Murakami M, Greer MA (1986) Type-II thyroxine 5′-deiodinase in the rat pineal gland. Biochem Biophys Res Commun 137:863–868
Vaughan MK, Vaughan GM, Chenoweth PC, Menendez-Pelaez A, Kumar P, Chambers JP, Rodriguez C, Hoover P, Reiter RJ (1991) Harderian gland porphyrin, lysosomal, ATPase and type II 5′-deiodinase rhythms in Sprague-Dawley and Fischer-344 rats kept under long or short photoperiod conditions. Int J Biochem 23:919–924

The Pineal and the Harderian Glands: Evidence for Mutual Interactions

RUSSEL J. REITER[1] and BRUCE A. RICHARDSON[2]

[1]Department of Cellular and Structural Biology, The University of Texas Health Science Center at San Antonio, 7703 Floyd Curl Drive, San Antonio, Texas 78284-7762, USA
[2]Department of Biology, The University of Michigan-Flint, 303 Kearsly Street, Flint, Michigan 48502-2186, USA

1 Introduction

The pineal gland and the Harderian glands have a number of biochemical and physiological features which are similar. Both organs contain melatonin and associated enzymes required for its synthesis (Cardinali and Wurtman 1972; Bubenik et al. 1978; Balemans et al. 1980; Reiter et al. 1981). Additionally, both the Harderian glands (Guerrero et al. 1987) and the pineal gland (Guerrero et al. 1988) contain the 5'-deiodinase enzyme which converts the relatively inactive thyroid hormone, thyroxine, to its active form, triiodothyronine (T_3). In both the Harderian and the pineal glands, melatonin and T_3 synthesis are rhythmic. Finally, both organs also exhibit day/night rhythms in vasopressin and oxytocin (Gauquelin et al. 1988). Why these organs have these similarities is unknown and, indeed, astonishing considering their widely different embryological derivations, etc. Recently, it has become apparent that there may be a bidirectional physiologic interaction between these organs. The purpose of this chapter is to summarize the presumptive interrelationships.

2 Evidence for Harderian Gland Effects on the Pineal Gland

The first evidence which suggested an effect of the Harderian glands on the pineal was provided in two reports by Wetterberg and colleagues (1970a, b). Their studies were prompted by the observation that the pineal serotonin rhythm (Quay 1963) persists in blinded suckling rats, indicating the existence of extraretinal photoreceptors in these animals (Zweig et al. 1966). Furthermore, the presumptive photoreceptors were localized in the head since, if the blinded rats were hooded, the pineal serotonin response to light was abolished; however, Zweig and co-workers (1966) did not suggest that the Harderian glands were the site of the photoreceptive process. The reason Wetterberg et al. (1970a) tested the possible photoreceptor capabilities of the Harderian glands was apparently

Webb et al. (Eds).
Harderian Glands

because of their location in the orbital cavity, their anatomical association with the eye, and because they had observed (although unreported at the time) that the porphyrin content of the Harderian glands was influenced by the lighting conditions to which rats were exposed. To test the involvement of the eyes and/or Harderian glands in mediating the pineal serotonin rhythm, newborn male and female Long-Evans rats were kept with their mothers under a light:dark cycle of 12:12 (lights on at 06.00 h). When they were 9 days of age, the suckling rats were either blinded by bilateral orbital enucleation or they were both blinded and had their Harderian glands surgically removed. At the termination of the experiment, the completeness of Harderian ablation was verified by exposing the remaining contents of the orbital cavity to fluorescence at 366 nm (Harderian gland porphyrins fluoresce when exposed to this wavelength).

As had been previously shown, Wetterberg and colleagues (1970a) found that daytime (13.00 h) levels of pineal serotonin were much higher (three fold) than nighttime (23.00 h) values even when the young rats were blinded (Fig. 1). For two groups of blinded rats, the light was extended into the dark phase (to prevent the nighttime drop in pineal serotonin). However, in the rats that were blinded only (Harderian glands intact), the extension of light into the dark phase was apparently "seen" by the pineal gland since the serotonin content of the gland remained higher than values in dark-exposed controls. In the blinded, Harderian glandectomized rats, however, the drop in pineal serotonin content was complete as if the animals were totally incapable of light perception (Fig. 1).

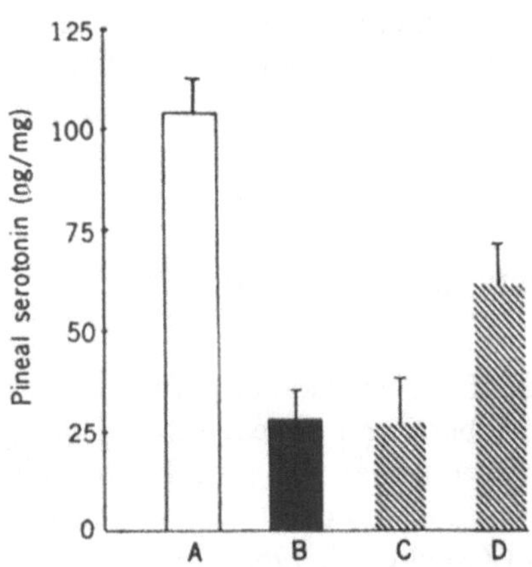

Fig. 1. Pineal serotonin content in Harderian glandectomized or Harderian gland-intact 12-day-old rats that were blinded (at 9 days of age) and exposed to either light or darkness at night. *A* Control rats killed at 13.00 h during the day; *B* control rats killed at 23.00 h, 5 h after darkness onset; *C* rats killed at 23.00 h in the light and subjected to Harderian glandectomy; *D* rats killed at 23.00 h in the light with Harderian glands intact. 12 rats/group; SEM are indicated by *vertical bars*. (Wetterberg et al. 1970a)

Thus, in these studies, removal of the Harderian gland from blinded suckling rats was similar to hooding of blinded rats in terms of its effects on pineal serotonin levels (Snyder et al. 1964, 1965; Zweig et al. 1966). The results are consistent with the idea that the Harderian glands, at least in pre-weanling rats, serve as extra-retinal photoreceptors involved in the mediation of the circadian rhythm of pineal serotonin. Wetterberg et al. (1970a) also mentioned the possibility that the destruction (by the combination of blinding and Harderian gland removal) of critical glandular or neural elements necessary for the daily changes in pineal serotonin may have been responsible for the observed changes. Assuming that the Harderian glands participate in the mediation of the effects of light on the metabolism of serotonin within the pineal gland, Wetterberg and coworkers (1970a) rendered no speculation as to how the orbital structures would signal, i.e., humoral, neural, etc., the pineal as to the photoperiodic status. Finally, the existence of a serotonin rhythm in the pineal of 12-day-old rat pups implies either the photoperiodic-dependent synthesis or metabolism of the monoamine in the gland at this age. The enzyme, serotonin N-acetyltransferase (NAT), which is of major importance in decreasing the nocturnal levels of serotonin in the pineal gland of the adult rat, does not exhibit a nighttime increase until rats are roughly 15 days of age; thus, the fluctuations in pineal serotonin in 12-day-old rats was presumably due to its metabolism via some other metabolic pathway, e.g., oxidation or methylation.

Soon after their original observation that the Harderian glands may be important in transducing photoperiodic information to the pineal gland, Wetterberg et al. (1970b) used another pineal end-point, i.e., the activity of the melatonin-forming enzyme hydroxyindole-*O*-methyltransferase (HIOMT), to assess the importance of the Harderian glands in pineal metabolic activity. At the time of their report, they relied on what is now known to be an inconsistent observation that pineal HIOMT activity exhibits a 24-rhythm with high values at night. In their study, Wetterberg and colleagues (1970b) again subjected Long Evans rat pups to surgical blinding and Harderian glandectomy. The pups and their respective dams were maintained under a light:dark cycle of 12:12 (lights on at 06.00 h). At 12 days of age, the pups were killed either during the day (13.00 h) or at night (23.00 h) and pineal HIOMT activity was estimated according to the procedure of Axelrod et al. (1965). Even in blinded pups, a day:night rhythm in pineal HIOMT activity was observed with the nighttime levels of enzyme activity being about twice those measured during the day (Fig. 2). In the blinded rats with intact Harderian glands, the extension of light into the normal dark period resulted in daytime HIOMT levels, suggesting that the pineal was still responding to light even though the animals were blinded; this suppression did not occur if the pups, in addition to being blinded, were also Harderian glandectomized (Fig. 2). Again, these findings were interpreted to mean that the Harderian glands mediate light effects on the pineal gland of suckling rat pups although the authors did present some other potential interpretations of the findings (Wetterberg et al. 1970b). Also, they implied that the porphyrins, which

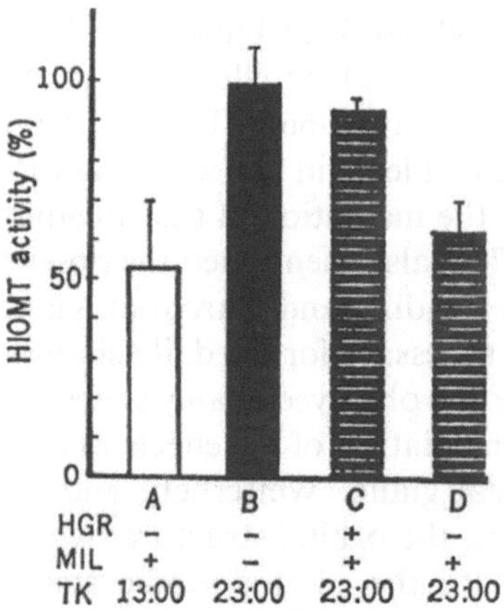

Fig. 2. Pineal HIOMT activity in 12-day-old rats exposed to either light or darkness and after Harderian glandectomy. *A* Control rats killed during the day (13.00 h); *B* control rats killed during darkness (5 h after onset) at 23.00 h; *C* rats killed at 23.00 h in light with Harderian glands removed; *D* rats killed at 23.00 h in light with Harderian glands intact. *HGR* Harderian glands removed; *MIL* maintained in light; *TK* time killed. Twelve rats/group and SEM are indicated by *vertical bars*. (Wetterberg et al. 1970b)

are in high concentration in the Harderian glands of rats (Klüver 1944), may somehow be involved in mediating light effects on the pineal gland.

The major problem with these findings is that the pineal HIOMT rhythm seems to be unreliable and, although a number of early studies suggested its existence (Wurtman et al. 1963; Axelrod et al. 1965), more recent observations on its supposed fluctuation generally suggest an absence of a rhythm (Reiter 1988). Early observations on the interactions of light, the Harderian glands, and the pineal gland are summarized in a brief review by Wetterberg et al. (1973). Also, a diagrammatic rendition of their proposed interactions is summarized in Fig. 3.

About the same time Wetterberg et al. (1970a,b) were examining light-Harderian-pineal gland interactions in suckling rats, Reiter and Klein (1971) were testing these potential relationships in adult rats. They exposed adult female rats to either continuous light or light:dark cycles of 14:10 (lights on at 06.00 h) for 67 days; additionally, some rats had their Harderian glands surgically removed. At the termination of the experiment, the animals were killed either during the day (12.00 h) or at night (24.00 h). Among other organs, the pineal gland of all animals was collected, weighed, and assayed for its activity of the two enzymes, i.e., NAT and HIOMT, involved in the conversion of serotonin to melatonin. As expected, continual light exposure significantly depressed the weights of the pineal gland but Harderian gland removal was without influence on this pineal parameter (Fig. 4). Pineal HIOMT activity was uninfluenced by either continuous illumination or loss of the Harderian glands. Also, in intact,

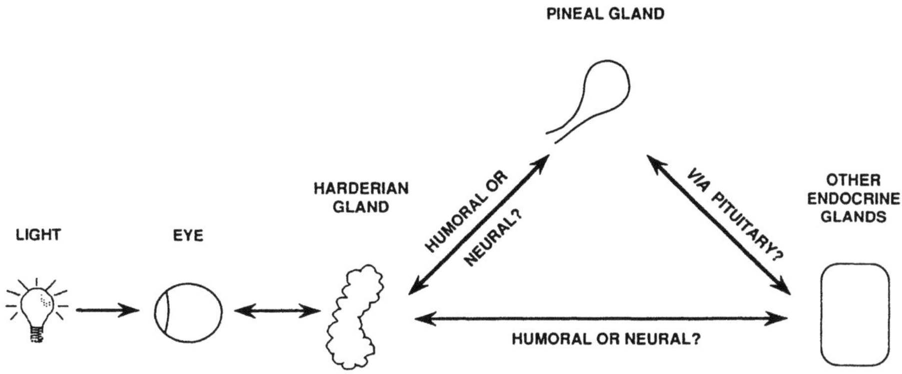

Fig. 3. Theoretical interactions among the photoperiod, the eyes, the Harderian glands, the pineal gland, and other endocrine organs. (After Wetterberg et al. 1973)

cyclic light-exposed rats, no day:night difference in pineal HIOMT activity was apparent. Pineal NAT activity exhibited the expected marked nighttime increase which was totally prevented by continual light exposure. On the other hand, Harderian gland removal was without effect on the pineal rhythm of serotonin N-acetylation (Fig. 4). Finally, surgical removal of the Harderian glands did not alter the weight decrease of either the ovaries, eyes, or retinas that were induced by continual light exposure. Interestingly, the Harderian glands, like the retinas (Noell et al. 1966; O'Steen and Lytle 1971), exhibited marked involutional changes as a consequence of light exposure for 67 days (Fig. 5), suggesting that these glands respond either directly or indirectly to light. This is consistent with the results of Wetterberg and colleagues (1970a, b), who concluded that, at least in suckling rats, the Harderian glands may be capable of direct photoreception.

The regressive responses of the Harderian glands to continual light exposure is not limited to observations in albino rats. Pigmented rats (Long Evans strain) as well respond to constant light exposure (for 65 days) with gross and cytological atrophy of their Harderian glands (Reiter 1973; Fig. 5). In these animals, as in the albino rats (Sprague-Dawley strain), loss of the Harderian glands is without significant effect on either the daytime or nighttime activity of HIOMT in the pineal gland.

That the pineal melatonin rhythm may be altered by bilateral removal of the Harderian glands in adult male Syrian hamsters was investigated by Panke and associates (1979a). In the study in question, hamsters were either left intact or they were subjected to Harderian glandectomy; thereafter they were maintained under a light:dark environment of 14:10 (lights on at 06.00 h). Two weeks after the surgery, pineal glands were collected from both intact and operated hamsters at 22.00 h (2 h after darkness onset), 04.00 h (8 h after darkness onset), and

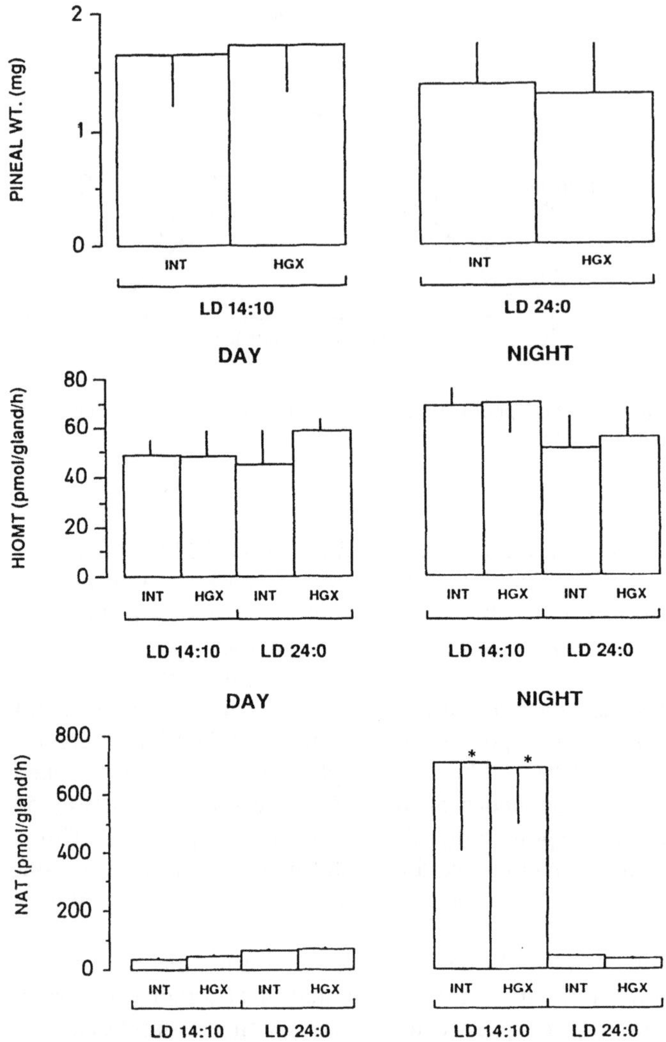

Fig. 4. Pineal weight, NAT, and HIOMT activities in albino rats kept in either cyclic (LD 14:10) or continuous (LD 24:0) light for 67 days. Additionally, some animals had been subjected to Harderian glandectomy (HGX). Continual light significantly depressed the NAT rhythm (*$P < 0.001$); however, removal of the Harderian glands had no effects on the pineal gland. (Data from Reiter and Klein 1971)

07.00 h (1 h after light onset). The 04.00 time point was of particular interest since in hamsters kept under these photoperiodic conditions, peak pineal melatonin levels normally occur at this time (Panke et al. 1979b). The findings reveal that loss of the Harderian glands in Syrian hamsters results in significantly lower

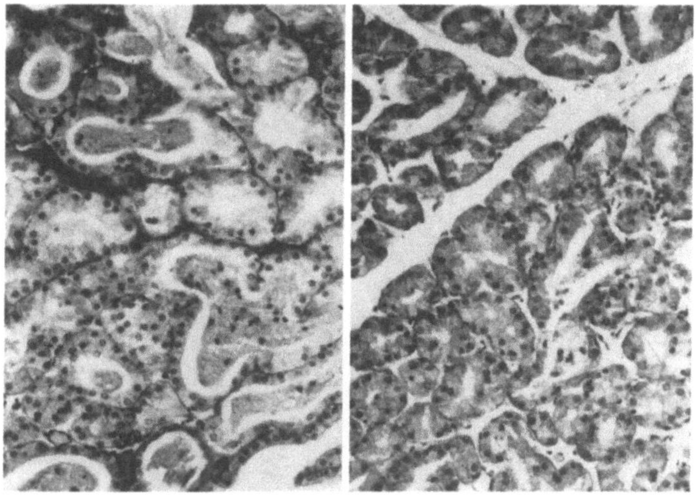

Fig. 5. Harderian glands of albino rats maintained in constant (*left*) or cyclic (*right*) light for 67 days. The glands of the constant light exposed rats weighed less than those in cyclic light and, as shown in the figure, they had numerous sloughed epithelial cells in their tubular lumina. (Reiter and Klein 1971)

melatonin levels 8 h after darkness onset (Fig. 6). However, the data from this study do not allow for the conclusion that peak pineal melatonin levels are blunted in Harderian glandectomized hamsters since it is possible that the melatonin rhythm was merely shifted and, thus, occurred at a time point not examined by Panke et al. (1979a).

The pineal melatonin rhythm is generally considered to be highly stable and its disruption by Harderian gland removal may be significant since there are few other surgical manipulations which are capable of doing this. However, these interactions have not been tested in any other mammal and have not been confirmed in the Syrian hamster. The study of Panke et al. (1979a) does prove that it is possible to surgically remove the Harderian glands without severe damage to either the eyes or optic nerves. Had this occurred, e.g., had the animals been blinded by surgical removal of the Harderian glands, the melatonin levels at 22.00 and 07.00 h would have had a wider distribution representing the free running melatonin cycle seen in blinded, adult rodents (Reiter et al. 1971). The perceived difficulty of visual system damage by Harderian glandectomy may be a major reason why so few studies have examined Harderian-pineal interactions using this procedure. In reality, the surgical procedure is rather straightforward when the orbital cavity is entered posteromedially for removal of the glands (Reiter and Klein 1971). The Harderian glands are, however, large (often larger than the eyes themselves) and it is possible to inadvertently leave behind portions of the glands when they are

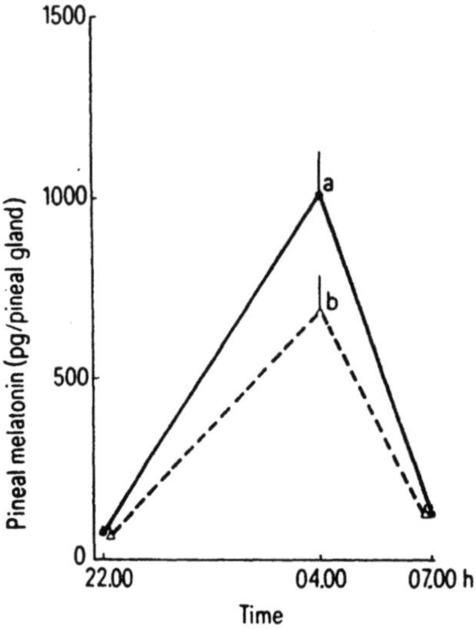

Fig. 6. Pineal melatonin levels in intact Syrian hamsters and in hamsters from whom the Harderian glands had been removed bilaterally. The results indicate that the nocturnal peak of melatonin may be blunted (*a* versus $b = p < 0.05$) in hamsters lacking their Harderian glands. (Panke et al. 1979a)

removed. Whether these residual fragments remain functional is unknown. Also, paired organs, e.g., the kidney, usually exhibit compensatory anatomical or functional changes when one of the pair is removed (Reiter 1972). Whether a remaining Harderian gland exhibits compensatory changes of any type after unilateral organ removal has not been tested.

3 Evidence for Pineal Gland Effects on the Harderian Glands

About the same time investigators began to examine the potential interactions of the Harderian glands with pineal function, other investigators were reporting on the reciprocal interrelationships, i.e., the effects of pineal manipulations on the morphology and function of the Harderian glands. Roughly two decades ago, Hoffman (1971) and Clabough and Norvell (1974) uncovered evidence that the morphological appearance of the Harderian glands was related to the functional status of the pineal gland. According to Hoffman (1971), surgical removal of the pineal gland from Syrian hamsters kept under long days (light:dark cycles of 16:8) was without effect on the cell types present in the Harderian glands of either males or females. However, bilateral orbital enucleation was shown to change the porphyrin content of the Harderian glands (Hoffman 1971; Wetterberg 1972). These changes implied that the pineal gland may have some effect on the Harderian glands since, at this time, it was already known that the pineal gland of Syrian hamsters is maximally stimulated by treatments which reduce the duration of the daily light period (Hoffman and Reiter 1965a; Reiter and Hester 1966). However, stimulation of the pineal gland leads, in Syrian hamsters, to gonadal involution and a resultant suppression of associated hormones. This prompted Hoffman (1971) to conclude that the apparent effects of the pineal gland on Harderian morphology were, in fact, a result of changes in either circulating pituitary gonadotrophins or gonadal steroids. This conclusion was also reached by Clabough and Norvell (1974) when they observed that blinding in female Syrian hamsters changed the structure and porphyrin content of the Harderian glands. They observed that 11 weeks after blinding, the hamsters had involuted uteri and diminished amounts of porphyrin pigments in their Harderian glands; when pinealectomy was combined with blinding, the Harderian glands were, in all respects, similar to those of intact females maintained under long days.

There certainly is ample evidence available showing that gonadal steroids in particular have a profound influence on the morphology and function of the Harderian glands, especially in the Syrian hamster (Marrufo et al. 1989; Menendez-Pelaez et al. 1990). Especially germane are the observations that steroid receptors have been identified in the Harderian glands of the Syrian hamster (Vilchis et al. 1987). Whereas the effects reported by Hoffman (1971) and Clabough and Norvell (1974) are easily explicable on the basis of an indirect action of the pineal gland on the Harderian glands via the gonads, direct interactions should not be totally ruled out in view of the recent demonstration that melatonin-binding sites have also been found in the Harderian glands (Lopez-Gonzalez et al. 1990). Melatonin is, of course, the chief hormonal product of the pineal gland (Reiter 1980) although it is also produced in the Harderian glands themselves (Bubenik et al. 1976; Pang et al. 1977).

Also in the Syrian hamster, Menendez-Pelaez et al. (1988a, b) have examined the functional impact of pineal manipulation on Harderian gland physiology. Particularly noteworthy in these studies was the finding that in pinealectomized female hamsters the changes in Harderian gland NAT and melatonin content that occur shortly after lights-on were disrupted. Normally, within 1 h after light onset, NAT levels rise while melatonin values fall in the Harderian gland of this species. After pinealectomy, the rise in NAT activity was delayed and, whereas the drop in melatonin occurred, when the levels were restored, the values were much higher than those measured at other time points (Fig. 7).

There are reports showing that pinealectomy also changes the melatonin synthesizing ability of the Harderian glands in both the rat (Reiter et al. 1983) and the pigeon (Vakkuri et al. 1985a). The first of these reports appeared in 1983 when Reiter and colleagues were investigating melatonin rhythms in extrapineal locations. They observed that the melatonin synthetic cycle in both the retina and the Harderian glands of rats was exaggerated in pinealectomized animals. In the study in question, male rats were either left intact or they were

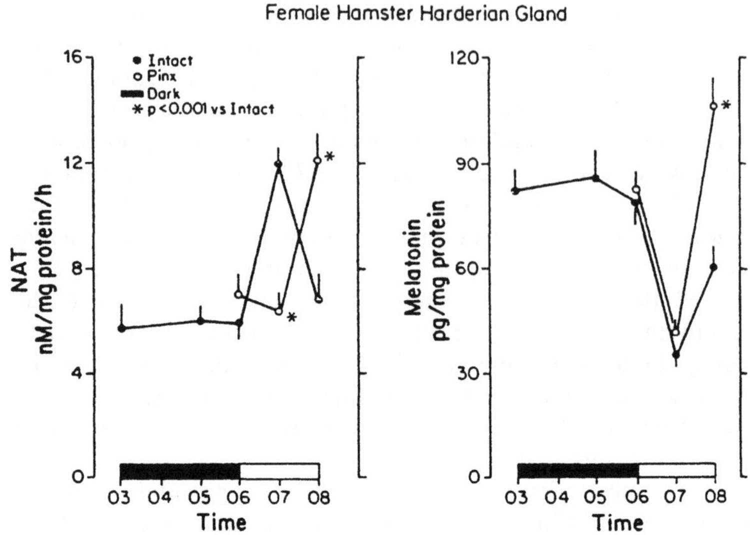

Fig. 7. NAT activity and melatonin levels in the Harderian glands of female Syrian hamsters during darkness (*black horizontal bar*) and shortly after light onset (*clear horizontal bar*). Animals were either intact or pinealectomized (*Pinx*). Pinealectomy disturbed the light phase levels of both NAT and melatonin. (Menendez-Pelaez et al. 1988b)

surgically pinealectomized; thereaftrer, they were maintained in a photoperiodic environment of 14:10 with lights on from 06.00 to 20.00 h daily. After 3 weeks of treatment, the Harderian glands, retinas, and pineal gland (from the intact rats) were collected at 4 h intervals over a 24-h-period and the melatonin content of the organs was determined by radioimmunoassay. In the intact animals, each of the three organs investigated exhibited a clear 24-h-cycle of melatonin production. After pinealectomy, these rhythms not only persisted in the Harderian glands and retinas, but they were exaggerated (Fig. 8). Thus, whereas pineal removal did not influence either the phasing of the rhythm or the minimal levels of melatonin in the glands, the peak values were higher than in the organs of intact rats. The authors theorized that the exaggerated melatonin response in the Harderian glands and retinas was a compensation for the loss of pineal melatonin production. Yu and co-workers (1981) had earlier shown a compensatory rise in retinal melatonin in the rat although the Harderian glands were not examined in this study.

In the pigeon (*Columbia livia*), the effects of pinealectomy in terms of the concentrations of melatonin in the Harderian glands may be even more dramatic (Vakkuri et al. 1985b). In this species, a day:night rhythm in the melatonin content of the Harderian glands exists (Vakkuri et al. 1985a). Two weeks after pinealectomy, the nighttime melatonin content of the Harderian glands was 100-fold greater than that in the glands of pineal-intact pigeons (Fig. 9). On the

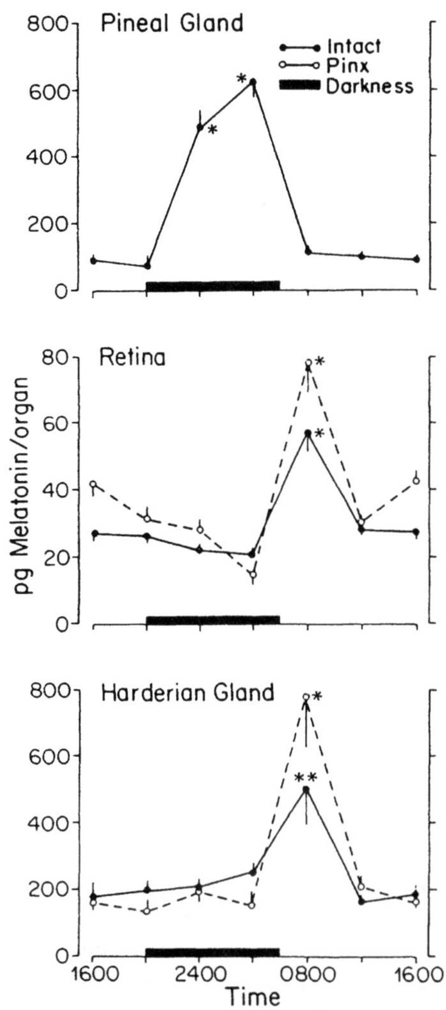

Fig. 8. Melatonin levels in the pineal gland, retina and Harderian glands over a 24-h-light:dark cycle in adult male rats. Also shown is the effect of pinealectomy (*Pinx*) on retinal and Harderian gland melatonin. Pinealectomy significantly increased peak melatonin levels in both organs. (Reiter et al. 1983)

other hand, pinealectomy decreased hypothalamic melatonin content and had no effect on the levels of the indoleamine in either the eyes or duodenum. On the basis of their results, Vakkuri et al. (1985b) concluded that the chief extra-pineal source of circulating melatonin was likely the Harderian glands.

The observations of Vakkuri et al. (1985b) on the pigeon were not confirmed by Cogburn and colleagues (1987) when single-comb White Leghorn cockerels (*Gallus domesticus*) were used as the experimental animals. In this study, young

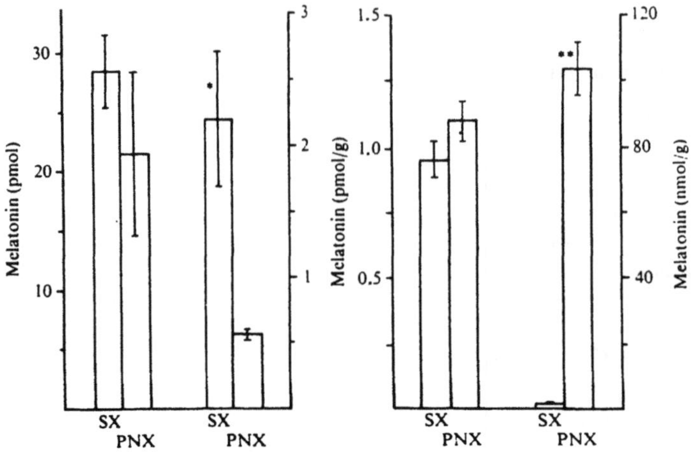

Fig. 9. *Left* to *right* Melatonin levels at midnight in the eyes, hypothalamus, duodenum, and Harderian glands of intact (*S*) and pinealectomized (*PNX*) pigeons 2 weeks following pinealectomy. $*P < 0.005$ and $**P < 0.001$ versus intact pigeons. (Vakkuri et al. 1985a)

cockerels were either unoperated, sham operated, or pinealectomized; the surgical procedures were carried out when they were 8–10 days of age. During the course of the study the birds were maintained in a light:dark environment of 12:12. When they were 25 days of age, the birds were killed at 4-h-intervals during a 24-h-period and their Harderian glands were analyzed for their content of melatonin using a radioimmunoassay. While a low amplitude rhythm (maximal levels at night) in Harderian gland melatonin was reported, neither the phasing of the rhythm nor the levels of melatonin were changed in the pinealectomized cockerels (Fig. 10). Thus, in this species, unlike in either pigeon (Vakkuri et al. 1985b) or rat (Reiter et al. 1983), the loss of the pineal gland did not cause a subsequent compensatory rise in Harderian gland melatonin levels. Besides the species difference, there is at least one other factor which could explain the difference among these studies. Whereas Reiter et al. (1983) and Vakkuri and co-workers (1985b) used adult animals, Cogburn and colleagues (1987) utilized sexually immature birds. How or whether this significantly impacted the results remains to be determined.

In the Indian jungle bush quail (*Perdicula asiatica*), a tropical zone bird, the weight of the Harderian glands exhibits a conspicuous annual rhythm which is positively correlated with the weight of the reproductive organs and inversely related to the size of the pineal gland (Haldar and Ghosh 1989; Fig. 11). Pinealectomy, of 2 month's duration, led to a significant decrease in the weights of the Harderian glands only in the recrudescence (March-cMay) and active (April-June) phases of the testicular cycle. The weight reduction was reflected in the histological appearance of the epithelial cells lining the follicles of the glands; in the case of the pinealectomized birds, these cells were reduced in their overall height.

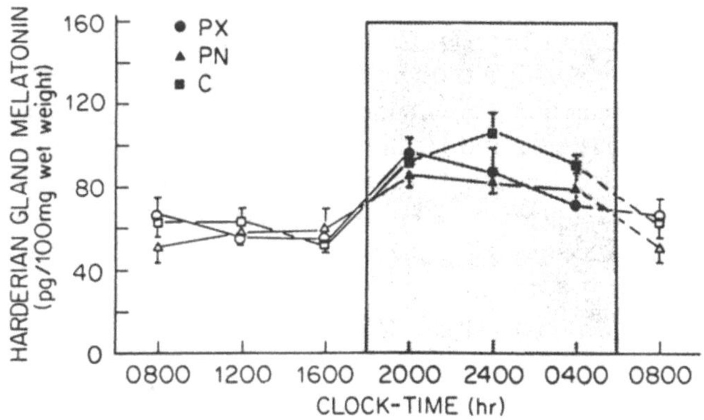

Fig. 10. Twenty four-hour melatonin rhythm in the Harderian glands of intact (*C*), pinealectomized (*PX*), and sham operated (*PN*) chickens. The *shaded area* represents the period of darkness. Each *point* represents the mean and SEM of six birds. (Cogburn et al. 1987)

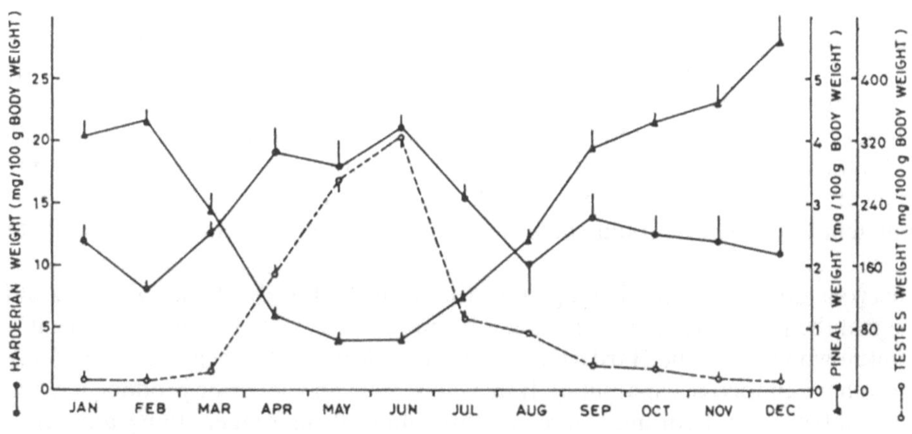

Fig. 11. Annual rhythms of weights of the Harderian glands, the pineal gland, and the testes in the Indian jungle bush quail. (Haldar and Ghosh 1989)

Melatonin administration by means of a subcutaneous implant markedly reduced the size of the Harderian glands in the quail during the developmental and active phases of their testicular cycle (Haldar and Ghosh 1989). As a result of their findings, these authors concluded that melatonin is an important factor in the control of Harderian gland function, but whether the relationship is a direct or indirect one was not determined.

Unlike the perception of surgical removal of the Harderian gland, pinealectomy in rodents is considered to be an easy, reliable, and relatively innocuous procedure (Hoffman and Reiter 1965b). However, in species where the pineal complex consists of deep and superficial pineal tissue, superficial pinealectomy (commonly known as pinealectomy) does not necessarily remove the deep part of the gland (Sheridan and Reiter 1970), although the remaining deep portion is generally considered to be nonfunctional after superficial pinealectomy because it is sympathetically denervated (Reiter and Hedlund 1976). Whereas superficial pinealectomy is assumed to destroy the ability of the deep pineal gland to signal the Harderian glands, this has not been definitively proven. There may be a hormone(s) released from the deep pineal gland which does not rely on the sympathetic innervation to the tissue.

4 Final Comment

The evidence establishing a functional link between the Harderian glands and the pineal is not compelling and those observations that have been made require confirmation. If the Harderian glands do in fact signal the pineal as to the photoperiodic conditions, it is necessary to explain how the signal processes are carried out, i.e., does it involve a neural or a blood-borne message? Both the Harderian glands (Norvell and Clabough 1972) and the pineal gland (Kappers 1960) have abundant innervations; however, the fibers seem to be exclusively efferent in function. If the Harderian glands convey a neural message to the pineal gland, the former organs would, by necessity, have to be innervated by afferent fibers which would transfer information from the Harderian glands to the brain and then eventually, presumably via the peripheral sympathetic nervous system, to the pineal gland. Alternatively, the Harderian glands could signal the pineal gland by means of a hormonal envoy. At the present time, it is unknown whether the Harderian glands, in addition to their exocrine functions, also have endocrine capabilities. If they do secrete an endocrine product, it could be that N-acetylserotonin serves as this humoral messenger (Pang and Allen 1986).

The observations suggesting a functional impact of the pineal gland on the Harderian glands seem to be somewhat more consistent than the observed reciprocal interactions. Loss of pineal tissue may, in fact, cause a compensatory rise in Harderian gland melatonin levels, and presumably synthesis. If this is the case it would seem likely that it is specifically the reduction in circulating melatonin due to pinealectomy which serves as the signal to the Harderian glands to change their melatonin synthetic machinery.

Of seeming interest to the overall field may be the extraorbital lacrimal glands (EOL). These structures, like the Harderian glands, seem to produce melatonin (Mhatre et al. 1988) and they undergo partial transformation into Harderian gland tissue, a process referred to as Harderianization, under certain conditions (Sashima et al. 1989). In the future it may be important to include the EOL with the Harderian/pineal/retinal complex.

References

Axelrod, Wurtman RJ, Snyder SH (1965) Control of hydrozyindole-O-methyltransferase activity in the rat pineal gland by environmental lighting. J Biol Chem 240:949–954

Balemans MGM, Pévet P, Legerstee WC, Nevo E (1980) Preliminary investigations on melatonin and 5-methoxytryptophol synthesis in the pineal, retina and Harderian gland of the mole rat (*Spalax ehrenbergi*) and in the pineal of the mouse "Eyeless". J Neural Transm 49:247–255

Bubenik GA, Brown GM, Grota LJ (1976) Immunohistochemical localization of melatonin in the rat Harderian gland. J Histochem Cytochem 24:1173–1177

Bubenik GA, Purtill RA, Brown GM, Grota LJ (1978) Melatonin in the retina and the Harderian gland. Ontogeny, diurnal variations and melatonin treatment. Exp Eye Res 27:323–333

Cardinali DP, Wurtman RJ (1972) Hydroxyindole-O-methyltransferase in rat pineal, retina and Harderian gland. Endocrinology 91:247–252

Clabough JW, Norvell JE (1974) Pineal influence on the Harderian glands of female golden hamsters. Anat Rec 178:119–125

Cogburn LA, Wilson-Placentra S, Letcher LR (1987) Influence of pinealectomy on plasma and extrapineal melatonin rhythms in young chickens (*Gallus domesticus*). Gen Comp Endocrinol 68:343–356

Gauquelin G, Gharib C, Ghaemmaghami F, Allevard A-M, Cherbal F, Geelen G, Bouzeghrane F, Legros J-J (1988) A day/night rhythm of vasopressin and oxytocin in rat retina, pineal and Harderian gland. Peptides 9:289–293

Guerrero JM, Puig-Domingo M, Vaughan GM, Reiter RJ (1987) Characterization of type II thyroxine 5'-deiodinase activity in the rat Harderian gland. Life Sci 41:1179–1185

Guerrero JM, Puig-Domingo M, Santana C, Menendez-Pelaez A, Reiter RJ (1988) Inhibition of pineal type II 5'-deiodinase does not effect the nocturnal increase of N-acetyltransferase activity and melatonin content in either euthyroid or thyroidectomized rats. J Pineal Res 5:513–520

Haldar C, Ghosh M (1989) Effect of pinealectomy and 5-methoxyindoles on Harderian gland activity of the Indian jungle bush quail *Perdicula asiatica*. Indian J Exp Biol 27:704–708

Hoffman RA (1971) Influence of some endocrine glands, hormones and blinding on the histology and porphyrins of the Harderian glands of golden hamsters. Am J Anat 132:463–478

Hoffman RA, Reiter RJ (1965a) Pineal gland: influence on gonads of male hamsters. Science 148:1609–1611

Hoffman RA, Reiter RJ (1965b) Rapid pinealectomy in hamsters and other small rodents. Anat Rec 153:19–22

Kappers JA (1960) The development, topographical relations and innervation of the epiphysis cerebri in the albino rat. Z Zellforsch Mikrosk Anat 52:163–215

Klüver H (1944) Porphyrins, the nervous system, and behavior. J Psychol 17:209–217

Lopez-Gonzalez MA, Calvo JR, Rubio A, Goberna R, Guerrero JM (1990) Characterization of melatonin-binding sites in the Harderian gland and median eminence of the rat. Life Sci 48:1165–1171

Marrufo B, Menendez-Pelaez A, Buzzell GR, Gonzalez-Brito A, Reiter RJ (1989) 5α-Dihydrotestosterone administration converts indoleamine metabolism and porphyrin content of the female Syrian hamster Harderian gland to the male type. Proc Soc Exp Biol Med 192:192–195

Menendez-Pelaez A, Reiter RJ, Howes KA, Puig-Domingo M, Vaughan MK, Troiani ME, Little JC (1988a) Harderian gland N-acetyltransferase activity in the male Syrian hamster: effects of gonadectomy, short photoperiod exposure or subcutaneous melatonin implants. Endocr Res 14:121–130

Menendez-Pelaez A, Reiter RJ, Guerrero JM, Santana C, Howes KA, Gonzalez-Brito A (1988b) N-acetyltransferase activity and melatonin concentrations in the Harderian glands of female Syrian hamsters: alterations following either pinealectomy of bilateral superior cervical ganglionectomy. Biochem Arch 4:231–238

Menendez-Pelaez A, Buzzell GR, Gonzalez-Brito A, Reiter RJ (1990) Androgenic control of N-acetyltransferase activity in the Harderian glands of the female Syrian hamster is mediated by 5α-dihydrotestosterone. J Cell Biochem 42:95–100

Mhatre MC, van Jaarsveld AS, Reiter RJ, (1988) Melatonin in the lacrimal gland: first demonstration and experimental manipulation. Biochem Biophys Res Commun 153:1186–1192

Noell WK, Walker VS, Kang BS, Berman S (1966) Retinal damage by light in rats. Invest Opthalmol 5:450–473

Norvell JE, Clabough JW (1972) Adrenergic and cholinergic innervation of the hamster Harderian gland. Science 178:1102–1103

O'Steen WK, Lytle RB (1971) Early cellular disruption and phagocytosis in photically-induced retinal degeneration. Am J Anat 130:227–234

Pang SF, Allen AE (1986) Extra-pineal melatonin in the retina: its regulation and physiological function. Pineal Res Rev 4:55–96

Pang SF, Brown GM, Grota LJ, Chambers JW, Rodman RL (1977) Determination of N-acetylserotonin and melatonin activities in the pineal gland, retina, Harderian gland, brain and serum of rats and chickens. Neuroendocrinology 23:1–13

Panke ES, Reiter RJ, Rollag MD, (1979a) Effect of removal of the Harderian glands on pineal melatonin concentrations in the Syrian hamster, *Mesocricetus auratus*. Experientia 35:1405–1406

Panke ES, Rollag MD, Reiter RJ (1979b) Pineal melatonin concentrations in the Syrian hamster. Endocrinology 104:194–197

Quay WB (1963) Circadian rhythm in rat pineal serotonin and its modification by estrous cycle and photoperiod. Gen Comp Endocrinol 3:473–480

Reiter RJ (1972) Compensatory growth of the ovaries, adrenal glands and kidneys in blinded, anosmic rats. Experientia 28:1492

Reiter RJ (1973) Comparative effects of continual lighting and pinealectomy on the eyes, the Harderian glands and reproduction in pigmented and albino rats. Comp Biochem Physiol 44A:503–509

Reiter RJ (1980) The pineal and its hormones in the control of reproduction in mammals. Endocr Rev 1:109–131

Reiter RJ (1988) Neuroendocrinology of melatonin. In: Miles A, Philbrick DRS, Thompson C (eds) Melatonin: clinical perspectives. Oxford University Press, Oxford, pp 1–42

Reiter RJ, Hedlund L (1976) Peripheral sympathetic innervation of the deep pineal gland of the golden hamster. Experientia 32:1071–1072

Reiter RJ, Hester RJ (1966) Interrelationships of the pineal gland, the superior cervical ganglia and the photoperiod in the regulation of the endocrine system of hamsters. Endocrinology 79:1168–1170

Reiter RJ, Klein DC (1971) Observations on the pineal gland, the Harderian glands, the retina, and the reproductive organs of adult female rats exposed to continuous light. J Endocrinol 51:117–125

Reiter RJ, Klein DC, Weller JC (1971) Pineal N-acetyltransferase activity in blinded and anosmic rats. Endocrinology 89:1020–1023

Reiter RJ, Richardson BA, Hurlbut EC (1981) Pineal, retinal and Harderian gland melatonin in a diurnal species, the Richardson's ground squirrel (*Spermophilus richardsonii*). Neurosci Lett 22:285–288

Reiter RJ, Richardson BA, Matthews SA, Lane SJ, Ferguson BN (1983) Rhythms in immunoreactive melatonin in the retina and Harderian gland of rats: persistence after pinealectomy. Life Sci 32:1129–1136

Sashima M, Hatakeyama S, Satoh M, Suzuki A (1989) Harderianization is another sexual dimorphism of rat exorbital lacrimal gland. Acta Anat 135:303–306

Sheridan MN, Reiter RJ (1970) Observations on the pineal system in the hamster. II. Fine structure of the deep pineal. J Morphol 131:163–178

Snyder SH, Zweig M, Axelrod J (1964) Control of the circadian rhythm in serotonin content of the rat pineal gland. Life Sci 3:1175–1179

Synder SH, Zweig M, Axelrod J, Fischer JE (1965) Control of the circadian rhythm in serotonin content of the rat pineal gland. Proc Nat Acad Sci 53:301–305

Vakkuri O, Rintamäki H, Leppäluoto J (1985a) Plasma and tissue concentrations of melatonin after midnight light exposure and pinealectomy in the pigeon. J Endocrinol 105:263–268

Vakkuri O, Rintamäki H, Leppäluoto J (1985b) Presence of immunoreactive melatonin in different tissues of the pigeon (*Columbia livia*). Gen Comp Endocrinol 58:69–75

Vilchis F, Hernandez A, Pérez AE, Perez-Palacios G (1987) Hormone regulation of the rodent Harderian gland: binding properties of the androgen receptor in the male golden hamster. J Endocrinol 112:3–8

Wetterberg L (1972) Increase in Harderian gland porphyrin content in castrated male hamsters dependent on light and visual function. Life Sci 11 (II):541–546

Wetterberg L, Geller E, Yuwiler A (1970a) Harderian gland: an extraretinal photoreceptor influencing the pineal gland in neonatal rats? Science 167:884–885

Wetterberg L, Yuwiler A, Ulrich R, Geller E, Wallace R (1970b) Harderian gland: influence on pineal hydroxyindole-O-methyltransferase activity in neonatal rats. Science 170:194–196

Wetterberg L, Ulrich R, Yuwiler A (1973) Light, the Harderian gland and the rodent pineal. In: Proc 4th Int Congr Endocrinology, Elsevier, New York, p 268

Wurtman RJ, Axelrod J, Philips LS (1963) Melatonin synthesis in the pineal gland: control by light. Science 142: 1071–1073

Yu HS, Pang SF, Tang PL (1981) Increase in the level of retinal melatonin and persistence of its diurnal rhythm in rats after pinealectomy. J Endocrinol 91:477–481

Zweig M, Snyder SH, Axelrod J (1966) Evidence for a non-retinal pathway of light to the pineal gland of newborn rats. Proc Nat Acad Sci 56:515–519

[illegible]

[illegible]

[illegible] Research [illegible]

[illegible]

[illegible]

The Harderian Gland as a Target for Steroid Hormone Action: Role and Characteristics of Intracellular Receptors

FELIPE VILCHIS, BERTHA CHÁVEZ, MARCO A. CERBÓN and GREGORIO PÉREZ-PALACIOS

Department of Reproductive Biology, National Institute of Nutrition S. Zubiran, Vasco de Quiroga No. 15, Tlalpan 14000, Mexico City, Mexico

1 Introduction

Evidence has been accumulated indicating that the Harderian gland in vertebrates is a target organ for hormone action. The effects of steroid hormones, thyroid hormones, growth hormone, and prolactin upon cellular growth, secretion, differentiation, and metabolic activities of the gland have been well documented. Futhermore, it has been shown that in some mammalian and non-mammalian species, the effects of steroid hormones of gonadal origin determine the sexual phenotype of the Harderian gland. Since most, if not all, of the direct activities of steroid hormones upon the gland are exerted at the genomic level, it was plausible to predict the presence of specific, high affinity, intracellular steroid binding sites in this intraorbital structure. Indeed, recent studies have provided evidence for the occurrence of androgen, estrogen, progestin, glucocorticoid, and thyroid hormone receptors in tissue preparations of Harderian glands from several species.

This chapter deals with the current concepts of the mechanisms involved in the mode of action of steroid hormones at target organs with particular focus on the Harderian gland. Emphasis is given to the binding kinetics, stereospecificity, and physicochemical features of the intracellular steroid hormone receptors of the Harderian gland in mammals, birds, and reptiles. Preliminary information involving an androgen-regulated protein in the hamster gland is also given.

2 Genomic Mechanisms of Steroid Hormone Action

Steroid hormones regulate cell growth and differentiation and many important metabolic events at the target organ level by the activation or repression of specific genes. These complex processes require the cooperation of specific hormone-responsive elements involved in the modulation of gene expression. Since an increase or a diminution in gene expression will ultimately determine the hormonal effects at the cellular level, it is important to understand the

Webb et al. (Eds).
Harderian Glands

mechanisms by which the steroid molecule is recognized by the target cell and all the subsequent intracellular steps required to achieve the hormone-genome interaction.

Steroid hormones are transported from their production site (endocrine glands) to the target organs by specific carriers (hormone-binding globulins) and enter the cells by diffusion. Once inside, the steroid is recognized, according to its chemical structure and configuration, by binder macromolecules of proteinic nature termed receptors. These intracellular receptors play a most important role in hormone action, since they specifically bind the steroid hormone and, after activation, specifically interact with certain regions of the genome.

Steroid receptors belong to a superfamily of DNA-binding proteins. The molecular cloning of the estrogen (Walter et al. 1985), glucocorticoid (Hollenberg et al. 1985), mineralocorticoid (Arriza et al. 1987), progestin (Misrahi et al. 1987), and most recently the androgen (Lubahn et al. 1988a, Chang et al. 1988;) receptor genes has allowed the characterization of their functional domains and primary amino acid sequences. Sequence comparisons have revealed structural features that are common to all groups of steroid hormone receptors as well as to other related regulatory proteins such as the vitamin D, thyroid hormone, and retinoic acid receptors.

The steroid receptor molecules possess a number of functionally independent domains, as depicted in Fig. 1. The E domain corresponds to the steroid binding region of the receptor, while the C domain functions as the DNA binding region of the molecule. Interestingly, the C domain exhibits a large percent homology (60 %) in terms of amino acid sequence among the different steroid receptors, and contains a highly conserved cysteine-rich region. This DNA-binding domain comprises two zinc finger motifs, each of which contains a zinc atom bound to cysteine residues (Fawell et al. 1990). In contrast, the steroid-binding domain (E) of the different steroid receptors presents less amino acid homology, a characteristic that allows specific steric recognition of the diverse steroid molecules. The D domain is a hydrophilic region of the receptor which serves as a hinge between the DNA-binding and the steroid-binding domains. The N-terminal region of the receptor molecule (domains A and B) appears to function in transcriptional activation processes.

The sex steroid hormone receptors are regulatory proteins which have similar structures but differ in size and composition. In addition, inter-species variations have also been observed; thus, the human estrogen receptor is a 66-KDa protein of 595 amino acids, while the rat and chick estrogen receptors have 600 and 598 amino acids respectively (Green et al. 1987). The androgen receptor in humans is a monomeric protein with a molecular weight of 98 999 Da and 919 amino acids, whereas the androgen receptor of the rat has a sequence of 902 amino acids and a molecular weight of 98 227 Da (Lubahn et al. 1988b). In the case of the progesterone receptor, several biochemical and molecular studies carried out in birds have disclosed the existence of two different subunits. The A subunit (79 KDa) of the receptor exhibits a high

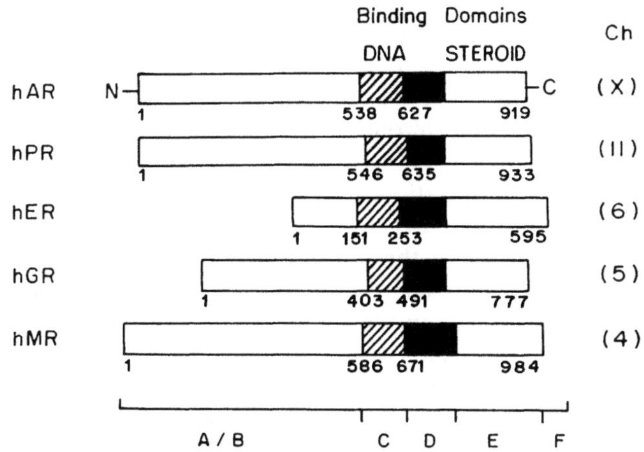

Fig. 1. Schematic representation of the structure of steroid hormone receptors. Diagram shows the main regions (*A-F*), the number of amino acid residues, as well as the DNA-binding and steroid-binding domains of the human steroid receptors. The localization of each receptor gene in the chromosomes (*Ch*) is also indicated (for details see the text)

affinity for DNA, while the B subunit (108 KDa) binds to DNA less effectively (Wei and Horwitz 1985). The progesterone receptor in humans is composed of 933 amino acids, while in the rabbit it contains 930 amino acids (Loosfelt et al. 1986). In summary, it can be said that despite the strong homology of the steroid receptors, the differences found between species and even among tissues may account for their different biological expression.

The intracellular steroid receptor exists in two forms: an inactive form (nontransformed state), which occurs prior to hormone interaction, and the activated form (transformed receptor) which is bound to the steroid. The inactive or native form of the receptor has an oligomeric structure and is associated to nonbinding "heat-shock" proteins, which are abundant and highly conserved marcomolecules (Joab et al. 1984). The native receptor undergoes transformation to its activated form upon binding to its specific hormone ligand. This transformation process implies separation of the hormone-binding component from the heat-shock proteins, as well as some structural changes. In some cases, receptor phosphorylation has been found to occur during the activation process (Logeat et al. 1985).

Recently, a number of studies have demonstrated that the activated receptor-hormone complexes interact with specific DNA sequences (steroid response elements) topographically located within or near the promotor region of hormone-regulated genes (Ryffel et al. 1988). A clearcut example is found in the progesterone-regulated uteroglobin gene in which the receptor-gene interactions have been well characterized (Bailly et al. 1986). It has also been suggested that other events such as dimerization of active receptor molecules may occur at the time of gene activation or repression (Fawell et al. 1990). From these data, it can be concluded that the primary action of the receptor-hormone complex is the modulation of specific gene expression at the transcriptional level. It must be pointed out, however, that effects of the steroid-receptor complex other than transcription have been well documented. These include changes in the rate of transcriptional elongation, mRNA stabilization, and protein turnover. Indeed, the androgen-receptor complexes in target organs stimulate intracellular RNA transport and migration of RNA through the nuclear envelope. Therefore it appears that steroid-receptor complexes exert their actions at both transcriptional and post-transcriptional levels. The intracellular location of the steroid receptors is a controversial and still unsolved issue. According to the "two-step" theory of hormonal mechanism of action originally proposed by Jensen et al. (1968), the native receptor resides in the cytoplasm and following "activation" (interaction with its corresponding hormone), it is transported into the nucleus where it binds to specific DNA regions. With the availability of polyvalent and monoclonal anti-receptor immunoglobulins and their use in immunocytochemistry, the "two-step" theory has been challenged, at least for some steroid hormones. Thus, in the case of estrogen, progestin, and androgen receptors, it has been strongly suggested that these macromolecules reside in the nuclear compartment (Husmann et al. 1990), while the glucocorticoid receptor appears to be located in both the cytoplasm and the nucleus (Picard et al. 1990). Even though a final answer to this intriguing question cannot yet be given, the general mode of action of steroid hormones remains essentially within the same conceptual frame. The intracellular steroid receptor, wherever it is originally located, represents the key molecule for initiating and directing early cellular events which result in the expression of hormonal effects. Accordingly, the evaluation of steroid hormone action in a given hormone-sensitive and/or hormone-dependent organ must include the assessment of steroid uptake, hormone-receptor binding kinetics, stereospecificity, receptor physicochemical and immunoreactive characteristics, northern blot analysis with cDNA probes, and if at all possible, the study of expression products of the hormone-regulated genes.

In the next section of this chapter, the current concepts and knowledge of the genomic mode of action of sex steroid hormones upon the Harderian gland in vertebrates are reviewed. Particular emphasis is given to the presence and characteristics of the intracellular steroid hormone receptors.

3 Androgen Receptors

The effects of androgens upon the Harderian gland in vertebrates give an adequate example of the mechanisms involved in the mode of action of steroid hormones at the target organs. Following the pioneer work of Christensen and Dam (1953), Woolley and Worley (1954), and Paule et al. (1955), who demonstrated that the gland of Harder presented a marked sexual dimorphism in the hamster and that gonadectomy in the male is accompanied by morphological changes in this orbital structure, the rodent Harderian gland has been used as a model to examine androgenic activity.

The effects of testosterone, the major testicular androgen, have been extensively studied in the Harderian gland of a number of species; other steroid hormones with androgenic activity including 5α-dihydrotestosterone (DHT), androstenedione, and dehydroepiandrosterone have been evaluated as well. However, it must be stressed that in most animal species the main biologically active androgens are testosterone and DHT. The formation of DHT at the target organs, which is mediated by the activity of an NADPH-dependent 5α-steroid reductase, appears to be critical for androgen-regulated cellular development and function. Both testosterone and DHT specifically bind with high affinity to a common intracellular receptor; the association of the androgen-receptor complexes with the nuclear acceptors allows the activation (or repression) of androgen-sensitive genes which regulate the synthesis of specific proteins. These androgen-regulated proteins will induce important cellular changes including growth, differentiation, and some metabolic events. These processes and events occurring at the cellular level can be recreated in the Harderian gland of vertebrates by the administration of androgens to castrated animals.

Even though a great number of observations have indicated that the Harderian gland is largely influenced by androgens, their mechanism of action has been studied only recently. The presence of androgen receptors in the Harderian gland was first suggested by Gustafsson and Pousette (1975), who demonstrated the specific retention of [^{3}H]-testosterone in the glands of castrated rats. Since increasing evidence clearly indicates that the Harderian gland of the golden hamster (*Mesocricetus auratus*) is extremelly sensitive to androgen activity, we decided to study the presence and characteristics of intracellular androgen-binding sites in this species using [^{3}H]-dimethylnortestosterone (DMNT) as the radioligand (Vilchis et al. 1987). The cytosolic androgen receptor exhibits an apparent dissociation constant (Kd) of $3.0 - 8.0 \times 10^{-10}$ M with a maximal binding capacity (number of binding sites) of 80-125 fmol/mg protein in the male and female glands of castrated hamsters. This receptor is highly specific for androgens since only natural and synthetic steroids possessing intrinsic androgenic potency (testosterone, DHT, DMNT, 19-nortestosterone and methyltrienolone) compete for androgen-binding sites, while other steroids such as progesterone, dexamethasone, tamoxifen and even DHA do not. Interestingly, the pheromonal steroid, 5α-16-androsten-3-one, does not interact with the androgen receptor of the hamster Harderian gland. The sedimentation properties of this receptor in linear sucrose gradients is 8-9 S in the glands of both male and female animals. When the androgen receptor is labeled with the natural androgen [^{3}H]-DHT, the steroid-receptor complex exhibits identical binding and physicochemical characteristics to those described for [^{3}H]-DMNT.

Cytosolic concentrations of the receptor are not affected in long-term castrated animals. Indeed, in spite of the distinct morphological changes, no differences in receptor concentration or in sedimentation coeficient are observed in glands removed after 3, 30, and 60 days after castration. Changes in the number of androgen-binding sites in the glands of animals under other experimental conditions have been recently reported (Stankov et al. 1989). The [^{3}H]-androgen-receptor complex is specifically retained in the nuclear fraction of the hamster Harderian gland where it activates (or represses) androgen-sensitive genes. Interestingly, a portion of androgen-receptor complexes are found associated to the nuclear matrix (unpublished observations).

Taking all these data together, it can be concluded that the androgen receptor of the hamster Harderian gland plays a pivotal role in the expression of those biological effects of androgens that require hormone-genome interaction. This mediating role of the androgen receptor is similar to that previously described for other androgen-responsive organs. In addition, the binding kinetics and physicochemical properties of the androgen receptor in the Harderian gland are similar to those found in other animal and human target tissues.

Once the androgen-receptor complex interacts with androgen-responsive elements in the nuclear compartment of the Harderian gland cells, a sequence of events is initiated which results in the transcription of specific genes with the subsequent synthesis of the corresponding proteins. Although it has been well established that androgens influence the concentration of a number of enzymes involved in the synthesis of porphyrins (Lin and Nadakavukaren 1981; Thompson et al. 1984), melatonin (Menéndez-Pelaez et al. 1988), and somatostatin (Puig-Domingo et al. 1988), very little is known of the transcriptional and post-transcriptional activities of androgens in the Harderian gland. The scarce information available on this issue is mainly due to the lack of knowledge of the molecular structure of the androgen receptor and of its specific interactions at genomic level.

Recently, two independent research groups reported the gene cloning of the human and rat androgen receptors (Chang et al. 1988; Lubhan et al. 1988a), thus making available cDNA probes as well as polyvalent and monoclonal antibodies against the androgen receptor. These important contributions will facilitate further research on the molecular mechanisms involved in androgen action in the Harderian gland. Nevertheless, to ascertain the biological function of the androgen receptor, we undertook a search for androgen-regulated proteins. Preliminary results have disclosed the presence of a cytosolic "redish" protein-like material sedimenting in the 10-11 S region in linear sucrose gradients of male hamster Harderian gland preparations. Further purification in denaturating polyacrylamide gel electrophoresis revealed that this material corresponds to a single band with an estimated molecular weight of 150-160 kDa (Fig. 2). Interestingly, this soluble protein is not present either in glands from castrated male hamsters or in female glands. Furthermore, this protein can be induced in castrated male and female animals by the exogenous administration of

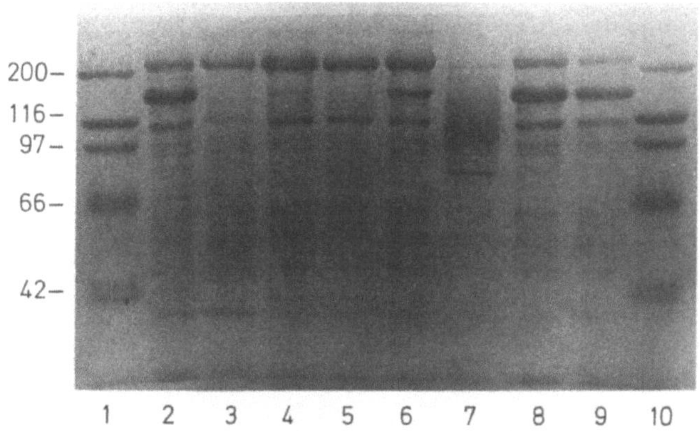

Fig. 2. Pattern of hamster Harderian gland proteins in SDS-PAGE. Total cytosolic proteins were separated in sucrose density gradients (20–35%) for 3 h. After centrifugation, gradients were collected in several fractions by tube puncture. Fraction IV of each group, (50 μg protein) was then separated in 7.5% SDS-polyacrylamide gels. *Lanes 1* and *10* molecular (kDa) weight markers (Bio-Rad); *lane 2* proteins from intact males; *lane 3* intact females; *lane 4* short-term castrated males (2 months); *lane 5* long-term castrated males (11 months); *lane 6* spayed females (2 months); *lane 7* prostate (control); *lane 8* castrated males treated s.c. with 5α-dihydrotestosterone (1 mg) for nine consecutive days; *lane 9* ovariectomized females treated as in lane 8. The gel was stained with Coomassie blue R-250. A 150-160-KDa protein was indentified in glandular preparations of intact males (*lane 2*) and androgen-treated animals (*lanes 8* and *9*)

testosterone, DHT and 3β-androstanediol as depicted in Fig. 3. This androgen-inducible protein in the hamster Harderian gland resembles that previously reported by Hoh et al. (1984) in gland homogenate preparations in the same species. The nature and functions of this specific hormone-regulated protein cannot yet be established, but a series of studies aimed at characterizing it are in progress in our laboratory.

The Harderian gland androgen receptor has also been characterized in other mammalian species. Indeed, we have studied the androgen receptor in the glands from rats (Urueta 1988), guinea pigs, rabbits, and several mouse strains. The binding characteristics of the androgen receptor in the glands of these species closely resemble those described for the hamster Harderian gland. Figure 4 shows the stereospecificity of the androgen receptor in the glands of castrated male rats, and Fig. 5 shows the binding kinetics of the Harderian gland androgen receptor in castrated male guinea pigs. A distinctive feature of the rabbit Harderian gland is that it is composed of two lobes, one pink and the other white, which present different anatomical characteristics (Kühnel 1971) and perhaps different functions. However, the number of androgen binding sites in the two lobes are essentially the same.

The biological effects of androgens on the Harderian glands of submammalian species have received only sporadic attention. In birds, Baba et al. (1988), reported that testosterone reduces lacrimal concentrations of immunoglobulins, presumably by direct action at the glandular level, and a recent report by

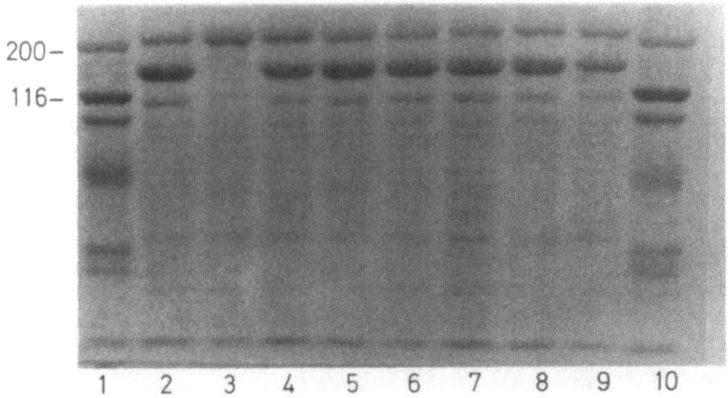

Fig. 3. SDS-PAGE of Harderian gland proteins from androgen-treated hamsters. Protein fraction IV (see legend fig. 4) were electrophoresed on 7.5% polyacrilamide gels. A major (150-160–kDa) protein was present only in the Harderian gland preparations of intact males and androgen-treated animals. *Lanes 1* and *10* molecular weight markers (kDa); *lane 2* intact males; *lane 3* intact females; *lane 4* castrated males plus testosterone (T); *lane 5* castrated males plus 5α-dihydrotestosterone (DHT); *lane 6* castrated males plus 3β-androstanediol (3β-diol); *lanes 7, 8* and *9* spayed females treated with T, DHT and 3β-diol, respectively. In all cases, steroids (1 mg/day) were administered s.c. for 10 consecutive days

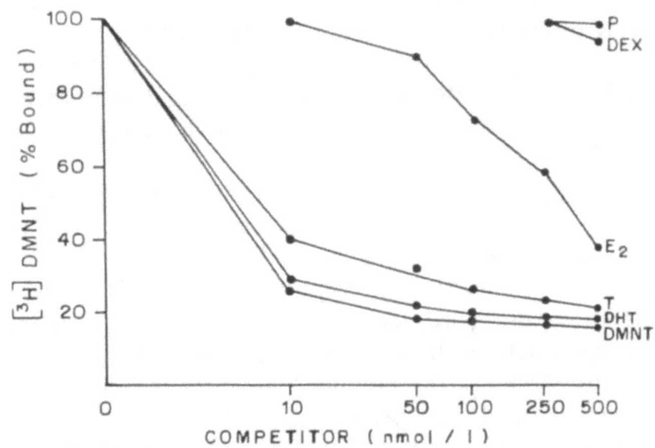

Fig. 4. Specificity of steroid binding to androgen receptors in the male rat Harderian glands. Cytosol samples were incubated to equilibrium with [^{3}H]-DMNT (1-2 nM) for 16 h at 4 °C with or without increasing concentrations of radionert steroids, including dimethylnortestosterone (*DMNT*), 5α-dihydrotestosterone (*DHT*), testosterone (*T*), 17β-estradiol (E_2), dexamethasone (*DEX*) and progesterone (*P*). Bound and free steroids were separated with DCC. Total binding in the absence of competitor was set at 100%

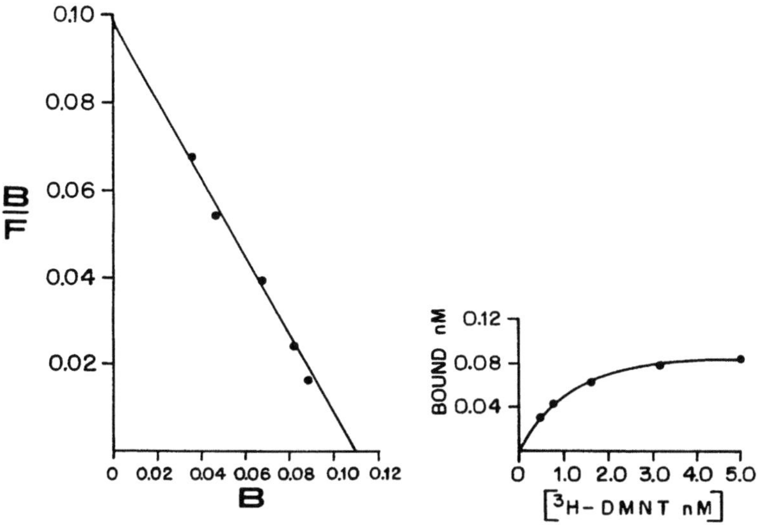

Fig. 5. Scatchard plot and saturation analysis of specific androgen binding in the Harderian gland of gonadectomized guinea pigs. Cytosol samples were incubated with increasing concentrations (0.5–5.0 nmol/l) of [^{3}H]7α, 17α-dimethyl-17β-hydroxy-4-estrogen-3-one (DMNT), in the presence or absence of radioinert DMNT. Specific binding was calculated by subtracting nonspecific binding from total binding. The apparent affinity of the androgen receptor in the male gland is 1.07 nmol/I

Minucci et al. (1989) claims the existence of a clear sexual dimorphism in the Harderian gland of the toad (*Bufo viridis*). We have recently demonstrated the presence of specific, high-affinity, 8 S androgen receptors in cytosol preparations of Harderian glands of ducks (*Anas platyrhynchos*) and chicks (*Gallus domesticus*). The characteristics of the androgen receptor in the avian gland are similar to those found in the mammalian Harderian gland. The cytosol androgen receptor seems to be more abundant in chick glands than in duck glands (Vilchis et al. 1991). d'Istria and his group (1989) have demonstrated the occurrence of androgen receptors in the Harderian gland of *Rana esculenta*. The receptor concentration was found to be higher in female as compared to male glands. Other nonmammalian vertebrates in which the androgen receptor of the Harderian gland has been characterized include the sea turtle (*Lepidochelys olivacea*) and lizards of the genus *Phrynosoma* (unpublished data).

In summary, the overall experimental findings point out that steroid hormones with androgenic potency exert a number of regulatory effects upon the morphology and physiology of the Harderian gland in vertebrates and underline the key role of the intracellular androgen receptor in mediating these effects. The data are in line with the concept that the androgen receptor in mammalian and nonmammalian Harderian gland exhibits similar binding and structural features, thus suggesting that this receptor is a well-conserved intracellular macromolecule.

4 Progesterone Receptors

Studies assessing the effects of gonadal steroid hormones with progestational activity upon the Harderian gland have yielded controversial results. This has been mainly due to the large interspecies variability that occurs in terms of progesterone responsiveness of the gland. Thus, whereas in some animal species adminstration of progesterone or related progestins alone exert important effects upon the maintenance of both structure and function of the Harderian gland (e.g. birds), in other species (rodents) these compounds appear to be completely devoid of significant biological activity. For example, administration of progesterone to adult female rats alters neither the weight of the gland (Lorincz and Lancaster 1957) nor the intraglandular concentration of porphyrins (Ulrich et al. 1974). Moreover, chronic administration of norethisterone, a potent synthetic progestin, does not have significant effects upon the rat Harderian gland. Although several investigators have stressed the importance of ovarian steroid hormones in the regulation of glandular function in hamsters (Moore et al. 1977; Payne et al. 1979; McMaster and Hoffman 1984; Spike et al. 1985, 1986) and mice (Shirama et al. 1981), there is no strong evidence for a clear function of progesterone in these species, and it appears that most of the effects observed after treatment with ovarian hormones are the result of the combined activities of estrogens and progesterone. However, the studies of Shirama et al. (1981) have suggested that porphyrin synthesis in the mouse (C3H) gland is progesterone-regulated. They demonstrated a significant increase in gland porphyrin content in this species following the administration of progesterone to adrenalectomized-ovariectomized animals and showed that this effect could be abolished by the simultaneous administration of androgens. On the contrary, progesterone has a well-defined activity in the avian Harderian gland. Gupta and Maiti (1983) have presented evidence which suggests that progesterone administration to female ducklings inhibits the secretory activity of the glands through the induction of drastic cell desquamation, while estradiol treatment increases gland mitosis but decreases cell size and mucosal substances.

The large difference found between Harderian glands of mammals and birds in regard to progesterone sensitivity is in line with the finding that the molecular structure of the intracellular progesterone receptor is different in these species (Wei and Horwitz 1985). Indeed, while the avian progesterone receptor is composed of two subunits (A, B) of different molecular weight, the mammalian progesterone receptor is a monomeric structure. In addition, these receptor macromolecules possess different immunoreactive characteristics. Another distinctive feature of the progesterone receptor of both mammals and birds is that there are two different molecular populations, one which is under strict estrogenic regulation and another which is estrogen-insensitive. The estrogen-inducible progesterone receptor is the more widely distributed in progesterone-sensitive organs, whereas the estrogen-insensitive receptor population appears to be confined to central nervous system structures, at least in mammals (MacLusky and McEwen 1978; Cerbón et al. 1989).

A series of experiments carried out in our laboratory have attempted to disclose the presence of the intracellular progesterone receptor in the Harderian gland and to assess its characteristics in several mammalian and avian species. Studies in the Harderian gland of castrated female and male hamsters using [^{3}H]-ORG2058 as radioligand revealed the absence of progestin-binding sites (Vilchis and Pérez-Palacios 1989). Priming with estradiol benzoate was completely ineffective in inducing progestin receptors, thus suggesting a lack of estrogen-inducible and estrogen-sensitive progestin-binding sites in the hamster Harderian gland. Since it has been previously reported that cholera toxin has a mitogenic effect upon the Harderian gland and that it can induce progesterone receptors in the hamster uterus (Hoffman and Alleva 1982; Alleva et al. 1983), we felt it was of interest to study the presence of progesterone receptors in the glands from cholera toxin-primed animals. The results support the idea of a lack of progesterone-binding sites in the hamster glands. Using a similar experimental approach, we have been unable to demonstrate the presence of either of the two intracellular progesterone receptor populations in the Harderian glands of mice (C3H/eB/Fe J, Swiss Webster, NIH, BALBc) rats, and guinea pigs. These data indicate that the Harderian gland of rodents does not possess functionally active progesterone receptors. Alternately, nongenomic mechanisms for progesterone action may explain some of the biological effects noticed in the mouse glands following progesterone treatment (vide infra). Preliminary results from our laboratory have revealed the presence of specific, high affinity, estrogen-inducible progestin binding in the rabbit Harderian gland. These progestin receptors are confined to the pink lobe of the rabbit gland but their functional role remains to be ascertained.

Very recently, we reported the presence and characteristics of the progesterone receptor in the chick and duck Harderian glands (Vilchis et al. 1991). The receptor of the avian gland exhibits a single type of binding site with an apparent Kd of $1.0 - 2.0 \times 10^{-9}$ M. The concentration of progestin receptors was similar in both species (30–60 fmol/mg protein), and they sedimented in linear sucrose gradients in the 7–8 S region (Fig. 6). Addition of an excess of radioinert ligand abolished the formation of [^{3}H]-ORG2058-receptor complexes, while cortisol and dexamethasone did not compete. When the progesterone receptor was studied in the glands of estradiol benzoate-primed animals, it became evident that most, if not all, progesterone-binding sites in the Harderian gland were estrogen-insensitive in both species. This striking result was in sharp contrast with the observation of a significant increase in progesterone receptor content in the oviduct and pituitary of the estrogen-primed animals. These data suggest that the intracellular progestin receptor in the avian Harderian gland mediates the genomic effects of progesterone. The estrogen-insensitive nature of these progestin receptors deserves further investigation.

Unpublished observations from our group have demonstrated the presence of a specific 7–8 S progestin receptor in the Harderian glands of the horned lizard. This progestin receptor has a nonspecific 3–4 S binding component (Fig. 6). Unfortunately, there is very limited information about the effects of progesterone upon the Harderian gland in reptiles and amphibians.

It is clear from these data that the effects of progesterone upon the Harderian gland of some species are exerted at the genomic level and therefore require an active functional role of the intracellular progestin receptors described herein. In those species in which the Harderian gland does not possess progestin receptors, the hormone action may be exerted through an alternate mechanism which uses a membrane-bound progestin receptor. In this regard, the specific binding of progesterone at the membrane level and its subsequent effects on adenylate cyclase activity have been demonstrated in *Xenopus laevis* oocytes (Baulieu 1983). Whether this mode of hormone action is operating in the Harderian gland has not been established and demands further experimental work. In addition, progesterone may exert its effects through its metabolic conversion products. In fact, some A-ring reduced progesterone derivatives, particularly those with a 5β configuration, have been directly related to the regulation of porphyrin

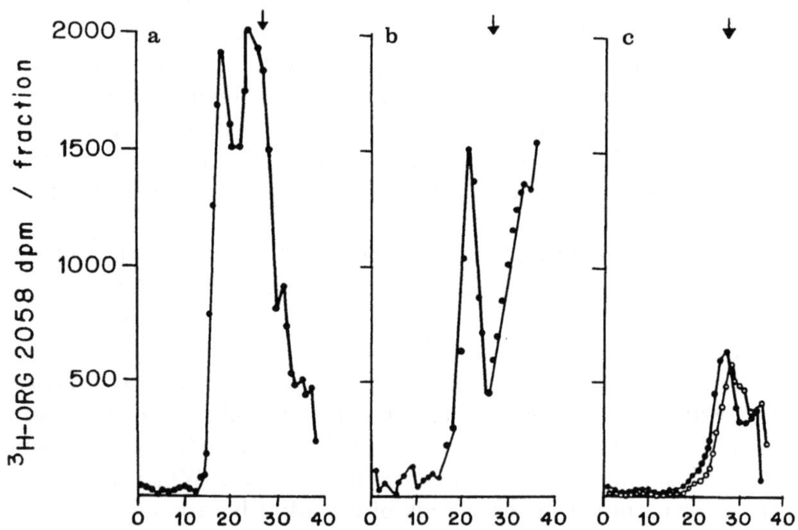

Fig. 6a–c Density gradient sedimentation profile of bound [^{3}H]ORG2058 complexes in Harderian glands of horned lizard (**a**), chick (**b**) and golden hamster (**c**). Cytosol preparations were labeled in vitro with 3-4 nmol [^{3}H]ORG2058/I in the absence (●–●–●) or presence (○–○–○) of an excess of radioinert ligand. Labeled cytosol samples were layered into linear sucrose gradients (20–35%) and centrifuged in a vertical rotor for 2.5 h at 4 °C. BSA-naphthol blue black (*arrows*) was used as sedimentation marker. No specific binding was detected in glands obtained from spayed estrogen-primed hamsters (**c**). A similar sedimentation profile was observed in glands from ovariectomized animals (not shown) indicating the absence of specific [^{3}H]-progestin-binding activity

biosynthesis in the liver (Granick and Kappas 1967). Whether this mechanism may be present in the Harderian gland has not been documented and will require additional studies.

5 Estrogen Receptors

Although estrogenic effects upon the Harderian gland were noticed early by Woolley and Worley, who in 1954 reported that estrogenic hormones produced marked changes in the glands of male hamsters, subsequent reports have been rather controversial. Indeed, Hoffman (1971) reported that in hamsters, neither ovariectomy alone nor in combination with estradiol-17α treatment modify the structural appearance of the gland. However, this observation must be taken with caution since the 17α-isomer of estradiol-17β is apparently devoid of biological activity, at least in mammals (Brecher et al. 1967). Furthermore, administration of estradiol-17β to ovariectomized rats does not alter the glandular concentration of porphyrins (Ulrich et al. 1974). Other reports indicate that ovarian hormones are necessary for maintaining the structure and activity of the Harderian gland; thus Spike et al. (1985, 1986) demonstrated that ovariectomy provokes changes in glandular structure and in porphyrin production. In addition, cyclic changes in glandular activity throughout the estrous cycle (Moore et al. 1977; Shirama et al. 1981), pregnancy, and lactation (Payne et al. 1979) have been reported. The studies of Gupta and Maiti (1981, 1983) revealed that estradiol-17β induces mitosis and also alters the secretions and cellular composition of the gland in male and female ducklings. These observations seem to indicate that the Harderian gland is a potential target for steroid hormones with estrogenic potency. Since estrogenic action in hormone-sensitive organs is mainly exerted through its genomic effects, it should be expected that the Harderian gland contained intracellular estrogen receptors. Support for this concept is provided by the data of Weaker et al. (1983), who, using an autoradiographic technique, demonstrated specific nuclear uptake of [^{3}H]-estradiol 17β in the periductal mucosal cells of the mucus-secreting lobules of the Harderian gland of armadillo (*Dasypus novemcinctus*). In general, very little information is available in terms of estrogen receptors in the Harderian gland of vertebrates. Studies carried out in our laboratory have disclosed the presence of specific, high-affinity binding sites for estrogens in the Harderian gland of several species. Estrogen receptors in the gland of hamsters bind [^{3}H]-estradiol 17β with high affinity and sediment at 7-8 S in low-salt sucrose density gradients (Vilchis and Pérez-Palacios 1989). Receptors with similar properties have also been identified in the glands of rat, mouse, guinea pig, and rabbit (unpublished observations). Interestingly, the concentration of estrogen receptors appears to be higher in the pink lobe than in the white lobe of the rabbit Harderian gland.

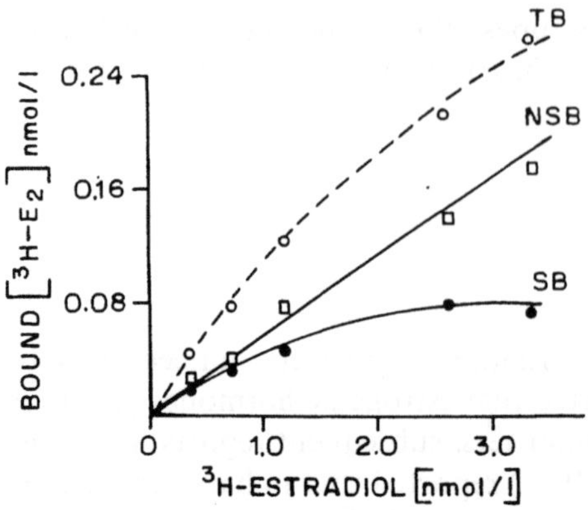

Fig. 7. Saturation binding analysis of estrogen receptor in duck Harderian glands. Cytosol fractions were incubated with different concentrations of [^{3}H]-estradiol for 18 h. Incubations were carried out at 4 °C in presence or absence of nonradioactive estradiol. Specific binding (*SB*) was calculated by subtracting nonspecific (*NSB*) from total binding (*TB*). Eadie-Hofstee analysis (not shown) of the data gave a Kd of 1.6×10^{-9} M

The marked changes of the gland induced by estradiol-17β in birds prompted us to study the characteristics of the estrogen receptor in two species (ducks and chicks) which present different glandular histology (Vilchis et al. 1991). Figure 7 shows the binding kinetics of the estrogen receptor of the duck Harderian gland, which possesses characteristics similar to those found in rodent glands. The estrogen-binding sites in these birds (Kd = 1-2 nM) are identical to those previously reported for other avian estrogen-sensitive organs. Although there is very little, if any, information about the functional role of gonadal steroids in the glands of amphibians and reptiles, we have demonstrated the existence of 4.6 S estrogen-binding sites in the Harderian glands of the horned lizard (*Phrynosoma*).

The presence of estrogen receptors in different species of vertebrates suggests that these intracellular binding macromolecules are involved in mediating some metabolic events of the Harderian gland.

6 Other Hormone Receptors

The Harderian gland in vertebrates seems to be sensitive to the action of hormones other than gonadal steroids, including those produced in the adrenals, the thyroid and even the pituitary (Boas and Bates 1954; Wetterberg et al. 1970b). In fact, adrenal steroids with glucocorticoid effects have been recognized as important modulators of glandular activity in several species; thus, administration of cortisol to newborn rats retards the onset of porphyrin synthesis in the gland (Wetterberg et al. 1970b), while adrenalectomy in adult female mice (C3H/Hc) results in a significant increase in the glandular content of porphyrins (Shirama et al. 1981). That glucocorticoids exert their effects in the Harderian gland through a receptor-mediated mechanism was suggested by the work of Butler et al. (1978), who demonstrated a large in vitro uptake of radioactive corticosterone by Harderian glands of the domestic duck, as well as its biotransformation to radioactive 11-dehydrocorticosterone, indicating the presence of an active 11β-hydroxysteroid dehydrogenase. Using the synthetic glucocorticoid, dexamethasone, as radioligand, we have recently demonstrated the presence of a specific, high-affinity, 8 S glucocorticoid receptor in the Harderian gland of adrenalectomized male and female golden hamsters (Vilchis and Pérez-Palacios 1989). Glucocorticoid-binding sites with similar characteristics are also present in the glands of guinea pigs, mice (Fig. 8), and rabbits (pink and white lobes). Our data favor the concept that the direct effects of adrenal steroids upon the Harderian gland are exerted througheir intracellular receptor. Additional studies, however, are required to establish their precise biological role in different species. To our knowledge, no information is available with regard to the effects of mineralocorticoids upon the gland.

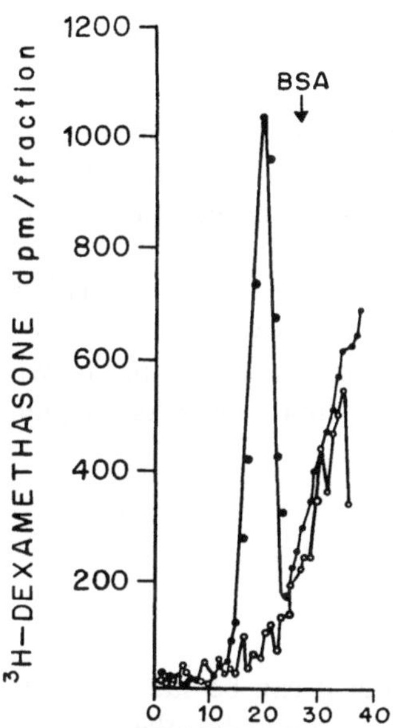

Fig. 8. Sedimentation profile of the glucocorticoid-receptor complexes from Harderian gland of ovariectomized-adrenalectomized mice using [^{3}H]-dexamethasone as radioligand. Cytosol aliquots (500 μl) preincubated with the [^{3}H] synthetic ligand were layered into linear sucrose gradients (20–35%) and centrifuged for 2.5 h at 370,000 *g*. Fractions incubated with (○–○–○) or without (●–●–●) an excess of unlabeled dexamethasone are shown

Among the nonsteroidal hormones that modulate the function of the Harderian gland, the thyroid hormones deserve special mention. Indeed, thyroxine plays an important role in the regulation of development, differentiation, and some metabolic events of the gland in several species. The early observations of Boas and Bates (1954) pointed out the effects of the thyroid in the maintenance of the rat Harderian gland; Shirama et al. (1982) demonstrated that thyroidectomy in tadpoles of *Buffo* and *Rana* precludes development of the Harderian gland, whereas replacement therapy with thyroid hormones reverses this effect. In rodents, thyroid hormones are also involved in regulation of porphyrin synthesis and metabolism (Wetterberg et al. 1970b; Hoffman 1971). The actions of thyroid hormones at their target cells are exerted using mechanisms similar to those described for steroid hormones. Their main actions are mediated by intracellular receptors that are located predominantly in the nuclear compartment. These thyroid hormone-binding macromolecules share a structure similar to that of the steroid receptors and present a high sequence homology in their hormone-binding and DNA-binding domains. In fact, they are considered as members of the same superfamily of regulatory proteins (Evans 1988). Furthermore, gene regulation of thyroid receptors appears to be similar to that of steroid receptors. Recently we have demonstrated the presence of a specific nuclear [^{125}I]-triiodothyronine, 4 S, high-affinity receptor in the Harderian glands of male and female golden hamsters (Vilchis et al. 1990a). Although the circulating levels of thyroid hormones are similar in both sexes, preliminary results indicate that gonadectomy diminishes the concentration of triiodothyronine-binding sites. These data strongly suggest that thyroid hormone effects upon the gland are mediated via their nuclear receptor and also indicate a functional inter-relation between the thyroid, the gonads, and the Harderian gland.

In addition to the effects of thyroid hormones, Shirama et al. (1981) have suggested specific developmental functions for the thyroid-stimulating hormone (TSH) of the anterior pituitary in amphibians. This observation is in line with the demonstration of functional TSH binding sites in the Harderian gland membrane of guinea pigs (Winand and Kohn 1971).

7 Concluding Remarks

From the experimental data reviewed in this chapter it can be stated that the Harderian gland in vertebrates is certainly a target organ for the action of gonadal and adrenal steroids.

The hormonal effects upon the gland appear to be exerted at genomic level, and therefore they are mediated by specific intracellular receptors. The binding and some structural properties of the androgen, progestin, estrogen, and glucocorticoid receptors have been characterized in the Harderian gland of a number of mammalian and nonmammalian species. The presence, concentration, and features of the receptor for a given steroid hormone in the Harderian gland vary from species to species; a finding which is in line with the variable responsiveness to steroids observed in the gland from different species. The study of gene expression products of steroid hormone-sensitive genes that could be used as biological markers will add a new dimension to our current knowledge of the mechanism of hormone action in the Harderian gland. In summary, because of their important role in mediating direct or indirect hormonal effects upon the gland, the steroid hormone receptors must be considered as key regulatory proteins in the mechanisms involved in steroid hormone action. Nevertheless, in some particular circumstances, steroid hormones may exert their effects through alternate non-genomic modes of action.

Acknowledgments. The experimental work described in this chapter was partially supported by grants from The Rockefeller Foundation (New York), The Special Programme of Research, Training and Research Development in Human Reproduction of the World Health Organization (Geneva), and The National Council of Science and Technology (Mexico City). The authors wish to thank Mr. José Luis Ramírez for his assistance in preparing the manuscript.

References

Alleva JJ, Kenimer JG, Jordan AW, Lamanna C (1983) Induction of estrogen and progesterone receptors and decidualization in the hamster uterus by cholera toxin. Endocrinology 112:2095–2106

Arriza JL, Weinberger C, Cerelli G, Glaser TM, Handelin BL, Housman DE, Evans RM (1987) Cloning of human mineralocorticoid receptor complementary DNA: structural and functional kinship with the glucocorticoid receptor. Science 237:268–275

Baba T, Masumota K, Nishida S, Kajikawa T, Mitsui M (1988) Harderian gland dependency of immunoglobulin A production in the lacrimal fluid of chicken. Immunology 65:67–71

Bailly A, LePage C, Rauch M, Milgrom E (1986) Sequence-specific DNA binding of the progesterone receptor to the uteroglobin gene: effects of hormone, antihormone and receptor phosphorylation. EMBO J 5:3235–3241

Baker AR, McDonnell DP, Hughes M, Crisp TM, Mangelsdorf DJ, Haussler MR, Pike JW, Shine J, O'Malley BW (1988) Cloning and expression of full-length cDNA encoding human vitamin D receptor. Proc Natl Acad Sci USA 85:3294–3298

Baulieu EE (1983) Steroid-membrane adenylate cyclase interactions during Xenopus laevis oocyte meiosis reinitiation. In: Bardin CW, Milgrom E, Mauvais-Jarvis P (eds) Progesterone and Progestins. Raven, New York, pp 91–105

Boas NF, Bates RW (1954) Role of the thyroid and anterior pituitary glands in the maintenance of the Harderian glands of the rat. Endocrinology 55:601–612

Brecher PI, Vigersky R, Wotiz HS, Wotiz HH (1967) An in vitro system for the binding of estradiol to rat uterine nuclei. Steroids 10:635–651

Butler DG, Wilson JX, Youson JH (1978) Transformation of corticosterone to 11-idehydrocorticosterone by nasal, Harderian and lacrymal glands of the domestic duck (*Anas platyhynchos*) in vitro. Gen Comp Endocrinol 34:475–479

Buzzell GR, Menendez-Pelaez A, Prokka-Heiskanen T, Pangerl A, Pangerl B, Vaughan MK, Reiter RJ (1989) Bromocriptine prevents the castration-induced rise in porphyrin concentration in the Harderian glands of the male Syrian hamster, *Mesocricetus auratus*. J Exp Zool 249:172–176

Cerbón MA, Martinez M, Pérez-Palacios G (1989) Oestrogen-insensitive progestin receptors in the central nervous system: physicochemical and immunorective characteristics. J Neuroendocrinol 1:291–298

Chang C, Kokontis J, Liao S (1988) Molecular cloning of human and rat complementary DNA encoding androgen receptors. Science 240:324–326

Christensen F, Dam H (1953) A sexual dimorphism of the Harderian glands in hamsters. Acta Physiol scand 23:333–336

d'Istria M, Chieffi-Baccari G, DiMateo L, Minucci S, Chieffi G (1989) Androgen receptors in the Harderian gland of *Rana esculenta*. Int. Sym Comp Endocrinol, Malaga, Spain, p 95

Evans RM (1988) The steroid and thyroid hormone receptor superfamily. Science 240:889–895

Fawell SE, Lees JA, White R, Malcolm GP (1990) Characterization and co-localization of steroid binding and dimerization activities in the mouse estrogen receptor. Cell 60:953–962

Granick S, Kappas A (1967) Steroid induction of porphyrin synthesis in liver cell culture. I. Structural basis and possible physiological role in the control of heme formation. J Biol Chem 242:4587–93

Green S, Kumar V, Kruts A, Chambon P (1987) The oestrogen receptor: structure and function. In:Moudgil VK (ed) Recent advances in steroid hormone action. Gruyler, Berlin, p 161

Gupta SK, Maiti BR (1981) Effect of estradiol on the Harderian gland in male domestic duckling. Mikroskopie 38:200–203

Gupta SK, Maiti BR (1983) Influence of sex-hormones on the Harderian gland activity in female domestic duckling. Mikroskopie 40:132–137

Gustafsson J-Å, Pousette Å (1975) Demonstration and partial characterization of cytosol receptors for testosterone. Biochemistry 14:3094–3101

Hoffman RA (1971) Influence of some endocrine glands, hormones and binding on the histology and porphyrins of the Harderian glands of golden hamters. Am J Anat 132:463–478

Hoffman RA, Alleva JJ (1982) Cholera toxin: mitogen for Harderian glands of golden hamsters. Comp Biochem Physiol 73:383–388

Hoffman RA, Johnson LB, Reiter RJ (1985) Harderian glands of golden hamsters: temporal and sexual differences in immunoreactive melatonin. J Pin Res 2:161–168

Hoh JH, Lin W-L, Nadakavukaren MJ (1984) Sexual dimorphism in the Harderian gland proteins of the golden hamster. Comp Biochem Physiol 77B:729–731

Hollenberg SM, Weinberger C, Ong ES, Cerelli G, Oro A, Lebo R, Thompson EB, Rosenfeld MG, Evans RM (1985) Primary structure and expression of a functional human glucocorticoid receptor cDNA. Nature 318:635–641

Husmann DA, Wilson CM, McPhaul MJ, Tilley WD, Wilson JD (1990) Antipeptide antibodies to two distinct regions of the androgen receptor localize the receptor protein to the nuclei of target cells in the rat and human prostate. Endocrinology 126:2359–2368

Jensen EU, Suzuki T, Kawashima T, Stumpf WE, Jungblut PW, De Sombre ER (1968) A two step mechansim for the interaction of estradiol with rat uterus. Proc Natl Acad Sci USA 59:632–638

Joab I, Radanyi C, Renoir M, Buchou T, Catelli M, Binart N, Mester J, Baulieu EE (1984) Common non-hormone binding component in non-transformed chick oviduct receptors of four steroid hormones. Nature 308:850–853

Kühnel W (1971) Strucktur und Cytochemie der Harderschen Drüse von Kaninchen. Z Zellforsch 119:384–404

Liao S, Chang C, Kokontis J, Popovich T, Hiipakka RA (1989) Structure and intranuclear dynamics of androgen receptors. In: (eds) Carlstedt-Duke J, Eriksson H, Gustafsson J-A The steroid/thyroid hormone receptor family and gene regulation. Birkhäuser Basel, p 83–92

Lin W-L, Nadakavukaren MJ (1979) The androgenic effect on the fine structure of the Harderian gland in the male hamster. Cell Tissue Res 198:119–127

Lin W-L, Nadakavukaren MJ (1981) Harderian gland lipids of male and female golden hamster. Comp Biochem Physiol 70B:627–630

Lin W-L, Nadakavukaren MJ (1982) δ-Amino levulinate synthase activity in the Harderian glands of male and female golden hamsters. Comp Biochem Physiol 71B:539–540

Logeat F, Le Cunff M, Pamphile R, Milgrom E (1985) The nuclear-bound form of the progesterone receptor is generated through a hormone-dependent phosphorylation. Biochem Biophys Res Commun 131:421–427

Loosfelt H, Atger M, Misrahi M, Guiochon-Mantel A, Meriel C, Logeat F, Benarous R, Milgrom E (1986) Cloning and sequence analysis of rabbit progesterone-receptor complementary DNA. Proc Natl Acad Sci USA 83:9045–9049

Lorincz AL, Lancaster G (1957) Anterior pituitary preparation with tropic acitivity for sebaceous, preputial, and Harderian glands. Science 126:124–125

Lubahn DB, Joseph DR, Sullivan PM, Willard HF, French F, Wilson EM (1988a) Cloning of human androgen receptor complementary DNA and localization to the X chromosome. Science 240:327–330

Lubahn DB, Joseph DR, Sar M, Tan J, Higgs HN, Larson RE, French F, Wilson EM (1988b) The human androgen receptor: complementary deoxyribonucleic acid cloning, sequence analysis and gene expression in prostate. Mol Endocrinol 2:1265–1275

Lubahn DB, Brown TR, Simental JA, Higgs HN, Migeon CJ, Wilson EM, French FS (1989) Sequence of the intron/exon junctions of the coding region of the human androgen receptor gene and identification of a point mutation in a family with complete androgen insensitivity. Proc Natl Acad Sci USA 86:9534–9538

MacLusky NJ, McEwen BS (1978) Oestrogen modulates progestin receptor concentrations in some rat brain regions but not in others. Nature 274:276–278

Maiti BR, Sahu A (1981) Effects of long-term treatment with norethisterone (a progestogen-only contraceptive) on the epidermis, the sebaceous glands, the meibomian gland and the Harderian gland in the cycling female rat. Mikroskopie 38:309–311

McMasters KM, Hoffman RA (1984) Harderian gland: regulation of sexual "type" by gonads and pineal gland. Biol Reprod 31:579–585

Menendez-Pelaez A, Reiter RJ, Guerrero JM, Puig-Domingo M, Howes KA(1988) Sexual dimorphism in N-acetyltransferase activity, hydroxyindole-o-methyltransferase activity, and melatonin content in the Harderian gland of Syrian hamsters: changes following gonadectomy. Proc Soc Exp Biol Med 187:287–291

Minucci S, Chieffi-Baccari G, DiMateo L, Chieffi G (1989) A sexual dimorphism of the Harderian gland of the toad, *Bufo viridis*. Basic Appl Histochem 33;299–310

Misrahi M, Atger M, d'Auriol L, Loosfelt H, Meriel C, Fridlansky F, Guiochon-Mantel A, Galibert F, Milgrom E (1987) Complete amino acid sequence of the human progesterone receptor deduced from cloned cDNA. Biochem Biophys Res Commun 143:740–748

Moore MR, Thompson GG, Payne AP, McGadey J (1977) Cyclical oscillations in the activity of δ-Aminolaevulinte synthase and porphyrin synthesis in the Harderian gland during the oestrous cycle of the golden hamster (*Mesocricetus auratus*). Biochem Soc Trans 5:1475–1478

Paule WJ, Hayes ER, Marks BH (1955) The Harderian gland of the Syrian hamster. Anat Rec 121:349–350

Payne AP, McGadey J, Moore MR, Thompson G (1977) Androgenic control of the Harderian gland in the male golden hamster. J Endocrinol 75:73–82

Payne AP, McGadey J, Moore MR, Thompson GG (1979) Changes in Harderian gland activity in the female golden hamster during the oestrus cycle, pregnancy and lactation. Biochem J 178:597–604

Payne AP, Mcgadey J, Johnston HS, Moore MR, Thompson GG (1982) Mast cells in the hamster Harderian gland: sex differences, hormonal control and relationship to porphyrin. J Anat 135:451–461

Picard D, Yamamoto KR (1987) Two signals mediate hormone-dependent nuclear localization of the glucocorticoid receptor. EMBO J 6:3333–3340

Picard D, Kumart V, Chambon P, Yamamoto KR (1990) Signal transduction by steroid hormones: nuclear localization is differentially regulated in estrogen and glucocorticoid receptors. Cell Regulation 1:291–299

Puca GA, Bresciani F (1969) Interactions of 6, 7-^{3}H-17β-estradiol with mammary gland and other organs of the C3H mouse in vivo. Endocrinology 85:1–10

Puig-Domingo M, Guerrero JM, Reiter RJ, Peinado MA, Menendez-Pelaez A, Santana C, Webb SM (1988) Androgenic control of immunoreactive somatostatin in the Harderian gland of the Syrian hamster. J Reprod Fertil 82:753–759

Ryffel GU, Klein-Hitpass L, Druege P, Doebbeling U, Heitlinger E, Cato ACB (1988) The estrogen responsive DNA element: structure and interaction with the estrogen receptor. J Cell Biochem 35:219–227

Shirama K, Furuya T, Takeo Y, Shimizu K, Maekawa K (1981) Influences of some endocrine glands and of hormone replacement on the porphyrins of the Harderian glands of mice. J Endocrinol 91:305–311

Shirama K, Kikuyama S, Takeo Y, Shimizu K, Maekawa K (1982) Development of Harderian gland during metamorphosis in anurans. Anat Rec 202:371–378

Spike RC, Johnston HS, McGadey J, Moore MR, Thompson GG, Payne AP (1985) Quantitative studies on the effects of hormones on structure and porphyrin biosynthesis in the Harderian gland of the female golden hamster. I. The effects of ovariectomy and androgen administration. J Anat 142:59–72

Spike RC, Johnston HS, McGadey J, Moore MR, Thompson GG, Payne AP (1986) Quantitative studies on the effects of hormones on structure and porphyrin biosynthesis in the Harderian gland of the female golden hamster. II. The time course of changes after ovariectomy. J Anat 145:67–77

Stankov B, Lucini V, Snochowski M, Cozzi B, Fumagalli P, Maccarinelli G, Fraschini F (1989) Cytosolic androgen receptors in the neuroendocrine tissues of the golden hamster: influence of photoperiod and melatonin treatment. Endocrinology 125:1742–1744

Sun CY, Nadakavukaren MJ (1980) Effect of testosterone on the female hamster Harderian gland pigmentation and ultra-structure. Cell Tissue Res 207:511–517

Thompson GG, Hordovatzi X, Moore MR, Mcgadey J, Payne AP (1984) Sex differences in haem biosynthesis and porphyrin content in the Harderian gland of the golden hamster. Int J Biochem 16:849–852

Ulrich R, Yuwiler A, Geller E, Wetterberg L (1974) Effects of sex hormones and environmental lighting on rat Harderian gland porphyrin. J Endocrinol 63:99–102

Urueta A (1988) Caracterización del receptor de androgenos en la glandula de Harder. Thesis, Universidad Nacional Autónoma de méxico

Vilchis F, Pérez-Palacios G (1989) Steroid hormone receptors and the sexual phenotype of the Harderian gland in hamsters. J Endocrinol 121:149–159

Vilchis F, Hernández A, Pérez-Palacios G (1987) Hormone regulation of the rodent Harderian gland: binding properties of the androgen receptor in the male golden hamster. J Endocrinol 112:3–8

Vilchis F, Langley E, Solis J, Pérez-Palacios G, Cerbón MA (1990a) Identification and characteristics of nuclear thyroid hormone receptors in the hamster Harderian gland. 72nd Annual Meet, Endocrine Soc, Atlanta. GA, USA, Abst 1199

Vilchis F, Chávez B, Pérez-Palacios G (1991) Steroid hormone binding in the Harderian gland of birds: characteristics of the androgen, estrogen and progestin receptors of *Anas platyrhynchos* and *Gallus domesticus*. Gen Comp Endocrinol 82:425–433

Walter P, Green S, Greene G, Krust A, Bornert J-M, Jeltsch J-M, Staub A, Jensen E, Scrace G, Waterfield M, Chambon P (1985) Cloning of the human estrogen receptor cDNA. Proc Natl Acad Sci USA 82:7889–7893

Weaker FJ, Villareal C, Sheridan PJ (1983) Cellular localization of estradiol in the Harderian gland of the female armadillo. J Exp Zool 225:5–7

Wei LL, Horwitz KB (1985) The structure of progesterone receptors. Steroids 46:678–695

Wetterberg L, Geller E, Yuwiler A (1970a) Harderian gland: An extraretinal photoreceptor influencing the pineal gland in neonatal rats? Science 167:884–885

Wetterberg L, Yuwiler A, Geller E (1970b) Harderian gland: development and influence of early hormonal treatment on porphyrin content. Science 168:996–998

Winand RJ, Khon LD (1972) The binding of [^{3}H] thyrotropin and a [^{3}H]-labeled exophthalmogenic factor by plasma membranes of retroorbital tissue. Proc Na-Acad Sci 69:1711–1715

Woolley GW, Worley J (1954) Sexual dimorphism in the Harderian gland of the hamster (*Cricetus auratus*). Anat Rec 118:416–417

Additional References

Unfortunately, space limitations have not allowed us the luxury of separately listing a complete bibliography on the Harderian glands nor could we include keywords to aid in its use. The following list of references, therefore, is comprised of only those which have not been referred to previously in the review chapters.

Between these two sources, however, we believe that most known published references to the Harderian glands can be found within this volume. With some exceptions, abstracts have not been included.

We hope that errors of omission or citation will be brought to our attention and also forwarded to Dr. Jan Hugo (see Foreword).

Roger A. Hoffman and Sherry Brooks

Abe J, Sugita A, Katsume Y, Yoshizuka M, Tamura N, Iwanga S, Nishida T (1980) Scanning electron microscopic observations of the Harderian gland in rat. Kurume Med J 27:239–246

Aitken ID, Survashe BD (1976) A procedure for location and removal of the lachrymal and Harderian glands of avian species. Comp Biochem Physiol 53:193–195

Albert DM, Frayer WC, Black HE, Massicotte SJ, Sang DN, Soque J (1986) The Harderian gland: Its tumors and its relevance to humans. Trans Am Ophthalmol Soc 84:321–341

Andrews AD, Dunn JD (1973) Effects of TSH on the chemical composition of orbital tissue in guinea pigs. Endocrinology 93:527–531

Aureli G, Carelli G (1956) Richerche morfologiche ed istochimiche sulla ghiandola di Harder del coniglio. Atti Soc Ital Sci Vet 10:441–446

Aureli G (1957) La ghiandola di Harder nel Maiale. Richerche morfologiche ed istochimiche. Riv Istochem Nor Patol 3:113–126

Baba T, Masumota K, Nishida S, Kajikawa T, Mitsui M (1988) Harderian gland dependency of immunoglobulin A production in the lacrimal fluid of chicken. Immunology 65:67–71

Ballantyne B, Fourman J (1967) The histology and histochemistry of the Harderian gland of the domestic duck. J Anat 101:194–199

Bellairs Ad'A, Boyd JD (1947) The lachrymal apparatus in lizards and snakes – I. The brille, the orbital glands, lachrymal canaliculi and origin of the lachrymal duct. Proc Zool Soc London 117:81–108

Böck P, Goldenberg H, Huettinger M, Kolar M, Kramar R (1974) Preparation and characterization of catalase positive particles ('micro peroxisomes') from Harder's gland of the rat. Exp Cell Res 90:15–19

Burns B, Pickwell GV (1972) Cephalic glands in sea snakes (*Pel amis*, *Hydrophis* and *Laticauda*). COPEIA 3:547–559

Burns RB (1979) Histological and immunological studies on the fowl lacrimal gland following surgical excision of Harder's gland. Res Vet Sci 27:69–75

Burns RB (1983) Antibody production suppressed in the domestic fowl, *Gallus domesticus*, by zinc deficiency. Avian Pathol 12:141–146

Collins ET (1930) The Harderian gland; xerophthalmia; vitamin A deficiency; keratomalacia. Ophth Soc Trans (UK) 50:201–230

Coujard M (1974) La "Hardérisation" de la glande de Loewenthal chez le Rat blanc. C R Soc Biol 168:231–233

Coujard M, Marchi N (1974) La glande de Loewenthal du Rat blanc en culture de tissus. C R Soc Biol 168:706–708

Coujard M (1975) Effet de la gangliectomie cervicale supérieure sur les glandes de Harder (Rat et Lapin) et de Loewenthal (Rat). C R Soc Biol 169:26–29

Da Costa JRV, Lopes RA, Iucif S (1977) Histochemical study of mucosubstance and proteins in the Harderian gland of *Metachirus nudicaudatus* (Marsupialia Mammalia). Acta Biol Par 6:45–49 (in Portuguese)

Da Costa JRV, Iucif S, Lopes RA (1978) Effect of hypervitaminosis A on the Harderian gland in rats: a morphologic and morphometric study. Int J Vitam Nutr Res 48:113–122

Da Costa JRV, Lopes RA, Piccolo AM, Campos SM, Petenusci SB (1980) Morphological and histochemical study of labial and Harderian glands of *Mastigodryas bifossatus bifossatus* (Ophidia, Colubridae). Rev Brasil Biol 40:349–355 (in Portuguese)

Damian E (1977) Extra-pineal gland sources of melatonin. Rev Roum Med Endocrinol 15:65–66

Danley DL (1973) The Harderian gland and site-finding by metacercariae of *Philophthalmus megalurus* (Trematoda: Digenea). J Parasitol 59:457–462

Dartt DA, Knox I, Palau A, Botelho SY (1980) Proteins in fluids from individual orbital glands and in tears. Invest Opthamol Vis Sci 19:1342–1347

Davelaar FG, Kouwenhoven B (1977) Influence of maternal antibodies on vaccination of chicks of different ages against infectious bronchitis. Avian Pathol 6:41–50

Davelaar FG (1980) A simple technique for surgical removal of the Harderian gland of 1-day-old chicks. Avian Pathol 9:591–595

Davelaar FG, Kouwenhoven B (1980) Effect of the removal of the Harderian gland in 1-day-old chicks on immunity following IB vacination. Avian Pathol 9:489–497

Davelaar FG Noordzij A, Van der Donk JA (1982) A study on the synthesis and secretion of immunoglobulins by the Harderian gland of the fowl after eye drop vaccination against infectious bronchitis at 1 day old. Avian Pathol 11:63–80

Del Cacho E (1990) Mechanismo de la respuesta inmune a nivel celular en la glándula de Harder de las aves. Doctoral Thesis, Universidad de Zaragoza

Di Majo V, Coppola M, Rebessi S, Bassani B, Alati T, Saran A, Bangrazi C, Covelli V (1986) Dose-response relationship of radiation-induced Harderian gland tumors and myeloid leukemia of the CBA-Cne mouse. J Natl Cancer Inst 76:955–966

Dimitrov DS, Nikiforov I, Kolev KK, Dimitrova D (1987) Histological special features of the intraorbital glands in pheasants, affected by tuberculosis. Vet Med Nauki 24:78–84 (in Bulgarian)

Durkó I, Juhász A (1986) Porphyrin synthesis in primary nervous tissue cultures from 10^{-3} M delta-aminolaevulinic acid in the presence of melatonin and neuropeptides. Neurochem Res 11:607–615

Eichhorn M (1988) The presence of carbonic anhydrase in orbital glands. An enzyme-histochemical study in cynomolgus monkeys and rabbits. Histochemistry 88:475–479

Eida K, Kikutani M (1969) Harderian gland. IV. Porphyrin formation from δ-aminolevulinic acid by the Harderian gland of rats. Chem Pharm Bull 17:927–931

Farrar PL, Wagner JE (1985) Bilateral removal of the Harderian gland and its effect on development of nasal dermatitis in the Mongolian gerbil. Lab Anim Sci 35:554

Farrar PL, Opsomer MJ, Kocen JA, Wagner JE (1988) Experimental nasal dermatitis in the Mongolian gerbil: effect of bilateral Harderian gland adenectomy on development of facial lesions. Lab Anim Sci 38:72–76

Figge FHJ, Davidheiser RH (1960) Influence of hypophysectomy on porphyrin synthesizing enzyme activity of rat Harderian glands and livers. Proc Soc Exp Biol Med 103:64–66

Fischer G (1970) Die Hardersche Drüse beim Hausschwein (*Susscropha domesticu*). Anatomische, morphologische und sub-strathistochemische Untersuchungen. Diss (Med Fak) Marburg

Fry RJM, Garcia AG, Allen KH, Sallese A, Staffeldt E, Tahmisian TN, Devine RL, Lombard LS, Ainsworth EF (1976) Effects of pituitary isografts on radiation carcinogenesis in mammary and Harderian glands of mice. In: Biological and environmental effects of low level radiation. Int'l Atomic Energy Agency, Vienna, pp 213–227

Gallego M; Glick B (1988) The proliferative capacity of the cells of the avian Harderian gland. So Dev Comp Immunol 12:157–166

Gardiner JE, Hellman K, Thompson JW (1962) The nature of the innervation of the smooth muscle, Harderian gland and blood vessels of the cat's nictitating membrane. J physiol 163:436–456

Giacomini C (1887) Annotazioni sulla anatomia del negro. Appen dice alle tre prime memorie. VIII. Esistenza della ghiandola d'Harder in un Boshchimane. Duplicate della cartilagine della plica semilunaris. Muscolo ciliare nei Negri. Distribuzione del pigmento del globo oculare. Atti D R Accad D Sc Di Torino XXII:693–711

Glick B, Rao S, Stinson R, McDuffie SC (1977) Immunoglobulin-positive cells from the gland of Harder and bone marrow of the chicken. Cell Immunol 31:177–181

Gupta SK, Maiti BR (1981) Action of prolactin on the Harderian gland of the chick. Gen Comp Endocrinol 45:289–292

Halder C, Ghosh M (1989) Effect of pinealectomy and 5-methoxy indoles on Harderian gland activity of the Indian Jungle bush squail *Perdicula asiatica*. Indian J Exp Biol 27:704–708

Hayes ER (1958) Comparative histochemical studies of the Harderian gland. Anat Rec 130:436

Hilton FK, Raque GH, Hilton MA (1981) Changes in concentrations of taurine in murine Harderian glands and retinas during postnatal development. J Exp Zool 216:493–496

Hoffman RA, Johnson LB (1985) Effects of photic history and illuminance levels on male golden hamsters. J Pineal Res 2:209–215

Holland JM, Fry M (1982) Neoplasms of the integumentary system and Harderian gland. In: Foster HL, Small JD, Fox JG (eds) The mouse in biomedical research, vol IV. Academic Press, New York, pp 521–528

Hughes MR (1974) Water content of the salt glands and other avian tissues. Comp Biochem Physiol 47A:1089–1093

Innes JR, Stanton MF (1961) Acute diseases of the submaxillary and Harderian glands (sialodacryoadenitis) of rats with cytomegaly and no inclusion bodies. Am J Path t38:455–468

Jacoby RO, Bhatt PN, Jonas AM (1975) Pathogenesis of sialodacryoadenitis in gnotobiotic rats. Vet Pathol 12:196–209

Jost U (1971) Vergleichende biochemische Untersuchungen über die Stoffwechselmöglichkeiten der Harderschen Drüse beim Schwein und Kaninchen. Inaug Diss (Med Fak) Marburg

Jost U (1972) Die Hardersche Drüse des Kaninchens. Eine erste biochemische Analyse, insbesondere die Charakterisierung der darin enthaltenen Lipide und die Strukturaufklärung von Glycerinäther/ester. Diss (Med Fak) Heidelberg

Kagan R, Marder J, Ikan R (1982) The rodent Harderian gland: a source of thermoregulatory pheromones? Phytoparasitica (Israel) 6:391–406

Kaltenbach JC, Harding CV, Susan S (1980) Surface ultrastructure of the cornea and adjacent epidermis during metamorphosis of *Rana pipiens*: a scanning electron microscopic study. J Morphol 166:323–336

Kamocki W (1883) Ueber die sogenannte Harder'sche Drüse der Nager. Biol Zentral 2:709–717

Karelin AA, Mardashev SR (1975) Guanidinoacetate-N-methyltransferase: location in mammalian retina and rat Harderian gland. Biochem 40:296–300

Kittel R (1962) Die postnatale Entwicklung der Gl. orbitalis externa und der Gl. infraorbitalis des Goldhamsters (*Mesocricetus auratus*, Waterhouse). Gegenbaurs Morphol Jahrb 105:484–496

Kue-Staniszewska A (1914) Cytologische Studien der Harderschen Drüse. Anat Anz 47:424

Kühnel W (1966) Enzymhistochemische Untersuchungen an der Harderschen Drüse des Kaninchens. Histochemie 7:230–244

Kühnel W, Wrobel KH (1966a) Die Histotopik von Aldolase und Alkohol-dehydrogenase in der Harderschen Drüse des Kaninchens. Histochemie 7:245–250

Kühnel W, Wrobel KH (1966b) Über die histochemisch cfassbare Ativitat der β-B-D-Glucuronidase und der β-B-D-Galaktosidase in der Harderschen Drüse des Kaninchens. Albrecht Graefes Arch Klin Exp Ophthal 171:173–183

Kühnel W, Burger G (1969) Cytochemische Localisation einiger Hydrolasen in drei Orbitaldrüsen des Kaninchens. Histochemie 20:87–98

Kühnel W, Beier HM (1972) Zur Morphologie der Harderschen Drüse der Hausgans. Verh Anta Ges 67:591–598

Lopes RA, Valeri V, Campos GM, Lopes OVP, De Faria RM (1973) Histochemical study of mucopolysaccharides of the cephalic glands of *Micrurus corallinus corallinus* (Wied) (Ophidea, Elapidea). Ann Histochim 18:131–139 (in Portuguese)

Lopes RA, de Oliveira C, Campos GM, de Barros EJM (1974) Morphological and histochemical study of the Harderian gland of some Brasilian reptilians: I. Ophidea. Mem Inst Butantan 38:41–50 (in Portuguese)

Lopes RA, de Oliveira C, Campos MNM, Campos SM, Birman EG (1974) Morphological and histochemical study of cephalic glands of *Bothrops jararaca* (Ophidia, Viperidae). Acta Zool 55:17–24

López JM, Tolivia J, Díaz C, Alvarez-Uría M (1990) Ultrastructural study of lamellar and nucleolus like bodies in the Harderian gland during postnatal development of the Hamster. Anat Rec 228:247-254

Lorincz AL, Lancaster G (1957) Anterior pituitary preparation with tropic activity for sebaceous, preputial, and Harderian glands. Science 126:124–125

Lucas J, Orten JM (1954) Dietary protein and glycine as precursors of porphyrins in the rat. J Nutr 52:89–97

Lütticken D, Rijke EO, Loeffen AHC (1987) Systemic and local antibody responses in chickens after infection and vaccination with infectious bronchitis virus. Prog Clin Biol Res 238:321–330

le Maire M, Wins P (1972) Some properties of particulate ATPases and phosphatases isolated from the Harderian gland of the duck. Arch Int Physiol Biochem 80:423–441

Mansikka A, Sandberg M, Granfors K, Toivanen P (1989a) B-cells in the chicken Harderian gland. In: Recent advances in avian immunology research. Alan R Liss, New York, p 59-65

Mansikka A, Sandberg M, Veromaa T, Vainio O, Granfors K, Toivanen P (1989b) B cell maturation in the chicken Harderian gland. J Immunol 142:1826–1833

Marty A, Evans MG, Tan YP, Trautmann A (1986) Muscarinic response in rat lacrimal glands. J Exp Biol 124:15–32

Mázló MA, Rohonyi B (1963) A study of the diameter distribution of the secretory lipid vesicles in the Harderian gland. Acta Biol Hung 14:191–198

McGee MA, Maronpot RR (1979) Harderian gland dacryoadenitis in rats resulting from orbital bleeding. Lab Anim Sci 29:639–641

Meek A (1983) On the occurrence of a Jacobson's organ with notes on the development of the nasal cavity, the lachrymal duct, and the Harderian gland in *Crocodilus porosus*. J Anat Physiol 27:151–160

Michael MI, Khalil SH, Matta CA, Rizk TA (1988) Normal development of the ocular glands of the mouse. Folia Morphol (Prague) 36:47–52

Miller ME, Habel RE (1951) Harder's gland in the dog. J Am Vet Med Assoc 118:155–156

Moissejeff E (1925) Zur Bedeutung der Harderschen Drüse im Cholesterinstoffwechsel. Z Gesamte Exp Med 47:359–368

Moore MR, Thompson GG, Payne AP, McGadey J (1980) Seasonal variation in δ-aminolaevulinate synthase and porphyrin content in the Harderian gland of the female golden hamster (*Mesocricetus auratus*). Int J Biochem 12:501–504

Mukai H (1926) Über die feinere Struktur der Harderschen Drüse beim Kaninchen. Albrecht von Graefes Arch Ophthal 117:243–272

Mykytowycz R (1966) Observations on odoriferous and other glands in the Australian wild rabbit, *Oryctolagus cunniculus* (L.), and the hare *Lepus europaeus*, P. III. Harder's, lachrymal and submandibular glands. CSIRO Wildl Res 11:65–90

Nash SM, Anderson BJ, Reed TL, Parrish JW, Davis SF (1986) Reward and non-reward odor cues: the role of the Harderian gland. Bull Psych Soc 24:141–144

Neumann U (1976) Die chirurgische Entfernung der Harderschen Drüse des Huhnes. Zentralbl Veterinaermed Reihe A 23:323–330

Neumann U, Kaleta EF (1977) Untersuchungen zur immunologischen Funktion der Harderschen Drüse des Huhnes. Zentralbl Vet erinaermed Reihe B 24:331–339

Oishi T, Lauber JK (1973) Photoreception in the photosexual response of quail. I. Site of photoreceptor. Am J Physiol 225:155–158

Oorjitham E, Norvell JE, Dunn JD (1985) Failure of Harderian gland removal to alter pituitary-adrenal function. Proc Okla Acad Sci 65:1–4

Parvari R, Ziv E, Lentner F, Tel-Or S, Burstein Y, Schechter I (1987) Analyses of chicken immunoglobulin light chain cDNA clones indicate a few germline V lambda genes and allotypes of the C lambda locus. EMBB J 6:97–102

Paule WJ (1951) The histology and histochemistry of the Harderian gland of the turtle. MS Thesis, The Ohio State Univ

Payne AP (1978) Attractant properties of the Harderian gland and its products on male golden hamsters of differing sexual experience. J Endocrinol 80:69–70

Payne AP, McGadey J, Johnston HS, Moore MR, Thompson GG (1978) Harderian gland histology and porphyrin content in the pre- and post-pubertal male hamster. J Anat 27:653–654

Pendergrass ML, Thiessen DD (1981) Body temperature and auto-grooming in the Mongolian gerbil, *Meriones unguiculatus.* Behav Neural Biol 33:524–528

Percy DH, Wojcinski ZW, Schunk MK (1989) Sequential changes in the Harderian and exorbital lacrimal glands in Wistar rats infected with sialodacryoadenitis virus. Vet Pathol 26:238–245

Peters A (1890) Beitrag zur Kenntnis der Harderschen Drüse. Arch Mikrosk Anat 36:192–203

Pévet, P, Balemans MGM, Legerstee WC, Vivien-Roels B (1981) Circadian rhythmicity in the activity of HIOMT in the formation of melatonin and of 5-methoxytryptophol in the retina, Harderian gland and pineal of the male hamster. Adv Biosci 29:201–206

Pévet, Heth G, Hiam A, Nevo E (1984) Photoperiod perception in the blind mole rat (*Spalax ehrenbergi,* Nehring): involvement of the Harderian gland, atrophied eyes, and melatonin. J Exp Zool 232:41–50

Phillips JA, Harlow HJ, McArthur NH, Ralph CL (1986) Epithalamus of the nine-banded armadillo, *Dasypus novemcinctus.* Comp Biochem Physiol 85:477–482

Piersol GA (1887) Beiträge zur Histologie der Harder'schen Drüsen der Amphibien. Arch Mikrosk Anat (Bonn) 29:594–608

Pochin EE (1945) Exophthalmos in guinea pigs injected with pituitary extracts. Clin Sci 5:75–91

Powell JR, Aitken ID, Survashe BD (1979) The response of the Harderian gland of the fowl to antigen given by the ocular route. II. Antibody production. Avian Pathol 8:363–373

Rausch RL, Lindsay JW (1974) Virus-associated tumours of the Harderian gland in laboratory-reared voles, *Clethrionomys rutilus* (Pallas). J Pathol 112:217–224

Reiter RJ, Guerrero JM, Puig-Domingo M, Santana C (1988) Type II iodothyronine deiodination: nyctohemeral rhythms in the cerebral cortex, the pineal gland, and the Harderian glands. In: Tietneld W (ed) Advances in the biosciences, vol 73. Pergamon Press, London, pp 123–136

Rock CB (1977) Harderian gland. In: Snyder F (ed) Lipid metabolism in mammals, vol II. Plenum Press, New York, pp 311–321

Saint-Girons H (1982) Comparative histology of the orbital glands of lepidosauria. Ann Sci Nat Zool Biol Anim 4:171–191

Santos R, Carlini EA, (1988) Red lacrimal secretion (Chromodacryorrhea) induced by cholinergic drugs in rats subjected to the watertank technique. Pharmacology 36:61–68

Satoh Y, Saino T, Ono K (1990) Effect of carbamylcholine on Harderian gland morphology in rats. Cell Tissue Res 261:451–459

Schneir ES (1950) The histochemistry of the Harderian gland of the rabbit. MS Thesis, The Ohio State Univ

Schoental R (1968) Selenium-75 in the Harderian glands and brown fat of rats given sodium selenite labelled with tselenium-75. Nature 218:294–295

Schramm U (1980) Lymphoid cells in the Harderian gland of birds. An electron microscopical study. Cell Tissue Res 205:85–94

Seely JC (1987) The Harderian gland (pathologically speaking). Lab Anim 21:33–39

Seyama Y, Otsuka H (1984) Biochemistry of lipids in Harderian gland. Seikagaku 56:1266–1271

Singh SP, McKenzie JM (1971b) ^{35}S-Sulfate uptake by mouse Harderian gland: effect of serum from patients with Graves' disease. Metabolism 20:422–427

Smelser GK (1943) Water and fat content or orbital tissues of guinea pigs with experimental exopthalmos produced by extracts of the anterior pituitary gland. Am J Physiol 140:308–315

Smelser GK, Ozanics V (1953) The effect of the Harderian and lacrimal glands upon regeneration of the corneal epithelium. Am J Ophthalmol 36:1545–1550

Smith SG, Belcher RV (1974) Distribution of some possible intermediates of the haem biosynthesis. Enzyme 17:1–10

Sonawane VD, More NK (1981) Mucopolysaccharide histochemistry of paraocular glands in passerine birds. Pavo 19:21–34

Spike RC, McGadey J, Johnston HS, Payne AP, Thompson GG, Moore MR (1985a) The effects of hormone manipulation on the porphyrin content and histology of the Harderian gland of the female golden hamster. Biochem Soc Trans 13:206–207

Spike RC, McGadey J, Johnston HS, Payne AP, Thompson GG, Moore MR (1985b) The effects of long term ovariectomy on the Harderian gland of the female golden hamster. Biochem Soc Trans 13:787–788

Stammer A (1964) Ein Beitrag zur Struktur und mikroskopischen Innervation der Harderschen Drüse der Vögel. Acta Univ Szeged 10:99–107

Stankov B, Lucini V, Snochowski M, Cozzi B, Fumagalli P, Maccarinelli G, Fraschini F (1989) Cytosolic androgen receptors in the neuroendocrine tissues of the golden hamster: influence of photoperiod and melatonin treatment. Endocrinology 125:1742–1744

Survashe BD, Aitken ID (1977) Removal of the lachrymal gland and ligation of the Harderian gland duct in the fowl (*Gallus domesticus*): procedures and sequelae. Res Vet Sci 22:113–119

Survashe BD, Aitken ID, Powell JR (1979) The response of the Harderian gland of the fowl to antigen given by the ocular route. I. Histological changes. Avian Pathol 8:77–93

Taddei D (1900) Contributo alla conoscenza istofisiologica dellaghiandola dell' Harder nel coniglio. Arch per esp e elin Milano 45:71–76

Takáts A, Faragó A, Antoni F, Fábián F (1974) Adenosine 3′,5′-monophosphate dependent protein kinase isolated from rat Harderian gland. Acta Biochim Biophys Acad Sci Hung 9:33–41

Tatro JB, Reichlin S (1987) Specific receptors for α-MSH are widely distributed in tissues of rodents. Endocrinology 121:1900–1907

Taub AM (1966) Ophidian cephalic glands. J Morphol 118:529–542

Thiessen DD (1980) Effects of water deprivation on the Mongolian gerbil, *Meriones unguiculus*: absence of Harderian gland involvement. Phys Psych 8:379–382

Thiessen DD, Pendegrass M (1982) Harderian gland involvement in facial lesions in the Mongolian gerbil. J Am Vet Assoc 181:1375–1377

Toolan HW (1983) Degeneration of lens and overgrowth of Harderian glands in hamsters neonatally injected with parvovirus MVM-i. Proc Soc Exp Bio Med 172:351–356

Tucker MJ (1979) Tumours of the Harderian gland. Inter Agency Res Cancer 23:135–140

van Jaarsveld , Mhatre MC, Reiter RJ (1989) Porphyrin levels in the Harderian glands of female and male Syrian hamsters: early changes following either gonadectomy or light deprivation and lack of a circadian rhythm. Biomed Res 10:1–8

Venable JH, Grafflin AL (1940) Gross anatomy of the orbital glands in the albino rat. J Mammal 21:66–71

Walcott B, McLean JR (1985) Catacholamine-containing neurons and lymphoid cells in a lacrimal gland of the pigeon. Brain Res 328:129–138

Werner HJ, Roberts JH (1951) The Harderian gland of the dog in chronic ectopia. J Am Vet Med Assoc 118:16–18

Wolters B (1976) Die Entwicklung der Harderschen Drüse des normalen und des bursektomierten Hühnchens (Licht- und elektron- enmikroskopische sowie morphometrische Untersuchungen). Inauguraldiss, Med Fak der Westf Wilhelms-Univ, Münster

Wolters B, Hultsch E, Niedorf HR (1977) Bursa-independent plasma cells in the Harderian gland of the chicken (*Gallus domesticus*). Histomorphometrical investigations. Beitr Pathol 160:50–57

Yoshie S, Ogawa T (1983) Occurrence of the myoid cell in the Harderian gland of the Japanese colubrid snake, *Rhabdophistigrinus*. Arch Histol Jpn 46:131–135

Yoshimura M, Sameshima M (1980) Mucosubstances of the acini of the rabbit Harderian gland: a histochemical and electron microscopic study. Acta Soc Ophthamol Jpn 84:101–112

Zawirska B (1963) Microfluoroscopic studies on porphyrin in Harder's gland and in tumours induced in it. I. Some morphological and physiopathological problems of the Harderian gland. Postepy Hig I Med Dosw 17:435–450 (in Polish)

Subject Index

MIX
Papier aus verantwortungsvollen Quellen
Paper from responsible sources
FSC® C105338

If you have any concerns about our products,
you can contact us on
ProductSafety@springernature.com

In case Publisher is established outside the EU,
the EU authorized representative is:
Springer Nature Customer Service Center GmbH
Europaplatz 3, 69115 Heidelberg, Germany

Printed by Libri Plureos GmbH
in Hamburg, Germany